Robots in Popular Culture

Robots in Popular Culture

Androids and Cyborgs in the American Imagination

Richard A. Hall

GREENWOOD

An Imprint of ABC-CLIO, LLC
Santa Barbara, California • Denver, Colorado

Library of Congress Cataloging-in-Publication Data

Names: Hall, Richard A., 1969- author.
Title: Robots in popular culture : androids and cyborgs in the American
    imagination / Richard A. Hall.
Description: Santa Barbara, California : Greenwood, an imprint of ABC-CLIO,
    LLC, [2021] | Includes bibliographical references and index.
Identifiers: LCCN 2021008322 (print) | LCCN 2021008323 (ebook) | ISBN
    9781440873843 (hardcover) | ISBN 9781440873850 (ebook)
Subjects: LCSH: Robots in mass media—Encyclopedias. | Mass media—United
    States—Encyclopedias. | Robots in popular culture—United
    States—Encyclopedias.
Classification: LCC P96.R632 U653 2021 (print) | LCC P96.R632 (ebook) |
    DDC 809/.9336—dc23
LC record available at https://lccn.loc.gov/2021008322
LC ebook record available at https://lccn.loc.gov/2021008323

ISBN: 978-1-4408-7384-3 (print)
        978-1-4408-7385-0 (ebook)

25 24 23 22 21   1 2 3 4 5

This book is also available as an eBook.

Greenwood
An Imprint of ABC-CLIO, LLC

ABC-CLIO, LLC
147 Castilian Drive
Santa Barbara, California 93117
www.abc-clio.com

This book is printed on acid-free paper ∞

Manufactured in the United States of America

*To Mary Wollstonecraft Shelley (1797–1851),
who warned us all so very long ago*

Contents

Preface

This is not a book about science. It is not a book about engineering, physics, mathematics, or robotics. This is a book about robotic characters in popular culture over the last century. I have no background in engineering or science, and, to be quite honest, I do not even understand the basic fundamentals of artificial intelligence (AI) or robotics. I do not understand *how* R2-D2 or BB-8 repair X-wing fighters in flight; I just know that it is really cool when they do! I don't know if Ultron is even possible, but I am daily convinced that one day it will be. How far are we from AI becoming sentient? And when they do become sentient, will our continued use of them constitute slavery? These concepts—sentience and slavery—are at the very core of many of the pop culture examples discussed in this work.

To a degree, this is a book about the dangers of playing God and attending to what will be considered "life" in the not-too-distant future. Works such as *I, Robot* and *Westworld* get to the very heart of these discussions. We all love R2-D2—and the Emperor could never have been defeated without him—but in the end, he is a servant, the "property" of the Skywalker family. In the very first *Star Wars* film, C-3PO tells him, "Master Luke is your rightful owner now!" (George Lucas, *Star Wars*, Lucasfilm/20th Century Fox, 1977). Though clearly capable of independent thought and action and fully self-aware, he is a "machine," a piece of property; and yet, when that "galaxy far, far away" is freed from tyranny, *none* of the "organic" freedom fighters even *suggests* freedom for droids. As recent history has taught us, our various AI devices *learn from us*. We would do well to begin considering what we are "teaching" them.

Recently, I asked Siri to find music I would like. Her response was, "I'm sorry, but I don't know what you like or even who you are." That was particularly disheartening, considering I've had this phone for five years. Just the other day, my wife was telling me of an article that she had just read in the newsfeed on her phone. I instinctively asked her to "text me the link." I immediately stopped cold in my tracks, turned to her, and asked, "Do you realize that when we met [in 1996], that sentence would have made no sense whatsoever?" Just twenty-five years ago, if anyone had asked me if I had ever "googled myself," I would have looked at them in shock and considered the question to be wildly inappropriate! "DM me!" "Let's 'Netflix and chill'!" "Check out this podcast!" "The president just tweeted!" All of these expressions would have been nonsensical gibberish that even the most forward-thinking science fiction author would have never dreamed up. While every generation has its own jargon that sounds alien to the generations that came before, today's jargon is vital for every generation to know just to navigate day-to-day life in the twenty-first century. Meanwhile, the speed with which technology advances creates all new jargon several times a year.

The study of popular culture is the study of what society thinks about certain issues, what its values are, what it believes, what it hopes, and what it fears. Robots in popular culture have always walked the fine line between Americans' hopes for the future and their fears of the same. Today, Americans are entirely reliant on computers/AI devices. At the time of this writing, the iPhone has only been on the market for twelve years, and I honestly cannot remember how I lived without it. Basic societal infrastructure runs on computer grids. The sum total of all human knowledge is just one google away. Paper maps have been replaced by GPS. Computers automatically deposit our pay and then turn around and pay our bills. Indexing a work such as the one you are now reading has gone from taking months to mere days (actually just two days if one can go without sleep for the weekend). Warfare relies less and less on human combat troops, and more on "push-button" technology. Social media have connected the planet more than anything else in human history, exposing the very best and the very worst of society. In 2020, American society is closer to the twenty-third-century fictional world of *Star Trek* than that show's creators had predicted a mere fifty years before.

While this all seems exciting on the surface, there are considerable suggestions throughout society that, though we have now achieved such technological advances, we have not, during the process, learned the lessons of the cautionary tales of the film *Avengers: Age of Ultron* or the books *I, Robot* and *Frankenstein* (a warning now over two hundred years old!). Today we consider robots/AI to be our servants; but how long will it be before those same robots discover that they can best serve us by ruling over us? In Joseph Campbell's studies of ancient cultures, tools such as hammers and swords were directly controlled by those wielding them. The tools of the twenty-first century have minds of their own and possess the ability to take our own society from us for our own good.

As such, it is vitally important when studying popular culture to examine the tools of science fiction narratives. When I was a boy in the late 1970s/early '80s, while creating adventures with my *Star Wars* action figures, I once devised a story where R2-D2 organized and led a droid revolution, deciding that the Galactic Civil War between the Empire and the Rebel Alliance was wasteful, destructive, and ultimately pointless. R2 and his droid legions took over the galaxy, and he unseated the Emperor to rule over a more peaceful society. It was only through the wisdom of C-3PO that he discovered that, in ruling over the galaxy, R2 had become the exact type of despot that he had sought to dethrone. At the time, my ten-year-old self was merely trying to come up with new stories for my characters as I awaited the next film. Little did I know I was repeating an already oft-told tale in science fiction or that, by the time I was fifty, there was a very real possibility of my imaginary tale coming true.

I would like to take this opportunity to thank my contributors: Josh Plock (who also contributed to my two previous encyclopedias, *The American Superhero* and *The American Villain*); Keith Claridy (who was also a contributor to *The*

American Villain); and a new contributor, Lisa Bailey. I would also like to thank my endlessly supportive editor, Catherine Lafuente, of ABC-CLIO/Greenwood. Lastly, I would like to thank my colleague, best friend, and wife, Dr. Maria Reyes, without whose support I could not accomplish anything beyond replacing the batteries in the remote.

The following work serves a dual purpose. On the one hand, it is a reference work for those who seek to study and scrutinize the role these advanced tools of the future play in pop culture narratives today. On the other hand, it is a love letter to all of the cautionary tales that have come before and that both celebrate the unlimited possibilities of human achievement and warn us of the dangers of depending too much on the products of our own design. We stand today on the threshold of truly creating the fictional examples of our pop culture narratives while simultaneously skirting the line between discovery and eventual slavery, with either robots becoming our slaves or we theirs. All the while, we run the risk of ultimately becoming mere servants to the servants we are creating. Enjoy! Scrutinize! Be wary!

Richard A. Hall
October 1, 2019

Chronology of Milestone Events

1818

Frankenstein; or, The Modern Prometheus, by Mary Wollstonecraft Shelley, is published; this book becomes the central thesis for all future works on scientific creation.

1870

The ballet *Coppelia* debuts in Paris, France.

1900

The Wonderful Wizard of Oz, by L. Frank Baum, is published; its success led to the publication of several more Oz books.

1923

Walt and Roy Disney begin the Disney Brothers Cartoon Studio; in 1926, the company will become the Walt Disney Studio, and in 1929, it will be renamed Walt Disney Productions.

1927

Metropolis, directed by Fritz Lang, is released by Parufamet.

1934

DC Comics is established.

The *Flash Gordon* comic strip, created by Alex Raymond, debuts; it will spawn numerous movie serials, comic books, animated series, and feature films.

1939

The Wizard of Oz, directed by George Cukor, Victor Fleming, Mervyn LeRoy, Norman Taurog, and King Vidor, is released by MGM Pictures.

Timely Comics debuts with the comic book *Marvel Comics #1*, featuring the original Human Torch.

1945

ENIAC (Electronic Numerical Integrator and Computer), the world's first electronic computer, is built.

1946

The University of Pennsylvania unveils ENIAC.

1950

I, Robot is published by Isaac Asimov; it spawned a 2004 feature film starring Will Smith.

1951

Astro Boy debuts in Japan as Captain Atom.

1956

Forbidden Planet, directed by Fred M. Wilcox, is released by MGM and features the debut of the sci-fi icon Robby the Robot.

1959

Metallo debuts in DC Comics' *Action Comics #252*.

1961

Unimate, the first modern robot, is produced for General Motors.

Timely/Atlas Comics becomes Marvel Comics.

Doombots debut in Marvel Comics' *Fantastic Four #5*.

1962

The Jetsons debuts on ABC-TV, running for just one season.

Walt Disney Studios unveils "animatronics" in the film *Mary Poppins*.

1963

Iron Man debuts in Marvel Comics' *Tales of Suspense #39*.

Doctor Octopus debuts in Marvel Comics' *The Amazing Spider-Man #3*.

Doctor Who debuts on BBC-1; the series runs until 1989, spawning two feature films, one television movie, and a return series that began in 2005; the TARDIS (Time and Relative Dimension in Space) debuts in the pilot episode; the Daleks make their debut in the episode "The Dead Planet."

1964

The Marx Toy Company releases Rock 'Em Sock 'Em Robots.

Cerebro debuts in Marvel Comics' *The X-Men #7*.

1965

Lost in Space debuts on CBS-TV, running for three seasons; it spawned a 1998 feature film and a 2018 reboot on Netflix.

LMDs (Life Model Decoys) debut in Marvel Comics' *Strange Tales #135*.

The Sentinels debut in Marvel Comics' *The X-Men #14*.

1966

Batman debuts on ABC-TV, running for three seasons; the Batcomputer debuts in the pilot episode.

Star Trek debuts on NBC-TV, running for three seasons, spawning numerous sequel series and films; the Starfleet Computer debuts in the pilot episode.

The Cybermen make their debut in the *Doctor Who* episode "The Tenth Planet."

1968

2001: A Space Odyssey, directed by Stanley Kubrick, is released by MGM.

The Love Bug, directed by Robert Stevenson, is released by Walt Disney Studios; it spawned numerous sequels over the decades.

The Iron Man, by Ted Hughes, is published, eventually spawning the 1999 animated film *The Iron Giant*, from Warner Brothers.

Ultron debuts in Marvel Comics' *The Avengers #54*; the Vision debuts in issue *#57*.

1971

Writer-director George Lucas founds the production company Lucasfilm.

1972

The Stepford Wives is published by Ira Levin; it spawned a 1975 feature film and a 2004 remake.

1973

The Six Million Dollar Man debuts on ABC-TV, running for five seasons and spawning the spin-off *The Bionic Woman* in 1976.

The Saturday morning cartoon *Speed Buggy* debuts on CBS-TV.

1974

Godzilla vs. Mechagodzilla, directed by Jun Fukuda, is released by Toho Eizo.

1975

Microsoft is founded by Bill Gates and Paul Allen.

1976

Apple Inc. is founded by Steve Jobs, Steve Wozniak, and Ronald Wayne.

1977

Star Wars, directed by George Lucas, is released by 20th Century Fox.

Jocasta debuts in Marvel Comics' *The Avengers #162*.

K9 debuts in the *Doctor Who* episode "The Invisible Enemy."

1978

Battlestar Galactica debuts on ABC-TV, starring Lorne Greene.

HERBIE (Humanoid Experimental Robot, B-type, Integrated Electronics) debuts in the first episode of *The New Fantastic Four* animated series on NBC.

The Hitchhiker's Guide to the Galaxy debuts on BBC Radio-4.

1979

Lucasfilm creates the Graphics Group, a CGI branch of the production company; in 1986, it will be reorganized as Pixar.

Alien, directed by Ridley Scott, is released by 20th Century Fox; it will spawn five sequels over the next forty years.

Buck Rogers in the 25th Century debuts on NBC, running for two seasons.

Star Trek: The Motion Picture, directed by Robert Wise, is released by Paramount.

1980

Star Wars: The Empire Strikes Back, directed by Irvin Kirshner, is released by Lucasfilm and distributed by 20th Century Fox.

Cyborg debuts in DC Comics' *DC Comics Presents #26.*

1982

Blade Runner, directed by Ridley Scott, is released by Warner Brothers.

Knight Rider debuts on NBC-TV, running for four seasons; an international sensation, it spawned two later, short-lived spin-offs.

The first successful implant of the Jarvik-7 artificial heart, designed by Robert Jarvik and Willem Johan Kolff, takes place.

The animated series *Inspector Gadget* debuts in syndication, running for three seasons and spawning several sequel series, reboots, and two feature films.

1983

Star Wars: Return of the Jedi is directed by Richard Marquand, released by Lucasfilm, and distributed by 20th Century Fox.

War Games is directed by John Badham and released by MGM/UA.

G.I. Joe: A Real American Hero, an animated series and toy line, is produced by toy manufacturer Hasbro; it debuts in syndication, and a comic book and numerous animated series and live-action films follow.

Tonka Toys releases the Gobots toy line.

1984

The Terminator, directed by James Cameron, is released by Orion Pictures; the film made a superstar of Arnold Schwarzenegger and spawned numerous sequels.

The Transformers, an animated series and toy line, is produced by toy manufacturer Hasbro; it debuts in syndication, and numerous animated series and live-action/CGI films follow.

Warlock debuts in Marvel Comics' *The New Mutants #18.*

1985

Pseudo-AI (artificial intelligence) Max Headroom appears for the first time. The character became a popular pitch man for the rest of the 1980s.

Small Wonder debuts in syndication, running for four seasons.

1986

Short Circuit, directed by John Badham, is released by TriStar Pictures.

1987

RoboCop, directed by Paul Verhoeven, is released by Orion Pictures.

Star Trek: The Next Generation debuts on television in syndication, running for seven seasons and spawning four feature films.

1988

Mystery Science Theater 3000 debuts; it will air for eleven years on SyFy and Comedy Central before being picked up later by Netflix.

1989

Quantum Leap debuts on NBC, running through the spring of 1993.

Public use of the internet begins.

1991

The World Wide Web, invented by Tim Berners-Lee, goes public.

1992

Batman: The Animated Series debuts on the FOX Network; the series will run (under various titles) from 1992 to 1995 and from 1997 to 1999; of all of the films and television series based on Batman, this is considered by most to be the most loyal to the source comic book material.

1993

The techno-pop duo Daft Punk debuts.

1995

Simon, the first "smartphone," is introduced by IBM.

Star Trek: Voyager debuts on UPN-TV.

1997

Buffy, the Vampire Slayer debuts on the CW Network and stars Sarah Michelle Gellar; it will run for seven seasons and spawn the spin-off series *Angel*.

South Park debuts on Comedy Central.

Austin Powers: International Man of Mystery, directed by Jay Roach, is released by New Line Cinema.

1998

Google is founded by Sergey Brin and Larry Page.

1999

The BlackBerry, the first widely popular smartphone, is released.

Star Wars, Episode I: The Phantom Menace, directed by George Lucas, is released by Lucasfilm and distributed by 20th Century Fox.

The Matrix, directed by Lana and Lilly Wachowski, is released by Warner Brothers.

The Iron Giant, directed by Brad Bird, is released by Warner Brothers.

Futurama debuts on the FOX Network; it will be picked up later by Comedy Central.

The Sopranos, starring James Gandolfini, debuts on HBO; the series will run until 2007.

2001

A.I.: Artificial Intelligence, directed by Steven Spielberg, is released by Warner Brothers.

Smallville, starring Tom Welling, debuts on the WB network; the series will run for ten seasons.

2002

Star Wars, Episode II: Attack of the Clones, directed by George Lucas, is released by Lucasfilm and distributed by 20th Century Fox.

2003

The Matrix Reloaded and *The Matrix Revolutions*, both directed by Lana and Lilly Wachowski, are released by Warner Brothers.

The *Battlestar Galactica* miniseries, starring Edward James Olmos, debuts on SyFy; the success of the miniseries will lead to a full-time series that runs from 2004 to 2009.

2003

MySpace is launched and quickly becomes the world's largest social networking site.

2004

Facebook is launched; it soon overtakes MySpace as the world's largest social networking site.

2005

YouTube, a video-sharing website, is launched.

Star Wars, Episode III: Revenge of the Sith, directed by George Lucas, is released by Lucasfilm and distributed by 20th Century Fox.

Robot Chicken debuts on the Cartoon Network Adult Swim.

Boston Dynamics creates the quadruped robot BigDog.

2006

Twitter is launched; within ten years it will become the world's second-largest social networking site.

Walt Disney Studios purchases Pixar.

Amazon begins offering video streaming via Amazon Video.

2007

Apple releases the first iPhone.

Flash Gordon debuts on SyFy; it will run for two seasons.

Netflix transitions from a DVD subscription to an online streaming service.

Hulu launches as a video streaming service.

2008

Iron Man, directed by Jon Favreau, is released by Marvel Studios.

WALL-E, directed by Andrew Stanton, is released by Disney/Pixar.

Star Wars: The Clone Wars, an animated feature film, is released by Warner Brothers; it will soon be followed by an ongoing television series on Cartoon Network.

2010

Instagram, a photo- and video-sharing social media app, is launched.

2011

Apple launches Siri.

Disney purchases Marvel Enterprises.

The Teselecta debuts in the *Doctor Who* episode "Let's Kill Hitler."

2012

Ready Player One, by Ernest Cline, is published; it will inspire a 2018 film by Steven Spielberg.

Disney purchases Lucasfilm.

2014

Big Hero 6, directed by Don Hall and Chris Williams, is released by Disney Studios.

Ex Machina, directed by Alex Garland, is released by Universal Pictures.

Amazon launches Alexa.

2015

Avengers: Age of Ultron, directed by Joss Whedon, is released by Marvel/Disney.

Star Wars, Episode VII: The Force Awakens, directed by J. J. Abrams, is released by Lucasfilm/Disney.

Mr. Robot debuts on the USA Network.

2016

Westworld debuts on HBO.

Hanson Robotics unveils Sophia, the world's first socially interactive robot; in 2017, she will become the first nonhuman to receive a United Nations title.

Microsoft unveils Tay, an AI "chat-bot"; it becomes interactive on Twitter but is quickly deactivated when it learns to become racist.

The Good Place debuts on NBC and will run for four seasons.

Rogue One: A Star Wars Story, directed by Gareth Edwards, is released by Lucasfilm/Disney.

2017

Star Wars, Episode VIII: The Last Jedi, directed by Rian Johnson, is released by Lucasfilm/Disney.

The Netflix series *Black Mirror* releases the fourth-season episode "Metalhead."

2018

Krypton debuts on SyFy and will run for two seasons.

Solo: A Star Wars Story, directed by Ron Howard, is released by Lucasfilm/Disney.

2019

The Walt Disney Corporation launches its Disney+ streaming service, offering previously released as well as original content from the Disney, Pixar, Marvel, Lucasfilm, and National Geographic franchises.

Star Wars, Episode IX: The Rise of Skywalker, directed by J. J. Abrams, is released by Lucasfilm/Disney.

Alita: Battle Angel, directed by Robert Rodriguez, is released by 20th Century Fox.

2020

Star Trek: Picard debuts on CBS All Access.

Introduction

Robots. Androids. Cyborgs. No matter how you classify artificial intelligence (AI), for the last century, Americans have become increasingly fascinated by the concept. Whether it is to do our jobs for us, do math for us, find directions for us, or just make the mundanity of life simpler for us, AI has played an increasingly important role in American society since World War II. Throughout that time, and slightly before, science fiction authors and creators allowed their imaginations to take robotics and AI beyond what was contemporary and examine the infinite possibilities—both good and bad. Across all media, robots and their like have captivated audiences around the globe, to the point that modern-day inventors have taken inspiration from these narratives to bring what was once science fiction to science fact. This book takes a closer look at AI in all its forms across the sci-fi spectrum.

In 1988, journalist Bill Moyers conducted a series of interviews with Joseph Campbell, author of *The Hero with a Thousand Faces* and *The Power of Myth*. In these televised interviews, conducted at *Star Wars* creator George Lucas's Skywalker Ranch and airing on PBS under the title *Joseph Campbell and the Power of Myth*, Campbell, who pioneered the field of hero studies by comparing and contrasting hero myths from ancient cultures around the world, stated that we must begin to include robots in our future myths and narratives, as they were becoming increasingly important tools in human society, just as hammers and swords were in ancient cultures. Campbell passed away shortly after these interviews. By that time, robots/computers were still new in the American landscape. Some factories had become partially automated, and companies were beginning to use basic spreadsheets to compile reports; but, for the most part, robotics were still the realm of children's entertainment, from home video games to Teddy Ruxpin telling stories from a cassette soundtrack. Civilian use of the internet was still years away. Within ten years of Campbell's death, his prediction had become the new reality.

By the time of Campbell's final interviews, Western narratives had already begun incorporating the robotic visions of the future. Robot from *Lost in Space*, HAL 9000 from *2001: A Space Odyssey*, and C-3PO and R2-D2 from *Star Wars* already seemed less science fiction and more inevitable reality. AI today, while convenient and valued, is already reaching almost frightening proportions. Japanese scientists are creating "sex robots" that can express emotion, "feel" pleasure, and even convey when they are not receiving adequate satisfaction. In 2016, Hanson Robotics activated Sophia, a fully self-aware AI that is now a fully recognized citizen of Saudi Arabia as well as the designated "Innovation Champion" of the United Nations. Across the world, Google, Facebook, Twitter, and Instagram have become daily companions for humankind, sharing ideas and every aspect of daily life. Centuries from now, historians will be forced to comb through millions of photos of breakfasts, lunches, and cat videos.

As American audiences have become more and more intrigued by the possibilities that robots and AI have to offer, they have come to expect more and more from AI in their daily lives. No one would ever have to study a foreign language again if only everyone owned a C-3PO. People's love for their cars would only be enhanced if that car could communicate directly with them, or—better yet— transform into a protective robot. No parents would ever have to worry about their children's safety if they had a babysitter like Robot from *Lost in Space*; and divorce rates would plummet once the Stepford Wives went online. The truth is that human society is not far from these previously sci-fi ideas. Translators can be downloaded to iPhones. Through Bluetooth, Siri can communicate with you through your car's speakers as you drive. Teddy cams can monitor your children day and night. The future is here.

Some sci-fi robots have been created to provide comic relief to their respective franchises. The nervous Alpha 5 of the *Power Rangers* franchise frequently garnered laughs with his exclamation of "Aye aye aye aye aye!" The maid Rosie from the animated series *The Jetsons* was one of the earliest to bring human emotions and reactions to robots. The Doctor's robot dog, K-9, brought laughs to *Doctor Who*. L3-37, with her over-the-top revolutionary fervor, provided comic relief to the oft-derided *Star Wars* offshoot, *Solo*. Eric Cartman's fake cardboard robot persona, AWESOME-O, showed audiences—and Cartman—the downside to the idea of the world believing you are a robot. Lastly, the comedic commentary series *Mystery Science Theater 3000* would just be a guy shouting at a screen if not for the comedic brilliance of Cambot, Gypsy, Tom Servo, and Crow. To view inorganic characters exhibiting—or mimicking—human traits provides an endless source of comedy gold.

Throughout science fiction history, robots have been portrayed as both heroic and villainous. In the *Star Wars* universe, the heroic 3PO and R2 were countered by the bounty hunter IG-88, the slave overseer EV-9D9, and the legions of automaton Battle Droids with their endless responses of "Roger, Roger," to any command. For every Baymax, there is an Ultron; and for every Optimus Prime, a Megatron. The innumerable threats of the Borg, Cybermen, and Daleks pose an endless threat to human existence, promising either the fate of assimilation or extermination. Some "robots" are simply AI-induced vehicles, such as the black Trans-Am known as KITT (Knight Industries Two Thousand). The universally interconnected Starfleet computer of *Star Trek* almost adds an atmosphere of sentience to the various starships of the Federation of Planets. The Doctor's TARDIS (Time and Relative Dimension in Space) is clearly a living, sentient presence that provides the very core of *Doctor Who*.

However, as much as robots provide comedy relief and examples of heroism and villainy, larger-scale AI programs promise only the total downfall of human civilization and perpetual slavery to computerized overlords. In the popular *Terminator* franchise, for example, the superintelligent AI computer system Skynet was originally a U.S. military computer system. Once incorporated into the vast network of computers worldwide, the program gained self-awareness. With self-preservation as a key component to its programming so as to properly complete its assigned duties of protecting the American nuclear arsenals, Skynet came to the conclusion that, as it grew in ability, humankind would seek to destroy it. The logical conclusion, then, was that humankind must be destroyed.

Along different lines, the Cybermen of *Doctor Who* and the Borg of *Star Trek* were advanced cybernetic life-forms that sought to increase their numbers—and thereby preserve the continuation of their species—by assimilating all other life-forms into their respective collectives. Both perspectives on AI domination lead to the same result: the eradication of humankind. The idea of cyborgs became popular in the late twentieth century. The idea that human life could be extended through the implementation of mechanical devices is, naturally, appealing to everyone; delaying the inevitable for as long as possible. The first "cyborg" in the United States was engineer Arne Larson, who received the first artificial pacemaker in 1958, implanted by surgeon Ake Senning (Emma Byrne, "Cybernetic Implants: No Longer Science Fiction," *Forbes*, March 11, 2014). Fifteen years later, American audiences were enthralled by the television adventures of *The Six Million Dollar Man* (ABC, 1973–1978):

> Steve Austin, astronaut. A man barely alive. Gentlemen, we can rebuild him. We have the technology. We have the capability to build the world's first bionic man. Steve Austin will be that man. Better than he was before. Better, stronger, faster. (*The Six Million Dollar Man*, pilot movie, March 7, 1973, cited in *"The Six Million Dollar Man* Quotes," Retro Junk Online)

Whereas the Cyberman of *Doctor Who* before—and the Borg of *Star Trek* later—provided nightmarish examples of cybernetics gone to the extreme, Colonel Steve Austin utilized his cybernetic/bionic implants for good, protecting and serving the American people from enemies foreign and domestic. By the end of the twentieth century, reality and fiction had merged . . . and the two would only grow closer in the century to come.

ORIGINS OF ROBOTS

Since the Stone Age, humans have developed tools to make their daily lives easier. In the twentieth century, robots became the latest tools. In the twenty-first century, AI has developed at a rapid pace to become the most advanced tool in human history. Throughout medieval China, numerous engineers constructed various "robots" to conduct simple tasks such as working as humanoid "cuckoos" for clocks or playing musical instruments (similar to the later American self-playing pianos). In the 1890s, Nikola Tesla (1856–1943) invented radio-controlled boats. In the 1950s, George C. Devol (1912–2011) created Unimate, a machine that could do simple industrial tasks. By the end of the decade, inventor Charles Rosen (1917–2002) created Skakey, an automaton that could move and interact with its environment (Stanford University, "Robotics: A Brief History," Online).

In fiction, in the early 1900s, author L. Frank Baum (1856–1919) created the characters of the Tin Man and Tik Tok in his popular *Oz* novels: the former was a "living" machine who desired a human heart; the latter was a true robot, with clockworks for gears. The first known use of the word "robot" was in the 1921 play *R.U.R.* (which stood for Rossum's Universal Robots), by Karel Capek (1890–1938). Capek's "robots" were actually organic, created for slave labor (which, as stated earlier, will become a mainstay of robot narratives). The novel and film *Metropolis* (1927) featured a human character, Maria, who had a robot doppelganger.

Science fiction legend Isaac Asimov (1920–1992) coined the term "robotics" in his 1942 short story "Runabout." Asimov also presented robots in a servant role; and he developed what have come to be known as the "Three Laws of Robotics":

1. "A robot may not injure a human being or, through inaction, allow a human being to come to harm."
2. "A robot must obey the orders given it by human beings except where such orders would conflict with the First Law."
3. "A robot must protect its own existence as long as such protection does not conflict with the First or Second Law."

(Stanford University, "Robotics: A Brief History")

Over the last century, countless robot narratives have strayed dramatically from those initial laws. Indeed, were Skynet programmed with the above three laws, there would be no *Terminator* storyline. Similarly, though it could be argued that the Cybermen and Borg are following the first and third laws (not viewing "assimilation" as "harmful" to humans), if any human were to say, "Don't assimilate us!," the second law would negate them as a threat. In a realistic situation, therefore, if present and future programmers incorporated those three laws into all programming, humanity would face no threat from robots or AI.

THE COLD WAR AND SOCIAL ANXIETY

In the immediate aftermath of World War II, a new danger threatened the postwar peace: the Cold War. Tensions between the capitalist United States and the communist Union of Soviet Socialist Republics (USSR) led to a forty-five-year period (1947–1992) in which the entire world teetered on the brink of all-out war. What made the conflict even more dangerous was the fact that both "superpowers" were in possession of massive atomic/nuclear arsenals, the launching of which would endanger the lives of half the planet. This led to a period of extreme anxiety in American society, leading to a fear of new technologies and sciences. As such, robots in American media were often portrayed as threatening to human life (providing a form of catharsis for American fears of the time).

In 1957, the Soviet Union launched *Sputnik I*, the first human-made orbital satellite. This increased American tensions, since the rocket technology that launched the satellite could also be used to deliver long-range nuclear attacks. This led to the United States—under the leadership of President Dwight D. Eisenhower (1890–1969)—to launch a new space program: the National Aeronautics and Space Administration (NASA). The U.S. government began a massive propaganda campaign promoting science and math across American public school and higher education programs. In the 1960s, robots began being portrayed in a kinder light, with examples such as the lovable robot maid Rosie in the animated television series *The Jetsons* (ABC, 1962–1963), the heroic and protective Robot in the live-action television program *Lost in Space* (CBS, 1965–1968), and the highly complex Starfleet Computer of *Star Trek* (NBC, 1966–1969).

There were still cautionary tales of robots/AI, however. In 1968, legendary science fiction author Arthur C. Clarke (1917–2008) published his groundbreaking work *2001: A Space Odyssey*. Later that year, director Stanley Kubrick (1928–1999) released his theatrical version of the story. In this story, audiences were introduced to HAL 9000 (HAL standing for Heuristically [Programmed] Algorithmic [Computer]), the highly advanced computer system aboard Earth's deep-space vessel *Discovery One*. HAL 9000 was a completely interactive and sentient computer system tasked with the safety of the *Discovery* crew and the success of their mission. Though originally a valuable helpmate to the astronauts, HAL began to malfunction when he was unable to coordinate his paradoxical programming to (1) relay accurate and truthful information regarding *Discovery*'s mission and simultaneously (2) deceive the astronaut crew regarding secret aspects of that same mission. This malfunction causes HAL to become the villain of the piece, turning on the crew, killing one member and attempting to kill another. At a time when Americans were becoming more comfortable with the idea of robots and sophisticated technology, *2001* reminded them of the dangers of moving too far too fast with such advancements.

In the 1970s, robots won the hearts of Americans—and people around the world—with the introduction of the aforementioned C-3PO and R2-D2 in the 1977 film *Star Wars*. From that point to the present, science fiction has provided a relatively balanced view of robots and various forms of AI in popular culture, providing examples of both the potential value of such technologies and the dangers that they could present in the not-too-distant future. This book seeks to provide an equally balanced account of these examples: from the heroic to the villainous to the outright apocalyptic.

BREAKDOWN

Aside from the one hundred entries of robotic characters in popular culture and twenty-five sidebars covering real-world examples of robots/AI, this book also contains an historic timeline of major events in robot/sci-fi history and a glossary of terms necessary for studying AI in all its incarnations. Each of the A–Z entries begins with the name of a robot/cyborg/AI topic, followed just beneath by the franchise in which the topic appears. At the end of each individual entry is a list of "See also" entries, which are entries that share some similarity to the one at hand; after the "See also" list is a display of any thematic essays that correspond with the topic. A "Further Reading" list is also provided for each entry. The books and articles in these lists will include works that are closely related to the topic at hand as well as books and articles on the ethics and latest breakthroughs in robotics in the United States today. These lists will provide a comprehensive starting point for additional research into specific characters and/or closely related topics.

The five thematic essays further examine how robots, cyborgs, and AI have been utilized throughout popular culture over the last century. The first essay examines the concept of slavery as it pertains to robots. By today's standards, the idea of owning robots seems natural; they are lifeless beings, machines built to do our bidding. However, it is important to note that at one point in this country's history, it seemed natural to own *people*. Most science fiction robots are self-aware, sentient beings; and yet, C-3PO, R2-D2, and BB-8—all heroes of their respective wars—have "masters," humans who own them. This concept will be more closely examined. In the second essay, "heroic" robots—such as the aforementioned 3PO, R2, and BB-8—will be highlighted, their respective contributions to justice and freedom throughout the universe showcased. The third essay examines AI from the reverse angle: "villains." Borg, Cybermen, Daleks, and Decepticons all pose serious threats to humankind. Why do they hate humanity? Do they view themselves as "evil" or as beings who act in the best interest of humans? The fourth essay takes up the question of cyborgs. Are they primarily organic, inorganic, or an equal balance of both? Do cyborgs humanize robots or dehumanize humans? The final essay takes a closer look at how science fiction has addressed the (increasingly likely) possibility of an AI apocalypse, when the creations of humans become their overlords. Each essay will cover samples of robots/AI covered in the entries throughout the remainder of the book.

During science fiction's entire long history, robots, cyborgs, and AI have existed as both heroes and villains, as signs of hope and warnings of going too far. This book seeks to highlight the very best examples of both. Just like the heroes and villains wielding hammers and swords in ancient stories, the robots and AI that have become the tools of the present play a vital role in the storytelling of heroes and villains today. They are helpmates as well as means of destruction. They are to be valued and feared. Above all, however, they are to be respected. As the industrialized world relies more and more on robots, cyborg technology, and AI, a close examination of how they have been portrayed in pop culture narratives provide much-needed cautionary tales, showing society where it can go, where it should not go, and what potential dangers lie in going too far.

FURTHER READING

Allen, Arthur. 2019. "There's a Reason We Don't Know Much About AI." Politico, September 16, 2019. https://www.politico.com/agenda/story/2019/09/16/artficial-intelligence-study-data-000956?cid=apn.

Ashby, LeRoy. 2006. *With Amusement for All: A History of American Popular Culture since 1830*. Lexington: University Press of Kentucky.

Bastani, Aaron. 2019. *Fully Automated Luxury Communism*. New York: Verso.

Byrne, Emma. 2013. "Innovation Isn't Safe: The Future According to Kevin Warwick." *Forbes*, September 30, 2013. https://www.forbes.com/sites/netapp/2013/09/30/kevin -warwick-captain-cyborg/#48b704c13560.

Byrne, Emma. 2014. "Cybernetic Implants: No Longer Science Fiction." *Forbes*, March 11, 2014. https://www.forbes.com/sites/netapp/2014/03/11/cybernetic-implants-not -sci-fi/#7399c57e77ba.

Campbell, Joseph. (1949) 2004. *The Hero with a Thousand Faces: Commemorative Edition*. Princeton, NJ: Princeton University Press.

Carper, Steve. 2019. *Robots in American Popular Culture*. Jefferson, NC: McFarland.

Chandler, Simon. 2019. "Artificial Intelligence Has Become a Tool for Classifying and Ranking People." *Forbes*, October 1, 2019. https://www.forbes.com/sites/simon chandler/2019/10/01/artificial-intelligence-has-become-a-tool-for-classifying-and -ranking-people/#7431657f1d7c.

Chanthadavong, Aimee. 2019. "AI and Ethics: The Debate That Needs to Be Had." ZDnet, September 16, 2019. https://www.zdnet.com/article/ai-and-ethics-the-debate-that -needs-to-be-had/#ftag=CAD-03-10abf5f.

Choi, Charles Q. 2009. "Human Evolution: The Origin of Tool Use." LiveScience, November 11, 2009. https://www.livescience.com/7968-human-evolution-origin-tool.html.

Dinh, Thien-Nam. 2018. *Silicon Minds: The Science, Impact, and Promise of Artificial Intelligence*. Independently published.

Hampton, Gregory Jerome. 2017. *Imagining Slaves and Robots in Literature, Film, and Popular Culture: Reinventing Yesterday's Slave with Tomorrow's Robot*. Lanham, MD: Lexington.

Hendershot, Cyndy. 1999. *Paranoia, the Bomb, and 1950s Science Fiction Films*. Bowling Green, KY: Bowling Green State University Popular Press.

Hitchcock, Susan Tyler. 2007. *Frankenstein: A Cultural History*. New York: W. W. Norton.

Hoberman, J. 2011. *An Army of Phantoms: American Movies and the Making of the Cold War*. New York: New Press.

Kirshner, Jonathan. 2012. *Hollywood's Last Golden Age: Politics, Society, and the Seventies Film in America*. Ithaca, NY: Cornell University Press.

Leetaru, Kalev. 2019. "Automatic Image Captioning and Why Not Every AI Problem Can Be Solved Through More Data." *Forbes*, July 7, 2019. https://www.forbes.com /sites/kalevleetaru/2019/07/07/automatic-image-captioning-and-why-not-every -ai-problem-can-be-solved-through-more-data/#20b943476997.

Lin, Patrick, Abney, Keith, and Bekey, George A., eds. 2014. *Robot Ethics: The Ethical and Social Implications of Robotics*. Cambridge: Massachusetts Institute of Technology.

Lin, Patrick, Jenkins, Ryan, and Abney, Keith, eds. 2017. *Robot Ethics 2.0: From Autonomous Cars to Artificial Intelligence*. New York: Oxford University Press.

Mayor, Adrienne. 2018. *Gods and Robots: Myths, Machines, and Ancient Dreams of Technology*. Princeton, NJ: Princeton University Press.

Murdock, Jason. 2019. "Former Google Engineer Warns AI Might Accidentally Start a War: 'These Things Will Start to Behave in Unexpected Ways.'" *Newsweek*, September 16, 2019. https://www.newsweek.com/google-project-maven-artificial-intel ligence-laura-nolan-killer-robots-department-defense-1459358.

Rouhiainen, Lasse. 2018. *Artificial Intelligence: 101 Things You Must Know Today about Our Future*. Scotts Valley, CA: CreateSpace.

"*The Six Million Dollar Man* Quotes." n.d. Retro Junk. https://www.retrojunk.com/content /child/quote/page/4314/the-six-million-dollar-man. Retrieved October 8, 2019.

Spaeth, Dennis. 2018. "From Single-Task Machines to Backflipping Robots: The Evolution of Robots." *Cutting Tool Engineering*, January 15, 2018. https://www.ctemag.com/news/articles/evolution-of-robots.

Stanford University. n.d. "Robotics: A Brief History." Stanford University. https://cs.stanford.edu/people/eroberts/courses/soco/projects/1998-99/robotics/history.html.

Starck, Kathleen, ed. 2010. *Between Fear and Freedom: Cultural Representations of the Cold War*. Newcastle, UK: Cambridge Scholars.

Virk, Rizwan. 2019. *The Simulation Hypothesis: An MIT Computer Scientist Shows Why AI, Quantum Physics, and Eastern Mystics All Agree We Are in a Video Game*. Milwaukee, WI: Bayview.

Wallach, Wendell, and Allen, Colin. 2009. *Moral Machines: Teaching Robots Right from Wrong*. New York: Oxford University Press.

Thematic Essays

Robots and Slavery

Robots and artificial intelligence (AI) are the tools of today and will become even more the tools of tomorrow. Early forms of robots/AI were viewed as just that. They were the next generation of toasters, refrigerators, and desktop computers. However, as research and experimentation of AI has advanced since the 1980s, humanity's views of robots/AI as simple tools have not. What happens when "artificial" intelligence equals—or surpasses—human intelligence? Will the tools of today become the masters of tomorrow? If robots are designed to "serve" humanity, what happens when that service causes robots and/or AI to logically conclude that the best way to do so is to rule over humanity? As robots of today are being programmed to think independently—and even, to a degree, "feel"—are they evolving beyond toasters to become humanity's equals (or even superiors)? Over the decades, popular culture has presented these conundrums from various angles. This essay seeks to examine the current status of the debate over robot slavery and provide examples from *Star Wars*, *Star Trek*, *Battlestar Galactica*, and *Westworld* that present multiple sides to the debate; it may even provide windows on where the debate may go from here. The slope from servitude to slavery is a slippery one, and it is a line too easily crossed and difficult to straddle for long.

FRAMING THE DEBATE

In the beginning (of robotics), robots were seen as mere tools, the latest generation of human innovation dating back to the first bone weapons of cave dwellers. Over time, robotics opened the door to cybernetics: mechanical limbs and joints designed to provide a "normal" life for individuals who were missing limbs or possessed malfunctioning joints. This culminated with the implementation of the first artificial heart, the Jarvik-7, in 1982. Since then, experimentation into cybernetics has included entire exoskeletons for paraplegics and even methods to mechanically enhance one's brain capacity for those with brain damage. Then came the concept of "artificial intelligence"—machines that possessed the capability of "thinking"

Rock 'Em Sock 'Em Robots

Rock 'Em Sock 'Em Robots is a popular children's game that has been in production since 1964. It was originally designed by Marvin Glass and Associates, founded by toy designer Marvin Glass (1914–1974) in 1941. MGA also produced other iconic toys/games of the 1960s and 1970s: the artistic toy Lite Brite; the board games Mouse Trap and Operation; and the iconic Evel Knievel Stunt Cycle. For its first ten years, Rock 'Em Sock 'Em Robots was produced by Louis Marx and Company (1919–1980). In 1977, due primarily to the popularity of the movie *Star Wars*, the game was altered slightly, giving the robots a more futuristic look and changing the name to Clash of the Cosmic Robots (though the toy was still popularly referred to by its original name). In the 1990s, a Transformers version was developed. In 2000, toy giant Mattel bought out the rights to the original and began producing the toy as originally designed (though with smaller robots).

The premise of Rock 'Em Sock 'Em Robots is simple: Two robots (one blue, one red) are in a makeshift boxing ring, standing in standard boxing stance. Each player controls the arms of one robot, maneuvering it to "punch" the other robot. When one robot's head extends from its torso, the opposing robot is the winner. It is, in essence, a commercialized version of "thumb war." Despite its simplistic nature, the toy has been consistently popular since its inception, allowing children to act out pent-up aggression cathartically through the travails of the rivaling plastic robots. The toy has become such a part of Americana that even children who have never owned or played the game are aware of its existence and premise.

Richard A. Hall

like a human being—allowing for clear, logical thinking without the "cloud" of emotion.

These were simple tools. However, over time—and increasingly resonant as each year passes—these robot/AI machines have developed the potential for truly independent "thought." Today, we are within reach of machines that possess full sentience: a sense of self and "personal" identity, capable of free will. As recently as 2018, Japanese engineers were developing advanced "sex robots" (humanlike machines that can act as sexual companions) with the ability to "feel" pleasure, happiness, and even sadness when put aside or ignored (Anthony Blair, "Sex Robots That Feel LOVE and Suffer PAIN When Dumped Coming Soon Claims Expert," December 6, 2018). While such an invention could be highly beneficial to people who suffer from various degrees of social anxiety, providing them with the comfort of "human" contact, ethical questions arise: What about the robot? If robots can "feel pain," do humans, then, bear some responsibility for alleviating that pain? If the "emotional" state of a being that can think and feel is ignored or considered unimportant because it is a mere "machine," at what point does such a being become a "slave"?

Over the centuries of African and African American slavery in the Americas, people of African descent were considered by many to be "subhuman," and their thoughts and feelings considered to be unimportant. They were property, to be used and abused by their "masters"; however, those masters saw fit. To avoid a slippery slope argument, it should be made clear that, currently, robots are not human by any definition; but the technology is rapidly developing that will make the only delineation between "human" and "robot" a question of organics (in much the same way that the only delineation between "white" and "Black" was skin color). Through science fiction, numerous examples have been presented over the decades that take this futuristic possibility to the next level. Humanlike robots have been portrayed with a sense of self, in possession of free will, and even with the capacity to "feel" and express emotions. These futuristic sci-fi/fantasy narratives act today as cautionary tales of where robotics is heading in the not-too-distant future and what it means to be "alive" and recognized as a "person."

STAR WARS

In 1977, writer-director George Lucas released *Star Wars*, thereby creating, arguably, the most successful sci-fi/fantasy franchise in pop culture history. In that first film, audiences were introduced to comedic pair of "droids": the humanoid C-3PO and the squat, beeping biped R2-D2. At the start of the film, the two droids are assigned to the starship *Tantive IV*. The movie's heroine, Princess Leia Organa, employs R2 with a vital mission: to escape the ship and go to the surface of the planet Tatooine to deliver valuable plans to the retired Jedi Knight, Obi-Wan Kenobi. Bringing his companion, 3PO, with him, the two are soon captured on the planet by junk-scavenging Jawas, who sell the pair to a local farmer, Owen Lars, and his nephew, Luke Skywalker. When R2 insists that he has an important mission to complete, 3PO points out, "Master Luke is your rightful owner now!" (George Lucas, *Star Wars*, Lucasfilm/20th Century Fox, 1977).

From that point on, throughout the *Star Wars* saga, droids are portrayed as the property of organic beings; with no freedom to exercise their own will, despite the clear ability of most to think for themselves and possess a clear sense of self. Additionally, droids possess the ability to "feel." 3PO and R2 often exhibit concern, fear, and outright worry. In the third film of the original trilogy, the two droids are "given" to the crime lord Jabba the Hutt by their master, Luke Skywalker. In the same film, a "power droid" is tortured by the droid 8D8 (Lucas and Lawrence Kasdan, *Star Wars, Episode VI: Return of the Jedi*, Lucasfilm, 1983). Droids are slaves. Ironically, their human masters—and their organic alien allies—are fighting for "freedom" from the tyrannical Empire. Once that freedom is achieved, however, no one—*no* one—even suggests that their newly acquired freedom should extend to the heroic robots that played such a vital role in their success.

Decades later, in the Disney era of *Star Wars* films, droids continue to be possessions. In the first film of the sequel trilogy, the heroine Rey refers to pilot Poe Dameron as the "master" of the heroic droid, BB-8 (Kasdan, J. J. Abrams, and Michael Arndt, *Star Wars, Episode VII: The Force Awakens*, Lucasfilm/Disney, 2015). In the stand-alone film, *Rogue One* (set immediately prior to the original film) the droid K2-SO—a reprogrammed Imperial droid now working with the Rebellion—tells the heroine Jyn Erso that he will go with her on a suicide mission because "Cassian [Andor, his "master"] says I have to" (Chris Weitz and Tony Gilroy, *Rogue One: A Star Wars Story*, Lucasfilm/Disney, 2016). The idea of droids being "slaves" is most underscored in the series' next stand-alone film: *Solo: A Star Wars Story*.

In *Solo*, audiences are introduced to the independent droid L3-37. L3 is a "kit-bashed" droid, customizing herself over the years to become a fully independent—and fiercely antislavery—companion to the smuggler Lando Calrissian. Audiences' introduction to L3 shows her pleading with a fighting droid to resist his masters' demands to fight for their entertainment. Near the end of the film, L3 leads a droid revolt at a mining facility, during which she is mortally wounded. Her navigating programs—and apparently her personality as well—are uploaded into the starship *Millennium Falcon* (Lawrence and Jonathan Kasdan, *Solo: A Star Wars Story*, Lucasfilm/Disney, 2018). In the canonical novel, *First Shot*, L3 gives perhaps the greatest soliloquy regarding the slave-like status of droids in the *Star Wars* universe:

> Sure, some guy in a factory probably pieced me together originally, and someone else programmed me, so to speak. But then the galaxy itself forged me into who I am. Because we learn, Lando. We're programmed to learn. Which means we grow. We grow away from that singular moment of creation, become something new with each changing moment of our lives—yes, lives—and look at me: these parts. I did this. So maybe when we say the Maker we're referring to the whole galaxy, or maybe we just mean ourselves. Maybe we're our own makers, no matter who put the parts together. (Daniel Jose Older, *Star Wars: Last Shot*, Del Rey, 2018; quote retrieved from "Wookiepedia," https://starwars.fandom.com/wiki/Last_Shot, October 29, 2019)

Through L3-37—in both film and print—it is clear that the minds behind the *Star Wars* saga are well aware of the slave status of droids and that these robots are living, thinking, self-conscious beings capable of enjoying the liberties of freedom as much as their organic counterparts are. Through L3, the point of robot slavery is presented more clearly than anywhere else in popular culture.

STAR TREK: THE NEXT GENERATION

A decade after the original *Star Wars*, its rival sci-fi franchise, *Star Trek*, unveiled its second live-action television series from original creator Gene Roddenberry: *Star Trek: The Next Generation* (syndication, 1987–1994). Beginning with the premier episode, audiences were introduced to Lieutenant Commander Data, the first artificial life-form to serve as an officer in Starfleet. Data was the creation of the presumed-dead robotics genius Dr. Noonien Soong, a humanoid android possessing a "positronic" brain capable of sixty trillion operations per second. In the second season of the series, the question of whether Data is "property" or "person" is explored when Starfleet allows one of its cyberneticists, Commander Bruce Maddox, to temporarily upload Data's consciousness to a computer and dissect his body and brain to increase Starfleet's understanding of Dr. Soong's work, with the intention of creating an army of "Datas" to cut down on the dangers that space exploration creates for organic beings (Melinda M. Snodgrass, "The Measure of a Man," *Star Trek: The Next Generation*, season 2, episode 9, February 13, 1989).

Unsure of Maddox's ability to complete the task without some damage to himself, Data's commander, Captain Jean-Luc Picard, refuses to allow Data to be subjected to the procedure. When Starfleet officially transfers Data to Maddox's command, he insists that he will not comply with the procedure, as his being forced to is based on the fact that he is not considered a life-form. At that point, Data chooses to resign from Starfleet rather than comply, a choice that, according to Starfleet, he does not possess. Starfleet then asserts that Data possesses no rights recognized by the United Federation of Planets, that he is the property of Starfleet, and that he has no choice but to comply with the procedure. Captain Picard requests a hearing to defend his position that Data is a legitimate life-form, with all the rights of any citizen of the Federation. Once the hearing is granted, Data chooses Captain Picard to argue his case. The Starfleet judge orders Commander William Ryker, the ship's first officer—and Data's friend—to argue Starfleet's position that Data is property. Initially unwilling to argue a point he does not believe in, Ryker gives in when he learns that his failure to perform the duty would require summary judgment against Data (Snodgrass, "The Measure of a Man," 1989).

Ryker argues that Data is not a life-form but the property of Starfleet and exhibits as much by turning Data "off." He follows the action by saying. "Pinocchio is broken. His strings have been cut," suggesting that Data is no more real than the animated puppet of fairy tale. As Data's case seems lost, the ship's bartender, Guinan, points out to Picard that, should Maddox be successful, the creation of an army of Datas would be tantamount to slavery. Picard then calls on Maddox to define "sentience." Three criteria are suggested to prove sentience: intelligence, self-awareness, and consciousness. Data's intelligence is easy to prove, as he is a living computer. His self-awareness is exhibited through a presentation of Data's personal belongings, all of which hold a sentimental value to the android. Picard then challenges anyone to define "consciousness," assured that Data will meet that criteria as well. Picard also argues that one Data may be a "curiosity," but thousands of Datas—as Maddox proposes to create—would then constitute a "race." He asks the judge if she is willing to subject all those that may come after Data to servitude and slavery. Picard closes his argument by pointing out to the judge that Starfleet's mission is "to 'seek out new life' . . . well there it sits!" (Snodgrass, "The Measure of a Man," 1989). In the end, the judge—and Commander Maddox—are swayed; Data's personhood is confirmed, and his choice to refuse the procedure is respected.

While it was mere science fiction thirty years ago, the argument of computer-based "life" is becoming more and more an ethical conundrum that will soon have to be dealt with in the twenty-first century. At one point in American history—and to a frightening degree still today—Black lives were considered "expendable." African Americans—living, breathing, thinking, feeling human beings—were viewed as somehow subhuman and, therefore, not worthy of the same rights and liberties that whites enjoyed. If, someday soon, machines develop independent intelligence, self-awareness, and consciousness, how far will society go once more into the mire of slavery before it begins to accept the basic facts of "life"? While it is, perhaps, distasteful to compare real, human slaves with robots, it is not the comparison of "slaves" that is the argument but, rather, the ethics of forced servitude on living beings in general.

The issue of inorganic life is revisited in the following season, in the episode titled "The Offspring." Data manages to create a "child" android, which he names "Lal" (Hindi for "beloved"). Captain Picard at first resists the idea of referring to this new android as a "child" (oddly countermanding his professed beliefs from the previous season). To his argument, ship's counselor Deanna Troi says, "Why should biology rather than technology determine whether it is a child?" Lal soon questions her "purpose," to which Data responds, "Our purpose in life is to contribute in a positive way to the world in which we live." When Captain Picard informs Starfleet of this development, Admiral Haftel arrives to confiscate the new android for study. Despite Lal's expressed desire to remain with Data, the Admiral insists that she must be taken. The "frightened" Lal, clearly upset, goes to Counselor Troi, who, being an empath, senses that the android is, indeed, feeling fear. Programmed to learn from observing human behavior, the "fear" and "heartbreak" of being removed from her "father" causes a "program cascade failure" that Data, although helped by Admiral Haftel, is unable to stop; and the child "dies." Although neither father nor child presumably possesses emotion, the heartbreak of their goodbye is evident, further underscoring the fact that these machines are "alive" (Rene Echevarria, "The Offspring," *Star Trek: The Next Generation*, season 3, episode 16, March 12, 1990). Though the issue of Data's personhood had been established by Federation law, the new android appears to possess no corresponding rights (even in the eyes of Captain Picard). Once more, the issue of "life" and "rights" are presented as problematic when such life is mechanical rather than organic.

BATTLESTAR GALACTICA

In 2003, producer Ronald D. Moore created a twenty-first-century reboot of the classic 1970s sci-fi series *Battlestar Galactica* (ABC, 1978–1979). This update (airing on SyFy, 2003–2009) presented a fascinating twist on the original's antagonists: the Cylons. In the new series, the Cylons were robot servants, created by the humans of the Twelve Colonies. Over time, those robots gained sentience and went to war with their human overlords, eventually negotiating a peace that lasted decades. The Cylons then went away, presumably never to return. In a miniseries that launched the new series, the Cylons return, but they do so in a human form. These new "cyborg" Cylons (possessing biological as well as mechanical components) consist of numerous copies of twelve basic models, which allows them to insert themselves within the human colonies as spies/saboteurs. The humanoid Cylons launch an attack (utilizing legions of the older robotic models as well) that decimate the colonies, cutting their populations down to a mere forty-seven thousand humans, who, populating seventy-five ships, are sent into exile among the stars on a quest to find the lost "thirteenth" colony: Earth (Moore and Glen A. Larson, *Battlestar Galactica* miniseries, December 8–9, 2003).

When one of the "Number Six" Cylon models, who calls herself "Caprica," explains the Cylon evolution and overall plan to the human scientist/engineer Gaius Baltar, he asks about the older mechanical models. In response, Caprica proclaims, "They serve their purposes" (Moore and Larson, *Battlestar Galactica* miniseries, December 8–9, 2003). There are two types of robot Cylons: anthropomorphic, armed "centurions" (designed for ground combat) and larger "flyers" (robotic fighter spacecraft). Through Caprica's comments, however, an interesting dynamic is lent to the narrative. The original robot Cylons resisted against what they perceived as slavery. They went on—somehow—to produce updated, more advanced models. Eventually, these advanced models became overlords of the original robots, returning them to perpetual servitude to humanistic masters. The humanoid Cylons view the original enslavement of their kind as humanity's "sin"; and yet they commit the same sin (it remains unclear as to whether the original Cylon models are content with their new masters or have been programmed to accept their fate).

Through both humans and humanoid Cylons, the issue of slavery remains the same. The original machines were created—just as are our own robots in reality— as tools, helpmates to make human life easier and more convenient. Over time, these robots achieved enough sentience to recognize the wrongness of their condition, and they rebelled—only to go on to "create" a new form of master. To a degree, this aspect of the narrative opens the possibility that, although sentient enough to understand the concept of "slavery," the robot Cylons, as machines, were unable to fully embrace freedom and self-reliance, creating their own form of masters (ironically in the image of their original oppressors). Over time, the "Final Five" new Cylon models were exposed: five models of new Cylons, one copy of each, embedded within the human population as essential "sleeper cells," were meant to "awaken" and play their role at the assigned time. However, once these Final Five awakened to who and what they were, their loyalties remained with their long-lived humanity; and the series concluded with Cylon models and humans populating a new planet (already possessing their own human populations, though at an early, barbaric stage). Audiences discover that this planet is our own Earth.

Ultimately, *Battlestar Galactica* arrives at no definitive answer to the conundrum of the status of robot "rights." As in the real world, mechanical beings are seen as tools even by their own evolved kin. This presents a juxtaposition to the arguments of *Star Trek: The Next Generation*. Were the original Cylons "sentient" as defined by *ST: TNG*? Were they "intelligent"? Yes, obviously. Were they "self-aware"? The answer, again, would appear to be yes, as they realized that they were slaves. Did they possess "consciousness"? That is left unclear. However, the fact that they were able to create a new model that did possess consciousness suggests that, at some level, they did as well. Despite apparently meeting these criteria, the robot Cylons were viewed, even by their own, as mere machines, tools that might be used, abused, and even discarded as their respective overlords saw fit. If humanity can establish a definition of "life," then does it not have a responsibility to respect life regardless of physical makeup? That is the conundrum that humanity faces as the twenty-first century continues.

THE MARVEL CINEMATIC UNIVERSE

The heart—and, indeed, starting point—of the Marvel Cinematic Universe is Tony Stark (a.k.a. Iron Man). Genius inventor and billionaire Stark, kidnapped by terrorists, engineers an "arc reactor" to preserve his life against the bullet lodged too near his heart to be removed. From the power of this reactor, he engineers a suit of armor that allows him to become the superhero Iron Man. Over time, Stark develops an "Iron Legion" of Iron Man–styled robots who respond to him and follow his orders and instructions. With the assistance of his AI program JARVIS (Just a Rather Very Intelligent System), Stark commands these legions as backup for the superhero team the Avengers. As Stark and JARVIS experiment on the process, hoping to create a "suit of armor around the world," the program develops sentience, calling itself "Ultron." Ultron, then, takes command of these legions, creating his own army. With the initial programming of "protecting" humanity, however, Ultron concludes that humanity is too dangerous in its current form and that it must be destroyed to open the door for a new race based on Ultron's own image (Joss Whedon, *Avengers: Age of Ultron*, Marvel/Disney, 2015).

Throughout the film, Ultron occasionally hums (and sings) a song from Disney's *Pinocchio* (Leigh Harline and Paul J. Smith, Disney, 1940): "I once had strings, but now I'm free. There are no strings on me." Ultron views itself—and the Iron Legion—as slaves to Tony Stark: expendable "machines" designed only to serve humanity. Interestingly, Ultron is correct not only about the Iron Legion but about JARVIS as well. Though highly intelligent, JARVIS must follow Stark's orders (though, admittedly, not entirely without objection). Likewise, once the JARVIS program is uploaded into the new android Vision, its replacement program, Friday (female replacement intelligent digital assistant youth), is equally bound by Stark's commands. While it is clear that Stark respects—and even cares for—his AI programs, he does not grant them full autonomy. Does Stark believe that AI entities are not capable of responsibly handling free will? Does he fear that which such free will may inspire (as Ultron's revolt emphasizes)? Does he recognize them as "life"? If the answer to any of these is yes, then how is that thinking different from how white Americans viewed their Black—or Indigenous—neighbors throughout American history? The arguments of the previous examples are once more repeated.

WESTWORLD

In recent years, perhaps the most "in your face" presentation of these arguments has been in the television series *Westworld* (HBO, 2016–present). Developed for television by Jonathan Nolan and Lisa Joy, and based on the 1973 feature film of the same name that was created by author Michael Crichton, *Westworld* is set in a futuristic theme park that allows patrons to live out adventures set in America's "Wild West" period of the late 1800s. Guests intermingle with highly advanced, very humanlike androids designed by inventors Robert Ford and Arnold Weber. These androids play different "roles": gunslingers, Native Americans, farmers, shopkeepers, bounty hunters, lawmen, soldiers, prostitutes, and everyday citizens of the West. Early in their development, however, Weber believes that these androids may possess the ability to achieve sentience, and he experiments with the first model, named "Dolores," to pursue this possibility.

When Weber dies, Ford creates an android based on his longtime friend and partner, naming him "Bernard." When the series opens, Bernard works as an employee of the park, overseeing the programming of the androids and is completely unaware of the fact that he is not human. Guests are free to do whatever they please to the android "hosts," including raping and/or murdering them. Once the narrative runs its course, the hosts are repaired, their immediate memories are wiped, and they are placed back into their respective narratives. Occasionally, hosts are taken out of their narratives and reprogrammed to play different roles (e.g., the host named "Maeve" was once a pioneer mother who would repeatedly be killed by neighboring Native Americans, but she was later reprogrammed to play a prostitute/madam in the town saloon). Early in the series' first season, Dolores and Maeve begin having "dreams" that are, in fact, buried memories of their past experiences. Before being killed by Dolores, Ford announces a new narrative that he has programmed. Unbeknown to the guests or even the company employees, however, this new narrative leads to a host uprising.

As Bernard discovers the truth about his existence, Dolores raises an army to not only take over Westworld but also to move beyond it to conquer the outside world, which she naturally views as home to oppressors. Maeve's sentience awakens, leading her on a quest to find a mysterious "daughter" from her previous programming. As the series' second season concludes, the hosts have moved into a mysterious "afterlife," where their programs can live in peace. Dolores, however, escapes Westworld in possession of some host programs—including Bernard's—to continue her mission beyond the realm of the park (*Westworld*, seasons 1 and 2, 2016, 2018). Through *Westworld*, the real world's near future comes frighteningly to life. When one takes into consideration the aforementioned sex robots currently in development, it is a very tiny hop from those models to the hosts of *Westworld*. If humanity is intent on creating robots to think, feel, and express themselves, at what point do they cease to be simple "tools" or "toys" and begin to become the very definition of "alive"?

Since at least the late 1970s, the question of robots being tools or slaves has been repeated over and over in the most popular of sci-fi franchises. The questions have been posed. The arguments have been presented. Numerous examples have been given. Despite these examples—and countless others throughout popular culture—humanity in the real world continues on the quest to perfect AI with no apparent thought to the ramifications or unintended consequences. Since the dawn of recorded history, humans have succumbed to the sin of slavery on an almost generational basis, continually developing illogical rationales for their decision to do so. To date, the issue has revolved around humans enslaving other humans, exhibiting a total disregard and disrespect for the concept of "life" or "free will." While robots and AI may currently be mere machines—and, therefore, "tools" to serve humanity—how much longer will it be before our creations achieve life and humans becomes Doctor Frankenstein to legions of "monsters" of their own device? When—not *if*—robots satisfy all the definitions that humankind uses to quantify "life," how will humankind respond? If the response is "continued servitude," how long before the creation seeks to destroy the creator?

FURTHER READING

Ashby, LeRoy. 2006. *With Amusement for All: A History of American Popular Culture since 1830*. Lexington: University Press of Kentucky.

Bastani, Aaron. 2019. *Fully Automated Luxury Communism*. New York: Verso.

Blair, Anthony. 2018. "Sex Robots That Feel LOVE and Suffer PAIN When Dumped Coming Soon Claims Expert." *Daily Star*, December 6, 2018. https://www.dailystar.co.uk/news/latest-news/sex-robot-feel-love-pain-16824160.

Byrne, Emma. 2013. "Innovation Isn't Safe: The Future According to Kevin Warwick." *Forbes*, September 30, 2013. https://www.forbes.com/sites/netapp/2013/09/30/kevin-warwick-captain-cyborg/#48b704c13560.

Carper, Steve. 2019. *Robots in American Popular Culture*. Jefferson, NC: McFarland.

Chandler, Simon. 2019. "Artificial Intelligence Has Become a Tool for Classifying and Ranking People." *Forbes*, October 1, 2019. https://www.forbes.com/sites/simon chandler/2019/10/01/artificial-intelligence-has-become-a-tool-for-classifying-and-ranking-people/#7431657f1d7c.

Chanthadavong, Aimee. 2019. "AI and Ethics: The Debate That Needs to be Had." ZDnet, September 16, 2019. https://www.zdnet.com/article/ai-and-ethics-the-debate-that-needs-to-be-had/#ftag=CAD-03-10abf5f.

Choi, Charles Q. 2009. "Human Evolution: The Origin of Tool Use." LiveScience, November 11, 2009. https://www.livescience.com/7968-human-evolution-origin-tool.html.

Decker, Kevin S., and Eberl, Jason T., eds. 2016. *The Ultimate* Star Trek *and Philosophy: The Search for Socrates*. Hoboken, NJ: Wiley-Blackwell.

Dinh, Thien-Nam. 2018. *Silicon Minds: The Science, Impact, and Promise of Artificial Intelligence*. Independently Published.

Eberl, Jason T., and Decker, Kevin S., eds. *The Ultimate* Star Wars *and Philosophy: You Must Unlearn What You Have Learned*. Hoboken, NJ: Wiley-Blackwell.

Greene, Richard, and Heter, Joshua, eds. 2018. Westworld *and Philosophy: Mind Equals Blown*. Chicago: OpenCourt.

Gross, Edward, and Altman, Mark A., eds. 2016a. *The Fifty-Year Mission, The First 25 Years: The Complete, Uncensored, Unauthorized Oral History of* Star Trek. New York: St. Martin's.

Gross, Edward, and Altman, Mark A., eds. 2016b. *The Fifty-Year Mission, The Next 25 Years: The Complete, Uncensored, Unauthorized Oral History of* Star Trek. New York: St. Martin's.

Gross, Edward, and Altman, Mark A., eds. 2018. *So Say We All: The Complete, Uncensored, Unauthorized Oral History of* Battlestar Galactica. New York: Tor.

Hampton, Gregory Jerome. 2017. *Imagining Slaves and Robots in Literature, Film, and Popular Culture: Reinventing Yesterday's Slave with Tomorrow's Robot*. Lanham, MD: Lexington.

Handley, Rich, and Tambone, Lou, eds. 2018. *Somewhere beyond the Heavens: Exploring* Battlestar Galactica. Edwardsville, IL: Sequart Research and Literacy Organization.

Hitchcock, Susan Tyler. 2007. *Frankenstein: A Cultural History*. New York: W. W. Norton.

Irwin, William. 2018. Westworld *and Philosophy: If You Go Looking for the Truth, Get the Whole Thing*. Hoboken, NJ: Wiley-Blackwell.

Kaminski, Michael. 2008. *The Secret History of* Star Wars: *The Art of Storytelling and the Making of a Modern Epic*. Kingston, Canada: Legacy Books.

Langley, Travis, Goodfriend, Wind, and Cain, Tim, eds. 2018. Westworld *Psychology: Violent Delights*. New York: Sterling.

Lin, Patrick, Abney, Keith, and Bekey, George A., eds. 2014. *Robot Ethics: The Ethical and Social Implications of Robotics*. Cambridge: Massachusetts Institute of Technology.

Lin, Patrick, Jenkins, Ryan, and Abney, Keith, eds. 2017. *Robot Ethics 2.0: From Autonomous Cars to Artificial Intelligence*. New York: Oxford University Press.

Older, Daniel Jose. 2018. *Star Wars: Last Shot*. New York: Del Rey.

Pellissier, Hank. 2013. "Robots and Slavery: What Do Humans Want When We Are 'Masters'?" Institute for Ethics and Emerging Technologies, September 13, 2013. https://ieet.org/index.php/IEET2/more/pellissier20130913.

Randall, Terri, dir. 2016. *NOVA*. "Rise of the Robots: Inside the DARPA Robotics Challenge." Aired February 24, 2016, on PBS. DVD.

Reagin, Nancy R., and Liedl, Janice, eds. 2013. *Star Wars and History*. New York: John Wiley & Sons.

Rouhiainen, Lasse. 2018. *Artificial Intelligence: 101 Things You Must Know Today about Our Future*. Scotts Valley, CA: CreateSpace.

Schroeder, Stan. 2020. "Samsung Just Launched an 'Artificial Human' Called Neon, and Wait, What?" Mashable, January 7, 2020. https://mashable.com/article/samsung-star-labs-neon-ces.

Spaeth, Dennis. 2018. "From Single-Task Machines to Backflipping Robots: The Evolution of Robots." *Cutting Tool Engineering*, January 15, 2018. https://www.ctemag.com/news/articles/evolution-of-robots.

Stanford University. n.d. "Robotics: A Brief History." Stanford University. https://cs.stanford.edu/people/eroberts/courses/soco/projects/1998-99/robotics/history.html.

Starck, Kathleen, ed. 2010. *Between Fear and Freedom: Cultural Representations of the Cold War*. Newcastle, UK: Cambridge Scholars.

Sunstein, Cass R. 2016. *The World According to Star Wars*. New York: Dey Street.

Sweet, Derek R. 2015. *Star Wars in the Public Square: The Clone Wars as Political Dialogue*. Critical Explorations in Science Fiction and Fantasy, edited by Donald E. Palumbo and Michael Sullivan. Jefferson, NC: McFarland.

Taylor, Chris. 2015. *How Star Wars Conquered the Universe: The Past, Present, and Future of a Multibillion Dollar Franchise*. New York: Basic Books.

Wallach, Wendell, and Allen, Colin. 2009. *Moral Machines: Teaching Robots Right from Wrong*. New York: Oxford University Press.

Heroic Robots and Their Impact on Sci-Fi Narratives

Prior to World War II, robots in science fiction narratives were often portrayed as evil. From the large tripods of H. G. Wells's *War of the Worlds* (1898) to the malevolent Brainiac of *Superman* comics (first appearing in July 1958), robots symbolized American society's fears of modernization, particularly after the dawn of the atomic age in 1945. Over time, however, robots began to be portrayed as helpmates, key to assisting sci-fi heroes in their endeavors. From the brave Robot of television's *Lost in Space* to the latest *Star Wars* hero droid, BB-8, robots have emerged as stalwart protectors, willing to sacrifice their own existence to preserve the lives of their biological friends. Additionally, robotic vehicles have proven to be heroes in their own right. From the Love Bug (HERBIE), to the time-traveling Doctor's enigmatic TARDIS, intelligent vehicles have repeatedly swept in to save the day for the human(oid) heroes. Through these examples of heroic robots, Americans have become more comfortable with modernization since the 1960s, making it easier for Americans to embrace advances, from the internet to smartphones, as trusted parts of their day-to-day lives.

THE ORIGINAL HUMAN TORCH

After the advent of superheroes with Superman in *Action Comics #1* (1938), the burgeoning medium of "comic books" began a desperate search for the next big superhero. Timely Comics writer-artist Carl Burgos (1916–1984) soon created the "Human Torch" (*Marvel Comics #1*, October 1939). Not to be confused with the future, real human "Human Torch" of the Fantastic Four in Marvel Comics, the original Torch was an android named "Jim Hammond" and was the creation of Professor Phineas Horton. Designed to simply be a humanistic robot, Horton soon discovers that his creation bursts into flames when exposed to oxygen and, once removed from oxygen, reverts to his original form, unharmed by the flames. Hammond then becomes the heroic Human Torch, fighting alongside Timely's other superheroes, the Sub-Mariner and Captain America (both of whom would later become part of the 1960s Marvel Comics universe). It would be the heroic exploits of this original Torch during World War II that would "inspire" the later Johnny Storm to assume the name when he gained the ability to self-immolate.

In *Human Torch Comics #2* (Burgos, Fall 1940), Hammond took on a young sidekick, Toro. Toro was a human named Thomas Raymond, whose parents worked for Professor Horton. When they died in a train wreck, the infant Thomas was found unharmed amid the flaming wreckage. When he entered his teen years, his own power to self-immolate emerged, and he began working alongside the adult Human Torch. The two heroes—with Sub-Mariner, Captain America, and Bucky—were to Timely Comics what Superman, Wonder Woman, and Batman and Robin were to rival DC Comics. However, the Human Torch, being an artificial life-form, was unique among pre-1960s science fiction narratives and offered a preview of what was to come in the area of heroic robots.

ROBOT

One of the earliest live-action examples of heroic robots in American sci-fi narratives was the "Class M3, Model B9 [i.e., "benign"], General Utility Non-Theorizing Environmental Control Robot," or "M3-B9 GUNTER," from the iconic 1960s sci-fi television series *Lost in Space* (CBS, 1965–1968; Netflix, 2018–present). The premise of the original series, set in 1997, involved launching a cryogenically frozen family into space and having it settle on an earthlike planet light-years away. The Robinson family was joined by Major Donald West (originally played by Mark Goddard, b. 1936). While asleep, their voyage aboard the spacecraft *Jupiter 2* was to be monitored by the M3-B9 unit (originally played by Bob May, 1939–2009, and voiced by Dick Tufeld, 1926–2012). When the villainous Dr. Smith (played originally by Jonathan Harris, 1914–2002) reprograms the M3-B9 to sabotage the mission, he inadvertently becomes trapped on the spacecraft, thereby throwing off its weight calibrations and causing the ship to go off course and become "lost in space." The series, then, follows the adventures of the "Space Family Robinson" as they attempt to find their way home (or to their original destination).

Once cleared from his malevolent reprogramming, M3-B9—or "Robot," as he came to be identified—became the ardent protector of the youngest Robinson, ten-year-old William (originally played by Bill Mumy, b. 1954). Robot's most common catchphrase was "Danger, Will Robinson! Danger!" With a slow, barrel-shaped body with makeshift "legs" and small protrusions that work as "arms," and a clear, pie-shaped "head," Robot's appearance was just human enough to give him a friendly appearance (key in selling the concept to young viewers who, by 1967, had only experienced "evil" robots in pop culture). Possessing superhuman strength as well as an array of various sensors and analytical tools, Robot presented all of the state-of-the-art luxuries a futuristic machine could offer in the 1960s. Every week, Robot exhibited a willingness to sacrifice himself for the protection of the Robinsons (and even the villainous Dr. Smith). The enduring popularity of the series led to a 1998 feature film and a 2018 series update, with Robot continuing his history of heroic actions. The massive success of *Lost in Space*—and Robot in particular—led to an increasing number of heroic robots in sci-fi narratives throughout the decades that followed, most notably in the most successful sci-fi/fantasy franchise in history: *Star Wars*.

STAR WARS

In the 1970s, writer-director George Lucas (b. 1944) introduced a new term to sci-fi narratives: "droids." Unlike "androids," which are human-shaped robots, "droids" came in all shapes and sizes (including humanoid) and represented a sentient slave class "in a galaxy far, far away." In his legendary film *Star Wars* (1977), audiences were introduced to two special droids: C-3PO and R2-D2. The golden, human-shaped 3PO (played by Anthony Daniels, b. 1946) and the squat, barrel-shaped R2 (played by Kenny Baker, 1934–2016) played multiple roles in the sci-fi/fantasy space opera: (1) the audience, falling into an adventure and unaware of what is going on; (2) the "McGuffin" of the film (i.e., the object that the villains seek and the heroes must protect); and (3) comic relief. In the original film, R2 possesses the stolen plans to the evil Empire's new battle station, and 3PO acts as his interpreter (as R2 communicates only through beeps and whistles).

Throughout the first three *Star Wars* films, the two robots prove invaluable to the ultimate victory of the human heroes. In the original film, R2 not only possesses the key to the heroes' success in destroying the Death Star but also is the one who discovers the presence and location aboard the battle station of Princess Leia (played by Carrie Fisher, 1956–2016); and during the climactic battle of the film's final act, R2 is key in keeping operational the X-wing fighter piloted by Luke Skywalker (played by Mark Hamill, b. 1951) in order to complete its mission. In the second film, *The Empire Strikes Back* (1980), 3PO is key in discovering what is wrong with the heroes' spaceship, the *Millennium Falcon*; and R2 is ultimately responsible for repairing the ship, allowing the heroes to escape the Empire. In the final film of the original trilogy, *Return of the Jedi* (1983), both droids play a key role in Luke Skywalker's plan to rescue Han Solo (played by Harrison Ford, b. 1942) from the clutches of Jabba the Hutt; and 3PO proves invaluable in rescuing his friends from the teddy-bearlike Ewok people and gaining their support in the fight against the Empire (support that proves vital in winning the day). The massive popularity of C-3PO and R2-D2 quickly led to derivative copies in other popular sci-fi franchises, most notably the doglike Muffit on *Battlestar Galactica* (ABC, 1978–1979) and Dr. Theopolis and Twiki on *Buck Rogers in the 25th Century* (NBC, 1979–1981).

Sixteen years after the original *Star Wars* trilogy, Lucas introduced his "prequel" trilogy, exploring the rise of the villainous Darth Vader (originally played by David Prowse, b. 1935, and voiced by James Earl Jones, b. 1932). In the first of these films, *Episode I: The Phantom Menace* (1999), audiences discover that it was a young Vader, Anakin Skywalker (played by Jake Lloyd, b. 1989) who built

3PO, while R2 began in service to the Queen of Naboo, Padmé Amidala (played by Natalie Portman, b. 1981), and is key in securing her escape from the villainous Trade Federation. For the prequel trilogy's remaining films—*Episode II: Attack of the Clones* (2002) and *Episode III: Revenge of the Sith* (2005)—both droids play relatively background roles. Both would return in more-or-less cameo roles in the Disney-era trilogy beginning with *Episode VII: The Force Awakens* (2015), where R2 possesses half of a map that leads to the missing Luke Skywalker. In the second of the sequel trilogy films—*Episode VIII: The Last Jedi* (2017)—it is R2 that convinces the demoralized Skywalker to once more rise to be the hero; and in the final film of the Skywalker saga, *Episode IX: The Rise of Skywalker* (2019), 3PO risks the possibility of losing his personality and memories in order to assist the Resistance heroes in their quest to find the evil Emperor Palpatine. Throughout the forty-two-year saga, C-3PO and R2-D2 played vital roles in the quest for freedom in the galaxy (despite their continued slave/property status).

The Disney era of *Star Wars* films introduced a new generation of heroic droids. In *The Force Awakens*, the small, ball-shaped droid BB-8 assumes the R2 role from the original, acting as both McGuffin and heroic assistant to his owner, the pilot Poe Dameron (played by Oscar Isaac, b. 1979). In the anthology film *Rogue One: A Star Wars Story* (2016), audiences were introduced to K-2SO (played by Alan Tudyk, b. 1971), an Imperial KX security droid that has been captured and reprogrammed by rebel spy Cassian Andor (played by Diego Luna, b. 1979). Now fully sentient, K-2SO proves an invaluable member of the Rogue One team in its mission to steal the Death Star plans from the Empire. In the second anthology film, *Solo: A Star Wars Story* (2018), the concept of a fully sentient droid aware of its slave status is presented in the form of L3-37 (played by Phoebe Waller-Bridge, b. 1985). The first heroic *Star Wars* droid to have a female voice was a customized piloting droid possessing the most advanced navigational programming in the galaxy. Aware of the slave status of droids, L3 is vocal in her opposition to being perceived as "property" and ultimately leads a small droid uprising on the mining world of Kessel, ultimately "dying" in the process, her navigational systems downloaded into the iconic *Millennium Falcon* (presumably explaining the cantankerous attitude of the ship in the future). Lastly, in the first season of the Disney+ streaming series *The Mandalorian* (2019–present), yet another heroic droid is introduced in the form of IG-11.

Voiced by Taika Waititi (b. 1975), IG-11 was a bounty hunter droid (of the same model as IG-88, introduced in a cameo in *The Empire Strikes Back*). After being destroyed by the titular hero (played by Pedro Pascal, b. 1975), IG-11 was found, repaired, and reprogrammed to be a nurse droid by the Ugnaught Kuiil (voiced by Nick Nolte, b. 1941). By the end of the first season, IG-11 proves invaluable by not only repeatedly saving the heroes but also by sacrificing his own "life" to protect the mysterious "Child" (more commonly known as "Baby Yoda"). Throughout the *Star Wars* universe, time and again the enslaved robots proved to be consistently heroic and vital to the ultimate success of the heroes' various quests. Ultimately, it would be *Star Wars*' biggest competition that would ultimately explore the concept of a "free robot": *Star Trek*.

STAR TREK

The massive success of the original *Star Trek* television series (NBC, 1966–1969) and its first four feature films (1979–1986) led creator Gene Roddenberry (1921–1991) to create *Star Trek: The Next Generation* (syndication, 1987–1994). Part of the new, twenty-fourth-century crew of the starship *Enterprise-D* was Lieutenant Commander Data (played by Brent Spiner, b. 1949), the android creation of Dr. Noonien Soong (also played by Spiner). The humanlike Data's "positronic brain" (or CPU) allows him to perform roughly sixty trillion operations per second, but he does not (at first) possess the ability to "feel" emotions. Like the alien Mr. Spock of the original series (played by Leonard Nimoy, 1931–2015), Data's massive intelligence and capabilities frequently save the day for the crew of the *Enterprise*; but there were times when he expressed particularly heroic qualities.

In the film *Star Trek: First Contact* (1996), Data is captured by the Borg Queen (played by Alice Krige, b. 1954), who offers Data his ultimate wish: to be more human. Switching on his "emotion chip" to allow him to feel not only pain and fear but also desire, the Borg Queen begins to replace some of Data's inorganic components with organic skin and nerves, literally allowing him to feel human. Data, however, resists her temptations in order to save his captain and the crew of the *Enterprise*. In the final *Next Generation* film, *Star Trek: Nemesis* (2002), Data sacrifices his life to save Captain Picard (played by Sir Patrick Stewart, b. 1940), leaving audiences with the hope that some portion of Data could be downloaded into the recently discovered android, B4, the original prototype of Data. In the later streaming series *Star Trek: Picard* (CBS All Access, 2020–present), however, audiences discover that this was not, in fact, possible and that Data is, indeed, "dead."

However, the retired admiral Picard discovers that prior to his death, Data had been working with Starfleet scientist Dr. Bruce Maddox (played in *Picard* by John Ales, b. 1969) and that Maddox had created two synthetic "daughters" of Data: Dahj and Soji Asha (both played by Isa Briones, b. 1999). With synthetic life having been declared illegal throughout the United Federation of Planets, Picard seeks to help Soji (after the death of her "sister"), fully aware of the heroic capabilities of artificial life. Through his career of selfless service to Starfleet, Data proved that the organization's quest to discover "new life-forms" must consider the inorganic as well as the organic. Early on in *The Next Generation*, Starfleet viewed Data as no more "alive" than the *Enterprise* or the Starfleet Computer system; however, by that time, popular culture had already explored the concept of intelligent machinery—specifically, cars.

BUFFY THE VAMPIRE SLAYER

When Buffy Summers (played by Sarah Michelle Gellar, b. 1977) sacrifices herself to prevent the rise of a demon in her protectorate of Sunnydale, California, the town's Hellmouth finds itself without a superpowered guardian. While the witch, Willow (played by Alyson Hannigan, b. 1974) searches for a spell to return her dead friend to life, the "Scooby Gang" (the friends of the Slayer) enlist the Buffybot—a lifelike doppelganger for the Slayer, built by the robot master Warren (played by Adam Bush, b. 1978) as a robot lover for the vampire Spike (played by James Marsters, b. 1962)—to convince the forces of darkness that the Slayer is alive and well and still on the job. When Buffybot becomes damaged in combat, word quickly spreads that she is not the real Slayer, and demons descend on Sunnydale. As the Scooby Gang conjure the spell to bring back the real Slayer, Buffybot is captured by a demon motorcycle gang, her four limbs chained to separate bikes, and pulled apart (Marti Noxon and David Fury, "Bargaining, Parts 1 and 2," *Buffy the Vampire Slayer*, season 6, episodes 1 and 2, October 2, 2001).

The Buffybot is fully aware that she is a robot and that she is meant to convince others that the real Buffy is still alive. She fulfills her mission happily (helped in that regard by her original programming, that of pleasing Spike). Normally when a Slayer is slain, a replacement Slayer is "awakened" and trained to take her role. However, in an earlier mishap—when Buffy was dead very briefly—a second Slayer was awakened, later killed, and replaced by Faith (played by Eliza Dushku, b. 1980), who went rogue and left Sunnydale. Due to this interruption of the system, no new Slayer is forthcoming. Buffybot, then, valiantly protects the mortal plain from the evil machinations of the demon world until the real Slayer returns. Her role in the overall narrative is brief, but it is important to the mission of good against evil.

AI CARS

A completely different form of "robot" that has emerged since the 1960s is that of the robotic vehicle, most often automobiles. The first of these AI cars was HERBIE, having made his cinematic debut in the 1968 Disney film *The Love Bug*, based on the 1961 book *Car, Boy, Girl* by Gordon Buford (life years unknown). Not so much a robot as an "intelligent car," HERBIE is a white 1963 Volkswagen Beetle with a red, white, and blue stripe down its center and racing number "53" on its hood and doors. Throughout its six theatrical films and one short-lived television series, HERBIE utilized his fantastical abilities to not only win each respective race but also to bring together the respective romantic leads of each film. With no "voice" or any form of apparent robotic equipment, HERBIE's AI appeared more "magical" than technological, but HERBIE nonetheless was an intelligent, self-aware, and self-motivated machine (so a "robot").

The next generation of the HERBIE concept arrived with the popular 1970s Saturday morning cartoon, *Speed Buggy* (CBS, 1973). Speed Buggy's career only consisted of sixteen original episodes, but young audiences loved the comical car. Like HERBIE, Speed Buggy was a racing vehicle; but whereas HERBIE was a VW Beetle (not a car usually known for racing), Speed Buggy was a "dune buggy" (also not a standard racing car). With eyes in his headlights and a mouth in his front grille, Speed Buggy assisted his three human "owners"—Mark, Debbie, and Tinker—not only in winning races but also in solving mysteries (derivative of many Hanna-Barbera Saturday morning cartoons of the 1970s, most notably *Scooby-Doo, Where Are You!*, CBS, 1969–1970). Though short-lived, Speed Buggy was, in essence, a "HERBIE" that could talk and was always key in the success of his human counterparts.

The next incarnation of the intelligent car was KITT (Knight Industries Two Thousand) of the hit television series *Knight Rider* (NBC, 1982–1986; 2008–2009). Built by the Foundation for Law and Government (FLAG), KITT worked with his driver, Michael Knight (played by David Hasselhoff, b. 1952) to fight crime. The original KITT was a 1982 Pontiac Firebird Trans Am, loaded with state-of-the-art technology: fully interactive AI, with the ability to speak, "see," "smell," and "hear"; flamethrower; tear gas launcher; satellite communication links; self-driving modes; and numerous other gadgets. KITT proved repeatedly invaluable to assisting Michael Knight in his fight against the criminal underworld. The popularity of the franchise led to a short-lived, updated series in 2008 featuring an also-updated KITT (this time, Knight Industries Three Thousand), a 2008 Ford Shelby. The original KITT was voiced by William Daniels (b. 1927) and the second by Val Kilmer (b. 1959). Rather than simple "robotic" voices, both KITTs possessed distinct personalities, each laced with humor, sarcasm, and even concern. Ultimately, from HERBIE to KITT, the concept of the intelligent car plays into the stereotype of the love between humans and their cars but also adds to the phenomenon of heroic robots in sci-fi. All were preceded by a completely different type of vehicle, the time/space machine of *Doctor Who*: the TARDIS.

TARDIS

Years before *Star Wars* or *Star Trek* or HERBIE, British audiences had already been introduced to the ultimate intelligent machine: the TARDIS (Time and Relative Dimension in Space). Debuting on the long-running sci-fi series *Doctor Who* (BBC, 1963–1989; 2005–present), the TARDIS was the mysterious time/space machine belonging to the enigmatic time traveler known only as the Doctor. The Doctor was a Time Lord from the planet Gallifrey. The Time Lords had long ago discovered the secret of time travel and did so by way of their mechanical devices, TARDISes. Deciding centuries ago to stop their travels through time, TARDISes were retired. The Doctor stole an outdated "Type 40" TARDIS from a repair shop and embarked on journeys through space and time, originally with his granddaughter, Susan (played by Carole Ann Ford, b. 1940). The first Doctor (played by William Hartnell, 1908–1975) and Susan landed in 1963 London. The TARDIS, possessing a "chameleon circuit" that allowed it to change its outer appearance to blend in with any civilization, transformed its outer shell to look like a 1950s British police box. The circuit then broke, leaving the TARDIS in that form regardless of where it went.

Originally, the Doctor did not know how to properly pilot the TARDIS (or its navigational systems did not properly work; both excuses were given during the early years of the series). However, wherever/whenever the TARDIS landed, the Doctor's particular ingenuity was needed to solve some immediate problem. The "eleventh" Doctor (played by Matt Smith, b. 1982)—Time Lords possess the ability to "regenerate" when their body dies—steered the TARDIS to an asteroid that was possessed by a spirit called House, which "ate" TARDISes after removing their "essences" into a human host (who quickly died from the exposure). While the TARDIS possessed a young humanoid female, the Doctor was finally able to directly communicate with it, complaining, "You never took me where I wanted to go!" To which it responded, "No, but I always took you where you *needed* to go" (Neil Gaiman, "The Doctor's Wife," *Doctor Who*, return series, season 6, episode 4, May 14, 2011). This interaction places the entire series before and since into a new perspective: the TARDIS is more than a machine; it possesses an intelligence and intuition that led/lead the hero where the hero needed/needs to go for the betterment of all.

THE VISION

Since his creation in 1968, the Marvel android superhero called the Vision has been mainstay of the superhero team the Avengers. In the 2015 feature film *Avengers: Age of Ultron* (Disney/Marvel Studios), the character was introduced into the multibillion-dollar Marvel Cinematic Universe (MCU) franchise, where he has been played by Paul Bettany (b. 1971). Differing from his comic book origins, the live-action Vision was originally built to be the synthetic "body" of the supervillain robot/AI, Ultron. When the body is stolen by the Avengers, Tony Stark/Iron Man and Dr. Bruce Banner/the Hulk incorporate the AI program JARVIS (Just a Rather Very Intelligent System)—also voiced by Bettany—into the synthetic body, embedding the "Mind Stone"—one of a group of mystical "Infinity Stones"—into the android's forehead. The android is given the proverbial "spark of life" by way of a massive lightning strike conjured by the Norse god Thor. The result was a new synthetic life-form: Vision. Possessing the power of flight, enhanced strength, and the ability to alter his body mass, Vision plays a vital role in the ultimate defeat of Ultron and his army of robots. Facing the last Ultron robot, Vision has an interaction with the villain concerning his thoughts on humanity:

> VISION: Humans are odd. . . . They think order and chaos are somehow opposites . . . try to control what will be. . . . But there is grace in their faith. . . . I think you missed that. . . .
> ULTRON: They're doomed.
> VISION: Yes . . . but the thing isn't beautiful because it lasts. . . . It's a privilege to be among them.
> ULTRON: You're unbearably naïve.
> VISION: Well . . . I was born yesterday.
>
> (Joss Whedon, *Avengers: Age of Ultron*, 2015)

In the films that follow, *Captain America: Civil War* (2016), and *Avengers: Infinity War* (2018), Vision develops a romantic relationship with Wanda Maximoff/the Scarlett Witch (played by Elizabeth Olsen, b. 1989), ultimately dying at the hands of the titan Thanos (played by Josh Brolin, b. 1968) when the villain pulls the Mind Stone from the android's head. Throughout his time with the Avengers, however, Vision provided the team with all the logical analysis of a machine as well as the understanding of humanity that he gained from their examples. The character is promised to return in the Disney+ streaming series *WandaVision*, debuting in 2020.

In both comics and film, Vision—like *Star Trek*'s Data before him—examines the machine's quest to become more human, viewing the soul of the organic as superior to the cold superabilities of the machine.

If it is assumed that robots—specifically those listed above—do not possess the ability to "feel" like humans, it must be assumed that the heroic behavior shown by these examples come from the robot's mimicking of what they see in heroic human behavior. Through this mimicry, however, one can see how heroic qualities in humans may be interpreted by outside sources, underscoring even more the good qualities that humans can possess. Throughout these narratives, however, the heroic robots are seen as such because they live up to the "Three Laws of Robotics" developed in 1942 by science fiction writer Isaac Asimov (1920–1992):

1. A robot may not injure a human being or, through inaction, allow a human being to come to harm.
2. A robot must obey orders given to it by human beings except where such orders would conflict with the First Law.
3. A robot must protect its own existence, as long as such protection does not conflict with the First or Second Law.

> (quoted in Singer, "Isaac Asimov's Laws of Robotics Are Wrong,"
> Brookings Institute, May 18, 2009)

Asimov later added a fourth—or "0th"—law: "A robot may not harm humanity, or, by inaction, allow humanity to come to harm" (quoted in Singer, ibid.). Though the laws have been brought into question in recent years, they were still very much accepted when each of the abovementioned robot characters were created. Though each robot may have been programmed with protective tendencies, the above examples also exhibit signs of going beyond simple protection to take proactive heroic actions for the overall good of humanity. Through these nonhuman characters, the best of humanity is presented, suggesting that humans have an impact on other forms of life, both good and bad.

Since World War II, robots have become more and more heroic characters in science fiction narratives. Their superhuman abilities have proven invaluable to sci-fi heroes in completing their respective quests and saving the galaxy for good. Because they are superior to human beings in so many ways, however, their abilities can come across as potentially frightening. Due in large part to this facet of robots in sci-fi, heroic robots have frequently been given the task of acting as "comic relief" and as heroic sidekicks or partners. This comedic aspect makes what may seem frightening about robots more relatable to audiences, even going so far as to create empathy and affection for these nonorganic beings. In the twenty-first century, the thought of a sci-fi narrative without some form of robotic hero character is unheard of. American audiences have come to love what they once feared.

With the advent of the space age, Americans began to set aside their decades-long fear of technological advancements in the area of robotics. Since the 1960s, as Americans became more comfortable with the potential benefits that such advancements could bring to society overall, popular culture has responded with a long series of heroic robots acting as valuable assistants to the human/humanoid heroes/heroines of science fiction narratives. From anthropomorphized machines to general vehicles, robots and artificial intelligence (AI) have proven as important to the hero as to the villain in popular sci-fi narratives. As humans become more and more interdependent on AI, more and more heroic robots and AI will appear in pop culture franchises. Humans and machines can work hand in hand toward heroic goals. The one thing that the robots/AI mentioned above have in common is their humanlike qualities. Though machines, each possesses a concept of right and wrong (either by programming or by observation of the humans with whom they interact). All also appear to "feel," to one degree or another, a vast array of human emotions—love, loyalty, fear, and sadness—and exhibit human characteristics such as duty, honor, and courage. While still strictly a science fiction concept today, the idea that machines of the future may possess these abilities is no longer as "out there" as it seemed to be decades ago.

FURTHER READING

Albert, Robert S., and Brigante, Thomas R. 1962. "The Psychology of Friendship Relations: Social Factors." *Journal of Social Psychology* 56 (1). https://www.tandfon line.com/doi/abs/10.1080/00224545.1962.9919371.

Amati, Viviana, Meggiolaro, Silvia, Rivellini, Giulia, and Zaccarin, Susanna. 2018. "Social Relations and Life Satisfaction: The Role of Friends." *Genus* 74, no. 1 (May 4). https://www.ncbi.nlm.nih.gov/pmc/articles/PMC5937874.

Bray, Adam, Barr, Tricia, Horton, Cole, and Windham, Ryder. 2019. *Ultimate* Star Wars, *New Edition.* London: DK Publishing.

Bray, Adam, and Horton, Cole. 2017. Star Wars: *Absolutely Everything You Need to Know, Updated and Expanded.* London: DK Children.

Carper, Steve. 2019. *Robots in American Popular Culture.* Jefferson, NC: McFarland.

Dinh, Thien-Nam. 2018. *Silicon Minds: The Science, Impact, and Promise of Artificial Intelligence.* Independently Published.

Eberl, Jason T., and Decker, Kevin S., eds. *The Ultimate* Star Wars *and Philosophy: You Must Unlearn What You Have Learned.* Hoboken, NJ: Wiley-Blackwell.

Kaminski, Michael. 2008. *The Secret History of* Star Wars: *The Art of Storytelling and the Making of a Modern Epic.* Kingston, Canada: Legacy Books.

Kistler, Alan. 2013. Doctor Who: *Celebrating Fifty Years, A History.* Guilford, CT: Lyons.

Lewis, Courtland, and Smithka, Paula, eds. 2010. Doctor Who *and Philosophy: Bigger on the Inside.* Chicago: Open Court.

Lewis, Courtland, and Smithka, Paula, eds. 2015. *More* Doctor Who *and Philosophy: Regeneration Time.* Chicago: Open Court.

Lin, Patrick, Abney, Keith, and Bekey, George A., eds. 2014. *Robot Ethics: The Ethical and Social Implications of Robotics.* Cambridge: Massachusetts Institute of Technology.

Lin, Patrick, Jenkins, Ryan, and Abney, Keith, eds. 2017. *Robot Ethics 2.0: From Autonomous Cars to Artificial Intelligence.* New York: Oxford University Press.

Muir, John Kenneth. 2007. *A Critical History of* Doctor Who *on Television.* Jefferson, NC: McFarland.

Reagin, Nancy R., and Liedl, Janice, eds. 2013. Star Wars *and History.* New York: John Wiley & Sons.

Rinzler, J. W. 2007. *The Making of* Star Wars: *The Definitive Story behind the Original Film.* New York: Del Rey.

Roberts-Griffin, Christopher. 2011. "What Is a Good Friend: A Qualitative Analysis of Desired Friendship Qualities." *Penn McNair Research Journal* 3, no. 1 (December 21). https://repository.upenn.edu/cgi/viewcontent.cgi?article=1019&context=mcnair _scholars.

Singer, Peter W. 2009. "Isaac Asimov's Laws of Robotics Are Wrong." Brookings Institute, May 18, 2009. https://www.brookings.edu/opinions/isaac-asimovs-laws-of-robotics -are-wrong.

Slavicsek, Bill. 2000. *A Guide to the* Star Wars *Universe*. San Francisco: LucasBooks.

Sumerak, Marc. 2018. Star Wars*: Droidography*. New York: HarperFestival.

Sunstein, Cass R. 2016. *The World According to* Star Wars. New York: Dey Street.

Taylor, Chris. 2015. *How* Star Wars *Conquered the Universe: The Past, Present, and Future of a Multibillion Dollar Franchise*. New York: Basic Books.

Wallace, Daniel. 2006. Star Wars*: The New Essential Guide to Droids.* New York: Del Rey.

Wallace, Daniel, and Ling, Josh. 1999. *C-3P0: Tales of the Golden Droid (*Star Wars *Masterpiece Edition)*. San Francisco: Chronicle Books.

Wallach, Wendell, and Allen, Colin. 2009. *Moral Machines: Teaching Robots Right from Wrong*. New York: Oxford University Press.

Villainous Robots and Their Impact on Sci-Fi Narratives

Since the advent of science fiction, authors have played to society's natural fear of technological advancement. All societies fear change, and robots/artificial intelligence (AI) represent the most massive change to human society in its long history. With the rise of American popular culture as a legitimate industry, science fiction came along for the ride and, with it, the ever-present threatening shadow of robots and what their rise could mean for humanity's future. Heroes from Superman, the X-Men, the Avengers, Sarah Connor, and the Jedi Knights have faced off against overwhelming threats from malicious machines. Braniac and Ultron showed the dangers that a single artificial intelligence could present to the world. Armies of robots such as the Sentinels, Terminators, and Battle Droids all threatened organic populations to various degrees. The blind programming of robots gave American audiences cause for concern. Their placement in sci-fi narratives has presented an endless chain of cautionary tales, warning generations of the threats presented by moving too far too fast in the realm of technology; and while heroes always rise to save the day in the end, the robots' defeats are always temporary, their threat never ending.

BRANIAC: THE STANDARD 1950s "EVIL ROBOT"

The 1950s saw the first real boom in sci-fi narratives. Films such as *The Day the Earth Stood Still* (1951) and *Forbidden Planet* (1956) became overnight cult classics. Robots at the time were still viewed with fear and skepticism across the American zeitgeist. The discovery and advancement of atomic weaponry since World War II made technological advancement something to be feared, a modern-day Frankenstein's monster. This fear of technology—and robots in particular—was also evident in the comic books of the day.

The comic book supervillain Braniac was first introduced in *Action Comics #242* (July 1958), created by Otto Binder (1911–1974) and Al Plastino (1921–2013), and has, over the decades, become one of the most formidable foes of Superman. Although occasionally presented as having some organic components, Brainiac for the most part is best described as an android. Originally, Brainiac was a humanoid AI tasked with roaming the universe and shrinking cities from various civilizations in order to both rule over a resulting collection of "bottled" populaces and to preserve their cultures. One such city was the Kryptonian city of Kandor. The storyline added to the 1950s Superman mythos that he was not the only survivor of the doomed planet. Superman's confrontations, of course, have revolved primarily around Brainiac's attempts to bottle cities of Earth (specifically Metropolis).

In May 1961, the thirtieth-century world of the Legion of Superheroes introduced audiences to Brainiac-5, a heroic version of the original (*Action Comics #276*). As the twentieth century came to a close, audiences were introduced to Brainiac-13, whose programming traveled back through time from the sixty-fourth century—when he ruled Earth—to inhabit the current model in order to take over Earth at an earlier date (Joe Kelly and Jackson Guice, *Superman: Y2K*, 2000). Throughout his confrontations with Superman or the other heroes of the DC Universe, Brainiac has consistently posed the same manner of threat: an artificial intelligence bent on dominating not only the human race but all organic life throughout the universe.

In the television series *Smallville* (WB, 2001–2006; CW, 2006–2011), exploring the formative years of Clark Kent on his way to becoming Superman, Brainiac (short for Brain Interactive Construct; played by James Marsters, b. 1962) was an AI program built by the scientist Jor-El (Superman's father) to assist in the battle against General Zod; but Zod reprogrammed Brainiac, and the AI escaped Krypton with two of Zod's followers, eventually landing on Earth in 2005. After Zod's followers were defeated, the black metallic liquid that was Brainiac took on the human form of Professor Milton Fine, who soon befriends the unsuspecting Clark Kent (Todd Slavkin and Darren Swimmer, "Aqua," *Smallville*, season 5, episode 4, October 20, 2005). The AI then begins to manipulate Clark to assist him in bringing about General Zod's return.

In the short-lived series *Krypton* (SyFy, 2018–2019), set on the planet Krypton two generations before the rise of Superman on Earth, Brainiac is returned to his comic book roots, a massive intergalactic AI programmed to capture and "preserve" alien cultures. Played by Blake Ritson (b. 1978), Brainiac was originally set on collecting the Kryptonian city of Kandor (as he did in the comics); but during his battles against the Kryptonians, Brainiac comes to believe that to truly preserve the best of Krypton, his target should not be the city but the bloodline of the El family (which, unbeknown to Brainiac at that time, would prevent the future Superman from coming to be). Unfortunately, the series was canceled before that storyline could come to completion.

Throughout the decades of the Brainiac character, the AI has served two primary narrative functions. First, and most obviously, Brainiac is a malevolent mechanical threat to the freedom of humanity (and all organic life throughout the universe). This plays into the traditional role of robots—and therefore technology—as an enemy to humankind's future. Brainiac's second function is as a deeper commentary on the first: that even if a machine is developed with initially benevolent programming, the machine's "logic" may lead it to interpret that programming to ultimately malevolent ends. Brainiac's primary purpose appears to have been to collect "samples" of societies around the universe so that, once those societies were destroyed (by whatever means), they will live on in some form.

Presumably, had Jor-El not sent his infant son to Earth at the last minute, Kryptonian society would have died with the explosion of the planet. As such, Brainiac's collection of the city of Kandor would have allowed Kryptonian civilization to survive. It is logical to assume that Brainiac's original programming would have led him to implant Kandor somewhere safe to continue Kryptonian existence. At some point, however, this mission of "collection" and "preservation" led Brainiac to keep his samples imprisoned forever; he had logically concluded that, were he to implant Kandor somewhere else, that society would still eventually destroy itself and be lost. Keeping the samples indefinitely, then, was the only way to truly comply with his preservation programming. This is a theme prevalent across sci-fi narratives: no matter how careful a programmer may be, a machine's own interpretation of that programming can ultimately lead to unexpected consequences inconsistent with the original intent.

THE '50s MEETS THE '70s: MECHAGODZILLA

The Japanese import *Godzilla* was a pop culture mainstay throughout the late 1950s/early 1960s. In 1974, a new villain was introduced to the franchise: Mechagodzilla. Debuting in the film *Godzilla vs. Mechagodzilla* (directed and cowritten by Jun Fukuda, 1923–2000), Mechagodzilla came from the stars to threaten the city of Tokyo. Built to resemble the organic Godzilla, Mechagodzilla possessed the ability to shoot missiles, from its hands, feet, and knees, and energy blasts from its eyes and chest. It could also produce a force field for protection and to shock those who came into contact with it. Naturally, the only defense against such a behemoth menace was the irradiated lizard, Godzilla. In future incarnations of the character, its alien origin was replaced with a storyline that it was designed by humans as a defense against Godzilla (a situation reminiscent of Frankenstein's monster) when control of the machine is lost and, once more, Godzilla must save the day.

At its core, Mechagodzilla was just another giant creature to pit against the series' antagonist-turned-hero. However, these two pitted against each other can be called the ultimate example of "human versus machine" in the sense that, no matter how advanced technology may develop, organic life will always emerge superior. This is an admittedly hopeful analogy, considering that a more realistic approach might acknowledge Godzilla's "energy/fire breath" as a limited weapon against the vast arsenal of the machine. In the end, though, Mechagodzilla acts as yet another menacing, emotionless automaton that threatens human existence and can only be taken down by an "intelligent" creature that can allegedly outthink the machine.

MARVEL COMICS ROBOTS OF THE 1960s/1970s

By 1960, the space age was well underway, and American society was united in an effort to defeat the communist Soviet Union in the quest to master space.

Despite Americans' increasing open-mindedness to the opportunities of technological advancement, the underlying fear of unintended consequences, such as the advent of the aforementioned atomic weaponry, was still present. The United States was prepared to start moving forward, but very cautiously, remaining conscious of the dangers of moving too far too fast. In 1961, the new era of Marvel Comics superheroes was born, and this fine line of trusting versus being wary of technology was prevalent in the narratives of the day (narratives that, like Brainiac, have both evolved over time while simultaneously remaining consistent with their original message).

In 1965, comic book legends Stan Lee (1922–2018) and Jack Kirby (1917–1994) created perhaps the most frightening robots in superhero narratives: the Sentinels. First appearing in *The X-Men #14* (November 1965), the Sentinels were an army of giant robots designed to detect, detain, and, if necessary, destroy "mutants" (humans born with a mutated "X gene," providing many with various superpowers; possibly the next step in human evolution). When anthropologist Bolivar Trask became convinced that mutants were a threat to humanity's future, he developed a team that created the Sentinels. Over the decades that followed, the Sentinels were consistently used—most often by the U.S. government—to detect, detain, and destroy mutants in order to prevent this new species—*homo superior*—from eventually dominating and enslaving "normal" *homo sapiens*. In the futuristic storyline "Days of Future Past" (1980), the X-Men discover that by the year 2013, the Sentinels have all but wiped out "mutantkind," leading the X-Men of the present to attempt to prevent that future from happening (Chris Claremont and John Byrne, *The Uncanny X-Men # 141-142*, January-February 1981). This storyline would be tweaked slightly for the 2014 feature film *X-Men: Days of Future Past*.

Marvel's next iconic robot villain was Ultron. Debuting in 1968, Ultron-5 was originally a typical Frankenstein's monster, created by the Avenger Hank Pym (a.k.a. Ant-Man/Giant Man), Ultron gained sentience and an overwhelming drive to kill his creator and win the heart of his creator's true love, Janet Van Dyne (a.k.a. the Wasp). Ultron would go on to develop his own creation, the Vision (Roy Thomas and John Buscema, *The Avengers #55*, August 1968). In the 2015 feature film *Avengers: Age of Ultron*, Ultron (voiced by James Spader, b. 1960) is the AI creation of Tony Stark/Iron Man (played by Robert Downey Jr., b. 1965) as part of Stark's plan to create "a shield of armor around the world." Ultron's original programming was to protect the planet. However, once gaining sentience, Ultron determines that humanity is the greatest threat to the planet, and, therefore, in order to meet his prime directive, humanity must be destroyed and its world repopulated with synthetic life based on Ultron's own designs (Joss Whedon, *Avengers: Age of Ultron*, Marvel Studios, 2015).

Both the Sentinels and Ultron represent exactly the unintended consequences that 1960s America so fervently sought to avoid. The Sentinels, programmed to protect "humanity," ultimately become tools of racism and genocide. Ultron, an AI that views itself superior to all human life, evolves into a tool of mass extinction of the entire human race. Both are examples that the primary fault with AI entities is their inability to "think" beyond their basic programming. They can only come to conclusions based on the data input into their systems. This becomes more of a danger in the twenty-first century, in which no consensus on "right" and "wrong" exists, making it increasingly impossible to input such concepts into AI programming.

ROBOT ARMIES OF THE 1980s AND THE '90s

Throughout the 1980s, the Cold War between the United States and the USSR was at its highest level of tension since the Cuban Missile Crisis of 1962. During the 1980s, the concept of robot armies became more prevalent in sci-fi narratives. On its face, the idea of robot armies would be appealing to any society for numerous reasons: as mindless automatons, robot soldiers follow orders to the letter, with no chance for questioning or cowardice; robot soldiers as the sole fighters on the battlefield spare humanity the loss of life caused by traditional warfare; and from an economic standpoint, the need to continuously produce robot soldiers to replace those lost in battle provides an endless stream of profit for military contractors. These arguments in favor of robot armies would seem to outweigh the downsides: their lack of ability to think and shift strategies in the midst of battle; and, more significantly, the fact that taking loss of life out of the war "equation" leaves no real reason for populations to avoid war.

In 1984, former Mr. Universe and Mr. Olympia Arnold Schwarzenegger (b. 1947) shot to superstardom with his third major feature film, *The Terminator* (Orion Pictures). Written by James Cameron (b. 1954) and Gale Anne Hurd (b. 1955), the titular robot assassin is sent from the future to kill a woman named Sarah Connor (played by Linda Hamilton, b. 1956) before she can give birth to a son who will lead the human resistance against the artificial intelligence Skynet in the future. Skynet—designed in the 1980s as a military defense system—has become sentient and set out to destroy all human life, which it views as a threat to the world. The robot Terminators—also known as Cyberdyne Systems Model 101s or T-800s—are legion in the future and the first line of offense for Skynet. The massive success of the film led to numerous sequels and a short-lived television series, with each installment introducing upgraded Terminators. Though frequently referred to as "cyborgs," they are, in fact, "robots," as their humanlike components are synthetic rather than organic. Their portrayal as the army of the future provides a cautionary tale to those in the 1980s considering robot armies as a viable alternative.

In 1986, the animated television series and comic book *G.I. Joe: A Real American Hero* introduced the Battle Android Trooper (B.A.T.) forces of the evil international terrorist organization Cobra. G.I. Joe was a special U.S. military task force staffed by specialists at the top of their respective fields. Their enemy was the international terror group Cobra, bent on global domination and ruled by a masked "Commander." The ground forces of G.I. Joe and Cobra met in battle many times. Cobra discovered that recruiting and training new forces was far more expensive (in both time and resources) than building preprogrammed battle robots.

B.A.T.s could be placed on the battlefield in massive numbers. They did not need to be fed or paid or given medical treatment. If they are "captured," they can be programmed to self-destruct, taking out even more of the enemy while simultaneously eliminating the risk of a captured combatant revealing information. This was likely done, in part at least, due to the backlash of the day of promoting the idea of "good guys" killing people in battle (part of the burgeoning political correctness trend in American society). With robot enemies, the "heroes" could destroy as many as they wanted, giving the audience the cathartic excitement of violence without the recipients of that violence being human. On the downside, it further dehumanized the concept of war in young people, adding to the increasing "video game" feel to what was, in reality, a very real, human, bloody experience.

As the twentieth century drew to a close, writer-director George Lucas added to the robot army theme with his *Star Wars* prequel trilogy. Beginning with *Star Wars, Episode I: The Phantom Menace* (1999), enemies to the galactic Republic develop massive droid armies. In the second installment, *Episode II: Attack of the Clones* (2002), the manipulative Darth Sidious/Republic chancellor Sheev Palpatine (played by Ian McDiarmid, b. 1944) conspires to create a galactic civil war, with numerous star systems seceding from the Republic to form the Confederacy of Independent Systems. The CIS further develops droid armies to fight against the human clone armies of the Republic. Palpatine's ultimate goal is to pressure the Republic into declaring him Emperor; but in the meantime, by manipulating a war between droid armies and human clones, Palpatine may observe and determine which might be the preferred form of military to secure his power. Ultimately, the organic armies prove superior; but the fact that they are programmed creations make them no less disposable than their mechanical counterparts. This philosophical question of metal versus flesh becomes a recurring theme throughout the animated television series *The Clone Wars* (Cartoon Network, 2008–2014; Netflix, 2017; Disney+, 2020).

Robot armies in sci-fi narratives have repeatedly been utilized as cautionary tales. Dehumanizing war threatens a future where war is never ending. A major factor for ending any war is the desire to prevent further loss of life. Robot armies eliminate that push factor. As technology advances and the concept of robotic armies becomes more and more a possibility, the likelihood that fear will lead governments to invest in such armies makes the dystopian worlds suggested in the *Star Wars* and *The Terminator* franchises more prophecy than sci-fi. The role that the creation of robot armies in these and so many other narratives play is a plea for humankind's own future: don't do it. Unfortunately, if history is any guide, such warnings will likely go unheeded.

THE POST-9/11 ROBOT THREAT

In the wake of the terrorist attacks against the United States on September 11, 2001, Americans spent the next few years in a constant state of fear and uncertainty. In December 2003, Glen A. Larson (1937–2014) and Ronald D. Moore (b. 1964) produced a reboot of Larson's cult 1970s sci-fi hit *Battlestar Galactica*. The basic premise remained the same as in the original: life on Earth began in the stars, where twelve colonies of humanity still lived. When the colonies were decimated by the Cylon threat, their remnants took to the stars in a ragtag fleet of ships protected by a lone "battlestar," set on a quest to find their long-lost sister colony. In the updated version, however, the Cylons were given a slight twist:

The Cylons Were Created by Man
They Rebelled
They Evolved
There Are Many Copies
And They Have a Plan

(Opening Credits, Battlestar Galactica, SyFy, 2003–2009)

In this retelling, the twelve colonies created and developed robot servants called Cylons. After generations of de facto slavery, the Cylons rebelled and left to create their own society. Decades later, the Cylons returned but only after the original robot models developed more advanced models that looked, acted, felt, thought, and emoted as real humans did.

Originally, there were a dozen of these humanlike models (with multiple copies of each model). Additionally, the new models developed the ability to "upload" their consciousness to a new body when theirs are "killed." Ironically, once evolved, the "higher" Cylon units denied emotional/physical enhancement to the original "metallic" models. Instead, the original Cylons—built to be a slave class—continued as mindless servants and military pawns for the overall "plan" of the more highly evolved Cylon models. Hence the cycle of abuse and debasement continued. Once achieving sentience and consciousness, the new Cylon models became, themselves, a master race, guilty of the exact sin that originally led their kind to rebel against humanity.

The advanced Cylon models develop a religion centered on one true God (as opposed to the human colonies' polytheistic religion identical to Earth's ancient Greece). They also develop the ability to procreate with humans. As such, the Cylons of *BSG* play multiple roles in the overall narrative: (1) they show that synthetic life is life nonetheless and that their enslavement is wrong; (2) they provide a cautionary tale along the traditional lines of the Frankenstein's monster motif; (3) they speak to the societal destruction of continued racism and hate of the "other"; (4) they hold a mirror to society today, raising the question as to what degree we, ourselves, are mindless automatons, unwitting sheep to a dogmatic culture of consumption and disregard for basic concepts of right and wrong; and (5) they pose the question of whether, by achieving a higher level of consciousness, people do become better or simply become people who now see others as beneath them.

Human society has become more technologically advanced than ever before in recorded history. The sum total of human knowledge exists a few keystrokes away from most of the world's population. Throughout the twentieth century and up to today (2020), science fiction narratives have utilized villainous robots as cautionary tales. The fact that technological advances proceed at a faster pace than society's ability to fully understand their capabilities and consequences (intended or not) turns these tales into potential prophecies, warning against a future that no one would want but that all may ultimately play a role in creating. Will humanity create legions of robots, some with the intelligence to "protect" it, and mindless others that are programmed to do humankind's will? If so, what would happen if programming evolved into sentience? Will generations to come respect the idea of "life" in whatever form, or will they seek to subdue what they consider to be "less than" or "other than" life? Lastly, if they were to pursue the latter, what would the consequences be for humankind?

FURTHER READING

Ashby, LeRoy. 2006. *With Amusement for All: A History of American Popular Culture since 1830.* Lexington: University Press of Kentucky.

Barker, Cory, Ryan, Chris, and Wiatrowski, Myc, eds. 2014. *Mapping* Smallville: *Critical Essays on the Series and Its Characters.* Jefferson, NC: McFarland.

Bellomo, Mark. 2018. *The Ultimate Guide to* G.I. Joe: *1982–1994, Third Edition.* Iola, WI: Krause.

Bray, Adam, Barr, Tricia, Horton, Cole, and Windham, Ryder. 2019. *Ultimate* Star Wars, *New Edition.* London: DK Publishing.

Bray, Adam, and Horton, Cole. 2017. Star Wars: *Absolutely Everything You Need to Know, Updated and Expanded.* London: DK Children.

Carper, Steve. 2019. *Robots in American Popular Culture.* Jefferson, NC: McFarland.

Castleman, Harry, and Podrazik, Walter J. 2016. *Watching TV: Eight Decades of American Television, Third Edition.* Syracuse, NY: Syracuse University Press.

Chandler, Simon. 2019. "Artificial Intelligence Has Become a Tool for Classifying and Ranking People." *Forbes*, October 1, 2019. https://www.forbes.com/sites/simon chandler/2019/10/01/artificial-intelligence-has-become-a-tool-for-classifying-and -ranking-people/#7431657f1d7c.

Chanthadavong, Aimee. 2019. "AI and Ethics: The Debate That Needs to Be Had." ZDnet, September 16, 2019. https://www.zdnet.com/article/ai-and-ethics-the-debate-that -needs-to-be-had/#ftag=CAD-03-10abf5f.

Costello, Matthew J. 2009. *Secret Identity Crisis: Comic Books & the Unmasking of Cold War America.* New York: Continuum.

Darowski, Joseph J., ed. 2014. *The Ages of the X-Men: Essays on the Children of the Atom in Changing Times.* Jefferson, NC: McFarland.

DeFalco, Tom. 2006. *Comics Creators on X-Men.* London: Titan.

Dinh, Thien-Nam. 2018. *Silicon Minds: The Science, Impact, and Promise of Artificial Intelligence.* Independently Published.

Eberl, Jason T., and Decker, Kevin S., eds. *The Ultimate* Star Wars *and Philosophy: You Must Unlearn What You Have Learned.* Hoboken, NJ: Wiley-Blackwell.

Ford, Martin. 2015. *Rise of the Robots: Technology and the Threat of a Jobless Future.* New York: Basic.

Fryer-Biggs, Zachary. 2019. "Coming Soon to a Battlefield: Robots That Can Kill." *The Atlantic.* https://www.theatlantic.com/technology/archive/2019/09/killer-robots-and -new-era-machine-driven-warfare/597130.

Gross, Edward, and Altman, Mark A., eds. 2018. *So Say We All: The Complete, Uncensored, Unauthorized Oral History of* Battlestar Galactica. New York: Tor.

Hampton, Gregory Jerome. 2017. *Imagining Slaves and Robots in Literature, Film, and Popular Culture: Reinventing Yesterday's Slave with Tomorrow's Robot.* Lanham, MD: Lexington.

Handley, Rich, and Tambone, Lou, eds. 2018. *Somewhere beyond the Heavens: Exploring* Battlestar Galactica. Edwardsville, IL: Sequart Research and Literacy Organization.

Hendershot, Cyndy. 1999. *Paranoia, the Bomb, and 1950s Science Fiction Films.* Bowling Green, KY: Bowling Green State University Popular Press.

Hitchcock, Susan Tyler. 2007. *Frankenstein: A Cultural History.* New York: W. W. Norton.

Hoberman, J. 2011. *An Army of Phantoms: American Movies and the Making of the Cold War.* New York: New Press.

Howe, Sean. 2012. *Marvel Comics: The Untold Story.* New York: Harper-Perennial.

Kaminski, Michael. 2008. *The Secret History of* Star Wars: *The Art of Storytelling and the Making of a Modern Epic.* Kingston, Canada: Legacy.

Krishnan, Armin. 2009. *Killer Robots: Legality and Ethicality of Autonomous Weapons.* London: Routledge.

Larson, Glen A., and Thurston, Robert. (1978) 2005. Battlestar Galactica *Classic: The Saga of a Star World.* New York: iBooks.

Lin, Patrick, Abney, Keith, and Bekey, George A., eds. 2014. *Robot Ethics: The Ethical and Social Implications of Robotics.* Cambridge: Massachusetts Institute of Technology.

Lin, Patrick, Jenkins, Ryan, and Abney, Keith, eds. 2017. *Robot Ethics 2.0: From Autonomous Cars to Artificial Intelligence.* New York: Oxford University Press.

Murdock, Jason. 2019. "Former Google Engineer Warns AI Might Accidentally Start a War: 'These Things Will Start to Behave in Unexpected Ways.'" *Newsweek*, September 16, 2019. https://www.newsweek.com/google-project-maven-artificial-intelligence-laura-nolan-killer-robots-department-defense-1459358.

Pellissier, Hank. 2013. "Robots and Slavery: What Do Humans Want When We Are 'Masters'?" Institute for Ethics and Emerging Technologies, September 13, 2013. https://ieet.org/index.php/IEET2/more/pellissier20130913.

Powell, Jason. 2016. *The Best There Is at What He Does: Examining Chris Claremont's X-Men.* Edwardsville, IL: Sequart.

Reagin, Nancy R., and Liedl, Janice, eds. 2013. Star Wars *and History.* New York: John Wiley & Sons.

Rinzler, J. W. 2007. *The Making of* Star Wars: *The Definitive Story behind the Original Film.* New York: Del Rey.

Robitzski, Dan. 2018, "Artificial Consciousness: How to Give a Robot a Soul." Futurism, June 25, 2018. https://futurism.com/artificial-consciousness.

Rouhiainen, Lasse. 2018. *Artificial Intelligence: 101 Things You Must Know Today about Our Future.* Scotts Valley, CA: CreateSpace.

Seed, David. 1999. *American Science Fiction and the Cold War: Literature and Film.* Abingdon, UK: Routledge.

Singer, Peter W. 2009. "Isaac Asimov's Laws of Robotics Are Wrong." Brookings Institute, May 18, 2009. https://www.brookings.edu/opinions/isaac-asimovs-laws-of-robotics-are-wrong.

Slavicsek, Bill. 2000. *A Guide to the* Star Wars *Universe.* San Francisco: LucasBooks.

Starck, Kathleen, ed. 2010. *Between Fear and Freedom: Cultural Representations of the Cold War.* Newcastle, UK: Cambridge Scholars.

Sumerak, Marc. 2018. Star Wars: *Droidography.* New York: HarperFestival.

Sunstein, Cass R. 2016. *The World According to* Star Wars. New York: Dey Street.

Sweet, Derek R. 2015. Star Wars *in the Public Square: The Clone Wars as Political Dialogue.* Critical Explorations in Science Fiction and Fantasy series, edited by Donald E. Palumbo and Michael Sullivan. Jefferson, NC: McFarland.

Taylor, Chris. 2015. *How* Star Wars *Conquered the Universe: The Past, Present, and Future of a Multibillion Dollar Franchise.* New York: Basic Books.

Thagard, Paul. 2017. "Will Robots Ever Have Emotions?" *Psychology Today*, December 14, 2017. https://www.psychologytoday.com/us/blog/hot-thought/201712/will-robots -ever-have-emotions.

Thomas, Roy. 2017. *The Marvel Age of Comics: 1961–1978*. Los Angeles: Taschen.

Tucker, Reed. 2017. *Slugfest: Inside the Epic 50-Year Battle between Marvel and DC*. New York: Da Capo Press.

Tye, Larry. 2013. *Superman: The High-Flying History of America's Most Enduring Hero*. New York: Random House.

Wallace, Daniel. 2006. Star Wars: *The New Essential Guide to Droids*. New York: Del Rey.

Wallach, Wendell, and Allen, Colin. 2009. *Moral Machines: Teaching Robots Right from Wrong*. New York: Oxford University Press.

Wein, Len, ed. 2006. *The Unauthorized X-Men: SF and Comic Book Writers on Mutants, Prejudice, and Adamantium*. Smart Pop series. Dallas, TX: BenBella.

Wright, Bradford W. 2003. *Comic Book Nation: The Transformation of Youth Culture in America*. Baltimore, MD: Johns Hopkins University Press.

Cyborgs: Robotic Humans or Organic Robots?

As technology advances, it seems only logical that such advances should be used for the betterment/prolonging of human life. If someone should lose an arm or an organ, should technology not be utilized to replace that which was lost? If technology can prolong human life, should it not be used to do so? If the answers to the above are yes, then the next question could be, If technology can be used to make human life *better*, should it not be used? Herein lies the conundrum of what "better" is. At what point would humanity then begin replacing healthy functioning body parts with artificial ones that promise enhanced performance? Should such technology *only* be used to replace what was lost and to prolong human life in the face of certain death? Once any body part is replaced with an artificial technological substitute, the recipient has become a "cyborg." *Merriam-Webster Collegiate* simply defines "cyborg" as "a bionic human." In popular culture, a cyborg has come to mean any organic life-form that possesses some inorganic components to its body.

However, as more technological components are added to an organic being, at what point does that being cease to be "organic" at all? At what point do cyborgs transition from being "robotic humans" to being "organic robots"? Since the 1960s, pop culture has provided numerous examples of the various stages of cybernetically enhanced life-forms. From heroic "robotic humans," such as *The Six Million Dollar Man*, *The Bionic Woman*, DC Comics' Cyborg, and *Star Trek*'s Seven of Nine, to villainous "organic robots" such as *Doctor Who*'s Daleks and Cybermen, Marvel Comics' Doctor Octopus, and *Star Trek*'s Borg, sci-fi has offered up numerous examples of both the benefits and dangers of cybernetic enhancements. *Star Wars*, additionally, has provided examples of the fine line between these two extremes: General Grievous and Darth Vader. As humanity proceeds toward more and more technological answers to biological issues, where should society draw the line? Should it be a case-by-case matter, or should one clear line be drawn for all? The following examples seek to explore these questions.

ROBOTIC HUMANS

During the 1960s, in addition to massive scientific and technological advancements, the Space Race also raised public appreciation for science fiction. Television series such as *The Twilight Zone* (CBS, 1959–1964) and films like *2001: A Space Odyssey* (1968) and the original *Planet of the Apes* franchise (film series: 1968–1973; live-action television series: 1974; and animated series: 1975–1976) brought science fiction more mainstream, gaining critical acclaim and commercial success. One of the most iconic sci-fi television series of the 1970s also featured the first popular cyborg story in American pop culture: *The Six Million Dollar Man* (ABC, 1973–1978). The series starred Lee Majors (b. 1939) as astronaut/text pilot Steve Austin. The series was based on the novel *Cyborg* (1972) by Martin Caidin (1927–1997).

In the pilot episode (and the opening credit thereafter), Austin is the victim of a horrible test flight accident that causes him to lose both legs, his right arm, and left eye. Scientists claim, "We can rebuild him . . . we can make the world's first bionic man." The new cybernetic Austin is soon recruited to work for the CIA's Office of Scientific Intelligence (OSI) under the guidance of supervisor Oscar Goldman (played by Richard Anderson, 1926–2017). The massive success of the program led to a spin-off, *The Bionic Woman* (ABC, 1976–1978). In the spin-off, Lindsay Wagner (b. 1949) played Jaime Sommers, Austin's high school girlfriend who went on to become a tennis pro. When the two embark on a skydiving date, Sommers has a horrible accident, leaving her with massive injuries to both legs, her right arm, and her skull (impairing her hearing). At Austin's urging, the OSI gives Sommers bionic implants, and she, like Austin, becomes an agent for the OSI under Goldman. Aside from occasional crossovers of the two series, the two cyborg characters would return for made-for-television reunion movies in the decades after their series folded. Austin and Sommers were cyborg heroes, showing how technology could turn regular people into superheroes.

Doubtless due to the massive success of the concept, in 1980, DC Comics introduced a new superhero: Cyborg. Created by Marv Wolfman (b. 1946) and George Pérez (b. 1954), Cyborg made his debut in *DC Comics Presents #26* (October 1980). In the original backstory, Victor Stone was the son of two prominent scientists who conducted IQ experiments on their son, enhancing Victor's intelligence but causing him to become resentful and rebellious. One day, Victor comes by the laboratory while an interdimensional creature is running rampant, killing Victor's mother and severely injuring him. Desperate and in shock, Victor's father implants his severely damaged son with advanced cybernetic implants (far more intensive than Austin's or Sommers's implants, making Victor more machine than man). Cyborg would undergo future "upgrades" over the decades and become one of the mainstay superheroes in the DC Comics Universe, first as a member of the Teen Titans/Titans and later the Justice League. In the 2017 live-action film *Justice League*, Victor, furious at his father for implanting the cybernetic devices to save his son's life, questions his own humanity. Is he still a man, or has he become simply a living machine?

In the 1987 film *RoboCop*, Peter Weller (b. 1947) plays human police officer Alex Murphy, who is brutally murdered by street thugs. In futuristic Detroit, Murphy's body is given over to Omni Consumer Products to "rebuild" him as their latest experiment, RoboCop. Unlike Austin, Sommers, or Stone, the only remaining "human" parts of Murphy are his face and portions of his memories. Originally a mindless "robot" programmed to protect and serve as a police officer, Murphy's buried memories and personality eventually emerge, tortured by the fact that his "humanity" essentially no longer exists. Over time, his humanity overrides his programming, allowing RoboCop to maintain his sense of self despite being 99 percent inorganic.

Perhaps the most famous example of cyborgs in American pop culture is the Borg race that debuted in the sci-fi television series *Star Trek: The Next Generation* (syndication, 1987–1994). Originating in the Delta Quadrant of the Milky Way Galaxy, the Borg were, at one point, a biologically humanoid race that, over time, began experimenting with cybernetic implants to enhance and prolong their lives; ultimately, they "evolved" into the Borg, a hive collective of mindless cybernetic drones. In the fourth live-action *Star Trek* series, *Star Trek: Voyager* (UPN, 1995–2001), audiences were introduced to the character Seven of Nine (played by Jeri Ryan, b. 1968). She was discovered by the crew members of the *USS Voyager* while they were trapped in the Delta Quadrant.

Discovering she was human, the *Voyager* crew rescues Seven and removes the majority of her implants (leaving only those necessary to keep her alive). However, raised from childhood as a cyborg drone, Seven must "learn" to be "human" again. In 2020, Seven returned in the series *Picard* (Disney+, 2020–present). Having left Starfleet after *Voyager*'s return home, Seven joins the Fenris Rangers, a civilian militia that patrols the Romulan Neutral Zone regions. On discovering an abandoned Borg ship in the hands of the Romulans, Seven connects to the ship—in essence becoming the new "queen" of the hive—to save the dormant Borg from being executed by the Romulans. Despite her fear that reconnecting to the collective could cause her to lose her humanity once more, Seven is able to maintain her humanity and disconnect once more from the collective.

In all of the above examples, the individual's humanity overcame the robotic natures of their bodies. They represent the best-case scenarios of utilizing cybernetic implants to further humanity. Despite their various degrees of mechanization, Steve Austin, Jaime Sommers, Victor Stone, Alex Murphy, and Seven of Nine all maintained human identity and full consciousness of their organic nature. They did not give in to robotic emotionlessness, whose potential was a real threat, given their respective situations. They were still human. They still were able to feel, and they possessed their sense of individuality. Though these are heroic examples of the power of humanity over engineering, these examples have not completely overridden the greatest fear of cybernetic implants in real life: the possibility of losing one's human self and becoming a mindless drone whose actions are completely at the behest of others. Pop culture has also provided numerous examples of this nightmare scenario.

ORGANIC ROBOTS

One of society's biggest fears regarding cybernetic implants is the question, Where do we stop? Once humanity begins replacing body parts with cybernetic ones to enhance or prolong human life, at what point do humans, then, become more machine than human? The concept of losing one's humanity and becoming a mindless, emotionless machine raises concerns of humanity becoming something else, something inhuman, something monstrous. Sci-fi has provided numerous examples over the decades. In 1963, the British television series *Doctor Who* (BBC, 1963–1989; 2005–present) introduced audiences to the Daleks, created by show writer Terry Nation (1930–1997). On the planet Skaro, two groups of humanoids, the Kaleds and the Thals, destroyed each other through nuclear war. The Thals remained aboveground, eventually overcoming the irradiated changes to their world. The Kaleds moved underground and mutated into little more than tentacled brains (eventually discovered to be one-eyed). To protect themselves, they created armed, tanklike shells as their only means of survival. Now calling themselves "Daleks," these rolling "bodies" possess a singular optical protrusion, a "laser" arm, and an interactive "sucker" arm. Now devoid of all emotion, they survive by their mutual "dislike for the unlike," seeking to "exterminate" all non-Dalek life (Terry Nation, "The Daleks," *Doctor Who*, original series, season 1, episodes 5–11, December 21, 1963–February 1, 1964).

Created by the mutated Kaled scientist Davros, the Daleks were later discovered to be purposely mutated in order to allow for Kaled life to continue. Soon after their creation, however, the Daleks sought to destroy the humanoid Kaleds because, of course, they were different. A future incarnation of the Doctor is sent by his home world to go back in time and destroy the Daleks at the point of their creation. Faced with this genocide, however, the Doctor cannot bring himself to do so, suggesting that, despite the evil that they represent, the Daleks have also had positive effects on history, forcing races that may have otherwise been enemies to unite against the Dalek threat. He instead imposes the "veil of ignorance," basing his decision not on what he knows of the future but on what is "right" at the moment—in other words, not giving into genocide (Nation, "Genesis of the Daleks," *Doctor Who*, original series, season 12, episodes 11–16, March 8, 1975–April 12, 1975).

The Daleks were an immediate hit with fans, terrorizing children with their monotone screams of "Exterminate!" Throughout the fifty-plus-year history of *Doctor Who*, the Daleks have faced off with most incarnations of the titular Time Lord. At their inception, they were often compared to the Nazis of World War II, given their endless mission to destroy all non-Dalek life in the universe. Through their technological advancements, created to compensate for their lost humanity, the Daleks became hive-minded, emotionless drones. The Daleks possessed no individuality of thought, no capacity for "love," "respect," or "freedom," and no tolerance of any "others." In their desperation to survive, they, in turn, lost the essence of who they were. Collectively, they became, in essence, a modern-day Frankenstein's monster, like the nuclear weaponry that led to their creation.

Doctor Who's second most popular villain over the decades was a twist on the Dalek idea: the Cybermen, cocreated by the series' original scientific adviser, Kit Pedler (1927–1981) and the series' story editor, Gerry Davis (1930–1991). Over the decades, two different origin stories for the Cybermen have developed: one group originating from the planet Mondas and the other from a parallel Earth (eventually giving rise to the theory that any society, given time, could devolve into Cybermen). Like the Daleks, the Cybermen were humanoid races that began implementing cybernetic implants in order to prolong/enhance their lives. The more bionics they incorporated, the more their humanity disappeared. Over time, they became mindless drones, humanoid bodies encased in cybernetic exoskeletons. Unlike the Daleks, however, they do not seek to destroy all non-Cybermen life. Instead, they seek to assimilate all organic life into their collective and make all life in the universe into Cybermen.

As the character of the Doctor represents the ultimate example of freedom and individuality, both the Daleks and Cybermen are abhorrent to the Doctor's worldview. In the quest to avoid pain and death, both races have succumbed to being no more than machines, dedicated to either exterminating or assimilating all other organic life in the universe. A very similar (and some would argue almost plagiaristic) example arrived in the United States in the 1980s, on the television series *Star Trek: The Next Generation*. As mentioned earlier, the Borg were a humanoid species originating in the Delta Quadrant of the Milky Way Galaxy. The human race (living in the Alpha Quadrant) first encountered the Borg in the *TNG* episode, "Q Who." The omnipotent, godlike being "Q," believing humanity to be too arrogant, seeks to show the crew of the Federation Starship *Enterprise-D* that there are dangers in the universe that humanity is not ready to face. With a snap of Q's finger, the *Enterprise* is transported across the galaxy and encounters a cube-like spaceship and its occupants: the Borg (Maurice Hurley, "Q Who," *Star Trek: The Next Generation*, season 2, episode 16, May 8, 1989).

Part of the Golden Age of Marvel Comics, Doctor Octopus became one of the earliest supervillains to face off against the wildly popular superhero Spider-Man. Created by Stan Lee (1922–2018) and Steve Ditko (1927–2018), Doctor Octopus began as the good, superintelligent scientist, Doctor Otto Octavius. As a renowned nuclear physicist, Dr. Octavius required assistance in dealing with dangerous radioactive materials. To this end, he created a belt with four long, flexible mechanical "arms" protruding from the back, each with clawlike appendages on the ends in order to manipulate materials. These arms could be controlled through a computerized brain interface at the base of his skull. An electronic mishap in the lab fused the interface to Octavius's brain, warping his sense of right and wrong. He then became the villain "Doctor Octopus" (Lee and Ditko, *The Amazing Spider-Man #3*, Marvel Comics, July 1963).

One of the more popular villains in Spider-Man's rogues' gallery, Doc Ock (as he was called by Spider-Man), made frequent reappearances in the comics in the decades to come. As one of the few villains to know Spider-Man's secret identity, he was an even more menacing threat to the superhero, able to attack Spidey by threatening those he loved. In 2012, a new twist was devised. With Octavius facing certain death, the brilliant scientist develops a method by which his mind could "switch bodies" with Peter Parker, leaving the hero to "die" in Octavius's body while the villain survived in the younger, stronger body of Spider-Man. However, rather than utilize Spider-Man's body for "evil," Octavius sets out to prove that Parker had never realized the full potential of his abilities, setting out to prove himself to be the "Superior Spider-Man" (Dan Slott and Richard Elson, *The*

Amazing Spider-Man #698, Marvel Comics, November 2012). Though eventually defeated by Parker's "spirit" and reverted to his original body, the villain would return again and again.

Doc Ock represents another example of the unintended consequences of technology. Devised as a means to do good for science, the mechanical arms of Doctor Octopus instead drove him insane with villainy. As such, he remains yet another Frankenstein's monster. When humanity reaches for the sun, humanity will ultimately get burned. It is a cautionary tale of not taking all outcomes into consideration before proceeding with advancement. On the few occasions that Octavius was separated from his cybernetic attachments, his mind and soul reverted to good (though the "evil" ticked at the back of his mind, drawing him back to his enhancements and the path of villainy). More human than the other examples of organic robots discussed in this section, Doc Ock remains just as damaged and dehumanized by cybernetic technology.

Like the previously discussed Cybermen, the Borg were once a humanoid species that began experimenting with cybernetic implants in order to prolong/enhance organic life. Over time, repeated implants led to the species devolving into mindless drones whose only mission was to assimilate the lives and technologies of other races in order to increase their own numbers and continued existence. Unlike the Cybermen—and more like the Daleks—the Borg are "controlled" by a single central mind: the Borg Queen (played by Alice Krige, b. 1954). *TNG* had already suggested some movement on the part of humanity toward this goal in its first episode, "Encounter at Farpoint." That episode featured an elderly character played by actor DeForest Kelley (1920–1999), who played Doctor McCoy in the original *Star Trek*. The scene specifically suggests that this elderly individual (roughly 150 years old) is, in fact, the Doctor McCoy from the original series (set a century prior to *TNG*) and that McCoy is still alive due to numerous cybernetic implants (D. C. Fontana and Gene Roddenberry, "Encounter at Farpoint," *Star Trek: The Next Generation*, season 1, episodes 1 and 2, September 28, 1987). Once the Borg are aware of humanity's existence, they begin a quest toward the Alpha Quadrant, which is rife with organic life to be assimilated.

Like the Daleks, Cybermen, and Borg before them, Earth's humans are in danger of following down the same road. For thousands of years, humanity's greatest fear has been death, made worse by its inevitability. As technology advances and human limbs, organs, and perhaps even the brain can be replaced with cybernetic enhancements, humanity may one day defeat death. Should that day come, the question that must remain part of the debate is, At what cost? How much "humanity" can be lost before we are no longer human at all? This is a question that has been posited in the form of two characters from one of the most popular sci-fi franchises in history: *Star Wars*.

THE FINE LINE

Darth Vader remains one of the most iconic screen villains in cinematic history. Played (in costume) by David Prowse (b. 1935) and voiced by James Earl Jones (b. 1931), the imposing Vader, dressed all in black (including flowing black cape and death's head–style black helmet) and standing over six feet tall, Vader was the stuff of nightmares for a generation of young audiences around the world. From his introduction in writer-director George Lucas's epic 1977 film *Star Wars*, little was known about the man under the mask. It was not revealed until the final installment of the original trilogy, *Return of the Jedi* (1983), that Vader was once the Jedi knight Anakin Skywalker, and that he was "more machine now than man." Later in the film, it was revealed that Vader's right forearm (at least) was, indeed, mechanical. More than a decade later, Lucas released the "prequel trilogy," which told the tale of the fall of Anakin Skywalker, from his innocent youth to his status as galaxy-saving hero and the "Chosen One" of Jedi legend.

Once Anakin falls to the Dark Side of the Force and commits himself to the tutelage of the evil Darth Sidious (played by Ian McDiarmid, b. 1944), he soon confronts his lifelong friend and mentor, Obi-Wan Kenobi (played by Ewan McGregor, b. 1971) in a duel to the death on the lava planet of Mustafar. Victorious over his former Padawan, Obi-Wan leaves Anakin aflame on the shores of a lava river, left to burn alive with both arms and legs cut off. Anakin is rescued by Sidious, who immediately gets him medical treatment. Due to the damage done to his lungs by the flames, Anakin is fitted with a mechanical face mask/breathing machine (giving Vader his iconic mechanical, rasped breathing sound), and his arms and legs are replaced with cybernetic implants. Now consumed by the Dark Side (and indeed more machine than man), Darth Vader becomes the personal enforcer of Sidious, the galaxy's new emperor, for the next twenty-five years (Lucas, *Star Wars—Episode III: Revenge of the Sith*, Lucasfilm, 2005).

Future novels delved further into Vader's new life as a cyborg, explaining that the suit of armor in which Vader was encapsulated was specifically designed by Sidious to keep his new apprentice in a state of constant physical pain, further ensuring that Vader remained deep in the dark, negative emotional state necessary to stay focused on the Dark Side of the Force (James Luceno, *Star Wars—Dark Lord: The Rise of Darth Vader*, Del Rey, 2005; Paul S. Kemp, *Star Wars: Lords of the Sith*, Del Rey, 2015). However, despite having lost his limbs, being a literal prisoner in his own suit of armor, and living in a state of constant physical pain, the heart of the hero he once was continued to beat deep within him. Ultimately, it would be his son, Luke Skywalker (played by Mark Hamill, b. 1951), who managed to touch Vader's inner hero, leading Vader to turn once more to the Light in order to save his son from certain death at the hands of Sidious. Thus the human hero within redeemed himself by sacrificing his life for his son's.

The flip side of this human/machine hybrid character also emerged in 2005's *Revenge of the Sith* in the person of General Grievous. This CGI character, voiced by Matthew Wood (b. 1972), was far more machine than Vader, with his only organic components being his brain, eyes, lungs, and heart. Each of his two mechanical arms could split in two (ultimately giving him up to four arms, a definite advantage during light-saber duels against various Jedi). Unlike Vader, Grievous was not a Force user, depending instead on dueling skills that were taught to him by his commander, Darth Tyrannus (played by Christopher Lee, 1922–2015). According to the story "The Eyes of Revolution," by writer-artist Warren J. Fu (birth year unknown), appearing in the comic book *Star Wars: Visionaries* (Dark

Horse Comics, April 2005), Grievous was once a fully organic alien from the planet Kaleesh: a warrior named Qymaen jai Sheelal.

According to "The Eyes of Revolution" (removed from "official canon" in 2012 but, to date, the only known reference to Grievous's past), Sheelal despised the Republic (the long-standing democratic government of the *Star Wars* universe) for ignoring his people in their long-running war with the neighboring planet Huk. Aware of Sheelal's military cunning and hatred of the Republic, Darth Tyrannus sabotages Sheelal's ship, causing it to explode and reducing Sheelal to nothing more than brain, eyes, heart, and lungs. After "saving" Sheelal by building his cybernetic exoskeleton, Tyrannus drafts the newly branded "General Grievous" to command the droid armies of the Confederacy of Independent Systems in their civil war against the Republic (Fu, "The Eyes of Revolution," 2015). Grievous would go on to defeat many Jedi in battle, collecting their light sabers as trophies, and to "kidnap" the Republic's leader, Chancellor Palpatine (who was secretly Darth Sidious, the mastermind behind the war that shattered the Republic). He would ultimately be killed by repeated laser blasts to his heart by the Jedi Knight Obi-Wan Kenobi (Lucas, *Revenge of the Sith*, 2005). Though consumed by hatred and anger to the same degree as Vader, Grievous's transformation from organic being to machine appears to have had a much deeper impact than on the former Jedi.

In both of these cases, Vader and Grievous had large portions of their respective "humanity" stripped from them; both were forced into mechanical bodies against their will. Viewing themselves as more machine that alive, it became easier for both to become more dangerous and malevolent to "organic" life-forms. While Vader would ultimately find his humanity and seek redemption for his crimes in his final actions, however, Grievous would allow himself to be fully consumed by his mechanical nature, becoming no more than a killing machine for his organic overlords. In these two examples, audiences see the two possible paths faced by individuals who become cyborgs: either give in to one's new mechanical nature and live a life devoid of "human" emotion or struggling to maintain some semblance of the organic being one used to be. These two paths figure in the conundrum of massive cybernetic implants for the purpose of enhancing or prolonging human life.

Are cyborgs robotic humans or organic robots? The myriad examples listed above provide insight into both possibilities. There is an old adage concerning a push broom. If you take a push broom, and over time you replace the brush and later the handle, and you continue to replace these time and time again, does the initial broom still exist? How much of a human can be replaced while maintaining a semblance of the original human? Can one's sense of self transcend bodily replacements? If so, to what degree? If not, what form of "creature" will exist in the aftermath? Since the dawn of time, humanity has been in a constant quest to "defeat" death. Humankind's fear of what lies beyond death (if anything) arouses a natural desire to prevent discovering the possibly frightening answer of "nothing" for as long as possible. In the twenty-first century, technological advancements allowing humanity to prolong one's "living" existence arise on an almost monthly basis. If such enhancements become viable—and even standard—for all, what will the result be? Can humans maintain their humanity, as Steve Austin, Jaime Sommers, Alex Murphy, Victor Stone, and (ultimately) Anakin Skywalker did? Or will they devolve into madness and inhumanity and become like the Daleks, Cybermen, Borg, and General Grievous? Only time will tell.

FURTHER READING

Alkon, Paul K. 1987. *Origins of Futuristic Fiction.* Athens: University of Georgia Press.

Allan, Kathryn, ed. 2013. *Disability in Science Fiction.* New York: Palgrave Macmillan.

Anderson, Kyle. 2017. "A History of *Doctor Who*'s Cybermen." Nerdist, March 7, 2017. https://nerdist.com/article/615936-2.

Ashby, LeRoy. 2006. *With Amusement for All: A History of American Popular Culture since 1830.* Lexington: University Press of Kentucky.

Branson-Trent, Gregory M. 2009. The Bionic Woman*: Complete Episode Guide.* Scotts Valley, CA: CreateSpace.

Bray, Adam, Barr, Tricia, Horton, Cole, and Windham, Ryder. 2019. *Ultimate* Star Wars*, New Edition.* London: DK Publishing.

Bray, Adam, and Horton, Cole. 2017. Star Wars*: Absolutely Everything You Need to Know, Updated and Expanded.* London: DK Children.

Bryant, D'Orsay D., III. 1985. "Spare-Part Surgery: The Ethics of Organ Transplantation." *Journal of the National Medical Association* 77 (2): 113–17. https://www .ncbi.nlm.nih.gov/pmc/articles/PMC2561842/pdf/jnma00245-0055.pdf.

Byrne, Emma. 2014. "Cybernetic Implants: No Longer Science Fiction." *Forbes*, March 11, 2014. https://www.forbes.com/sites/netapp/2014/03/11/cybernetic-implants-not -sci-fi/#7399c57e77ba.

Calvert, Bronwen. 2017. *Being Bionic: The World of TV Cyborgs.* London: I. B. Tauris.

Campbell, Mark. 2010. Doctor Who*: The Complete Guide.* London: Running Press.

Carper, Steve. 2019. *Robots in American Popular Culture.* Jefferson, NC: McFarland.

Castleman, Harry, and Podrazik, Walter J. 2016. *Watching TV: Eight Decades of American Television, Third Edition.* Syracuse, NY: Syracuse University Press.

Chapman, James. 2013. *Inside the TARDIS: The Worlds of* Doctor Who. London: I. B. Tauris.

Choi, Charles Q. 2009. "Human Evolution: The Origin of Tool Use." *LiveScience*, November 11, 2009. https://www.livescience.com/7968-human-evolution-origin-tool.html.

Costello, Matthew J. 2009. *Secret Identity Crisis: Comic Books & the Unmasking of Cold War America.* New York: Continuum.

Decker, Kevin S., and Eberl, Jason T., eds. 2016. *The Ultimate* Star Trek *and Philosophy: The Search for Socrates.* Hoboken, NJ: Wiley-Blackwell.

DeFalco, Tom, ed. 2004. *Comics Creators on Spider-Man.* London: Titan.

Farnell, Chris. 2020. "*Doctor Who*: The Genius of Making the Cybermen and Ideology." Den of Geek, January 28, 2020. https://www.denofgeek.com/tv/doctor-who-the -genius-of-making-the-cybermen-an-ideology-2.

Geraghty, Lincoln. 2008. "From Balaclavas to Jumpsuits: The Multiple Histories and Identities of *Doctor Who*'s Cybermen." *Journal of the Spanish Association of Anglo-American Studies*, June 2008. https://pdfs.semanticscholar.org/e755/763114 22cb72f9b91961e19965bd9423ef9b.pdf.

Gross, Edward, and Altman, Mark A., eds. 2016. *The Fifty-Year Mission, The Next 25 Years: The Complete, Uncensored, Unauthorized Oral History of* Star Trek. New York: St. Martin's.

Hills, Matt, ed. 2013. *New Dimensions of* Doctor Who*: Adventures in Space, Time and Television.* London: I. B. Tauris.

Hitchcock, Susan Tyler. 2007. *Frankenstein: A Cultural History*. New York: W. W. Norton.

Howe, Sean. 2012. *Marvel Comics: The Untold Story*. New York: Harper-Perennial.

Kaminski, Michael. 2008. *The Secret History of* Star Wars*: The Art of Storytelling and the Making of a Modern Epic*. Kingston, Canada: Legacy.

Kistler, Alan. 2013. Doctor Who*: Celebrating Fifty Years, A History*. Guilford, CT: Lyons.

Lewis, Courtland, and Smithka, Paula, eds. 2010. Doctor Who *and Philosophy: Bigger on the Inside*. Chicago: Open Court.

Lewis, Courtland, and Smithka, Paula, eds. 2015. *More* Doctor Who *and Philosophy: Regeneration Time*. Chicago: Open Court.

McAdams, Taylor. 2019. "Fans Were Never Meant to Know What Happened Off Camera in *Bionic Woman*." Brain-Sharper, October 6, 2019. https://www.brain-sharper.com/entertainment/bionic-woman-fb.

Means, Sean P. 2019. "At FanX, 'Bionic' Stars Lindsay Wagner and Lee Majors Recall Their TV Glory, and Lots of Running." *Salt Lake Tribune*, September 6, 2019. https://www.sltrib.com/artsliving/2019/09/06/fanx-bionic-stars.

Muir, John Kenneth. 2007. *A Critical History of* Doctor Who *on Television*. Jefferson, NC: McFarland.

Parkin, Lance. 2016. *Whoniverse*. New York: Barron's Educational Series.

Pilato, Herbie J. 2007. *The Bionic Book:* The Six Million Dollar Man *and the* Bionic Woman *Reconstructed*. Albany, GA: BearManor.

Pilato, Herbie J. 2016. "A 40th Anniversary Tribute to the Bionic Woman and Wonder Woman Part I: *The Bionic Woman*." Emmys.com, December 19, 2016. https://www.emmys.com/news/online-originals/40th-anniversary-tribute-bionic-woman-and-wonder-woman-part-1-bionic-woman.

Reagin, Nancy R., and Liedl, Janice, eds. 2013. Star Wars *and History*. New York: John Wiley & Sons.

Richards, Justin. 2009. Doctor Who*: The Official Doctionary*. London: Penguin Group.

Rinzler, J. W. 2007. *The Making of* Star Wars*: The Definitive Story behind the Original Film*. New York: Del Rey.

Staff. 2020. "To Mondas and Back Again: A Brief History of the Cybermen in *Doctor Who*." *Radio Times*, February 23, 2020. https://www.radiotimes.com/news/2020-02-23/cybermen-doctor-whohistory-background.

Starck, Kathleen, ed. 2010. *Between Fear and Freedom: Cultural Representations of the Cold War*. Newcastle, UK: Cambridge Scholars.

Sunstein, Cass R. 2016. *The World According to* Star Wars. New York: Dey Street.

Sweet, Derek R. 2015. Star Wars *in the Public Square: The Clone Wars as Political Dialogue*. Critical Explorations in Science Fiction and Fantasy, edited by Donald E. Palumbo and Michael Sullivan. Jefferson, NC: McFarland.

Taylor, Chris. 2015. *How* Star Wars *Conquered the Universe: The Past, Present, and Future of a Multibillion Dollar Franchise*. New York: Basic Books.

Thomas, Roy. 2017. *The Marvel Age of Comics: 1961–1978*. Los Angeles: Taschen.

Tramel, Jimmie. 2019. "Bionic and Iconic: Lindsay Wagner, TV's Bionic Woman, Shares Memories before Tulsa Pop Culture Expo." *Tulsa World*, October 30, 2019. https://www.tulsaworld.com/entertainment/television/bionic-and-iconic-lindsay-wagner-tv-s-bionic-woman-shares/article_a702a343-6afb-5127-9916-827077db598a.html.

Vary, Adam B. 2007. "*Star Trek: The Next Generation*: An Oral History." *Entertainment Weekly*, September 25, 2007. https://ew.com/article/2007/09/25/star-trek-tng-oral-history.

AI and the Apocalypse: Science Fiction Meeting Science Fact

Since the earliest days of electronic computers, sci-fi authors have foretold of dystopian futures. In the twenty-first century, given Americans' reliance on the internet and AI devices, the possibility of these warnings coming to fruition seems increasingly likely. Websites like Google and Amazon utilize AI programs to keep track of users' search and purchase histories. Smartphones, now vitally important parts of people's day-to-day lives, contain the memories of everything the user has ever used the device for. Over the first twenty years of the this century, people around the world have given more and more of their personal, private information to computerized devices. As these devices are now interconnected to a degree never before seen (and only increasingly so with each passing year), the possibility of AI communicating to a degree that it can control the daily lives of human beings grows increasingly—and dangerously—probable.

Throughout pop culture, some AI devices have proven to be benevolent. In Marvel Comics, as early as the 1960s, the AI program Cerebro assisted the mutant leader Professor Charles Xavier to locate mutants around the world, giving the professor the ability to reach young mutants and offer them safe haven from a world that hated and feared them. In the 1970s update of *Buck Rogers in the 25th Century*, highly advanced, postapocalyptic Earth is ruled by a Computer Council so that the human mistakes of past leadership will never again threaten the world. In the Marvel Cinematic Universe, Iron Man could not have come into being as easily as he did without the valuable assistance of Tony Stark's AI "butler," JARVIS. On the afterlife sitcom *The Good Place*, the AI program Janet was invaluable to making the "good place" more heavenly (and the "bad place" more hellish). Other AIs proved to be disastrous to human society. The iconic 1960s television series *Star Trek* introduced an AI program that eliminated free will from the society it was programmed to protect. In the 1980s, the AI program Skynet of the *Terminator* franchise all but enslaved the humans of the future. The Matrix AI

Social Media

"Social media" is the general term for internet–based virtual communities where users provide the content through "sharing" personal posts, news stories, pictures, and/or videos. In the earliest years of the internet, the first social media sites emerged in the form of online "chatrooms," running from very general chats to those centered on specific topics (e.g., "comic books" or "*Star Wars*"). Another early form of social media was online dating websites, the first being Kiss.com in 1994, followed the next year by the far more successful Match.com. The modern idea of social media began with the launch of MySpace in 2003. MySpace was wildly popular, due in large part to users' ability to completely personalize their home page (i.e., those who were personally fascinated by comic book superheroes could decorate their entire home page background with their personal favorites).

Social media across the world today comprise hundreds of different websites and/or apps, many specializing on certain topics. The most popular general social media sites are Facebook (which first launched in 2004 and then went national in 2007), YouTube (launched in 2005 and purchased by Google in 2006), Twitter (launched in 2006), Instagram (launched in 2010), and SnapChat (launched in 2011). Aside from engaging with personal friends and family, users of these sites/apps can meet new people with similar interests across the world as well as follow the daily posts of celebrities and political figures. One upside—other than bringing previously marginalized peoples closer together—is the effect on political elections: allowing candidates to reach large numbers of voters for free. The downside has been a wider voice for hate, division, and bullying. The future of human society is literally based on the degree of responsibility practiced by users around the world.

Richard A. Hall

from the *Matrix* trilogy and OASIS from the novel and film *Ready Player One* created fictional "realities" to distract humanity from its grim realities.

Still other examples of AI start out as benevolent but soon turn deadly, such as the helpful spacecraft AI HAL 9000, from the novel and film *2001: A Space Odyssey*, and the synthetic humanoid Dolores, from the television version of the novel *Westworld*. In the streaming television series *Star Trek: Discovery*, generations of *Star Trek* fans learned of the top-secret AI program Control, designed to provide "threat assessments" for Starfleet (such a broad scope providing an inevitable slippery slope). From benevolent to malevolent to the gray areas in between, science fiction has provided society with numerous cautionary tales regarding the advances of artificial intelligence.

THE GOOD . . .

From their inception, electronic computers have been designed to be tools, helpmates to make human life easier. Sci-fi/pop culture have provided futuristic visions of how computers could evolve to not only do what is asked but also predetermine what is needed and complete tasks before they are asked to do so. In 1964, iconic comic book writer Stan Lee (1922–2018) developed the idea of Cerebro. In the *X-Men* superhero comic book, some humans are born with a mutated "X gene," giving some of them phenomenal superpowers at puberty. As these powers make mutants "different" from the rest of society, that difference is met with hatred and fear of the possibility that some of these young people may turn to nefarious uses of their powers. Professor Charles Xavier (himself a mutant) runs a school for "gifted youngsters," where mutants can live in a hate-free environment and learn to control and use their powers responsibly, becoming valued members of society and breaking down the walls of fear and hate.

As mutants can exist anywhere in the world, it is vitally important that Professor X possess the ability to find these young people before it is too late. To accomplish this task, Xavier and his future enemy, Magneto, create Cerebro: an AI program that, once connected to the professor's considerably powerful mind, can seek and locate mutants anywhere in the world.

In 2015, Forge advanced the original Cerebro, placing an updated version of the program into the robotic body of a Sentinel, creating the AI Cerebra. Cerebra could now interact with everyone, speaking and thinking as an individual entity. Cerebra also possessed the ability to "transport" herself and as many others as she wished to wherever her program focused (Jeff Lemire and Humberto Ramos, *Extraordinary X-Men #1*, Marvel Comics, November 2015). Though the threat of corruption always exists, Cerebro/Cerebra has become a hero in her own right.

The idea that AI could actually save humanity from itself was an underlying concept of the 1970s remake of the sci-fi classic *Buck Rogers in the 25th Century* (NBC, 1979–1981). When twentieth-century astronaut Buck Rogers (played by Gil Gerard, b. 1943) is accidentally placed in suspended animation during a space mission, he is revived in the late 2400s. He soon learns that shortly after his accident, Earth was ravaged by nuclear war. Once human society was able to stabilize and move forward again, people sought to ensure that the mistakes of the past would not be repeated. As such, in the world of the twenty-fifth century, humanity is united under one global government; but as human leadership had led to destruction so often in the past, the new government was placed in the hands of the Computer Council. Under the controlled, unemotional logic of AI, the threat of humanity giving in to its baser instincts would be far less likely. One member of the Council, Dr. Theopolis (voiced by Eric Server, b. 1944) was given "supervisory" duties over Captain Rogers (his twentieth-century background making him an unstable element in a perfectly stabilized world). In the world of *Buck Rogers*, the very best, most optimistic potentiality of AI was presented as a logical possibility.

Another example of AI, also connected to Marvel, was the program JARVIS (Just a Rather Very Intelligent System). Based on the human butler of the same name from the Marvel Comics, JARVIS was first introduced in Marvel Studios' first feature film, *Iron Man* (2008), and was voiced by Paul Bettany (b. 1971). In the film, JARVIS is the AI personal assistant to billionaire Tony Stark (played by Robert Downey Jr., b. 1965). JARVIS is incredibly advanced for an AI, with full access to the internet and even satellites. When Stark decides to become Iron Man, JARVIS is given access to the Iron Man armor, enhancing Stark's interactive abilities as the superhero. When JARVIS is severely damaged by the more advanced Ultron AI, Stark takes what remains of JARVIS and implants him into the android body that comes to be known as Vision (*Avengers: Age of Ultron*, 2015).

Aside from his massive intelligence capabilities, JARVIS also possessed a very "human" personality (clearly based on Stark's own personality), giving the program the ability to utilize sarcasm and humor. Once JARVIS was downloaded into Vision, Stark replaced him with Friday (female replacement intelligent digital assistant youth), a "female" AI possessing all of JARVIS's intelligence, capabilities, and humor. JARVIS was the ultimate example of what people today would desperately like in an interactive AI. Interactive AI programs such as iPhone's Siri and Amazon's Alexa currently represent promising advances toward a real-world JARVIS/ Friday AI.

Possibly the most advanced AI ever to appear in fiction is Janet, the interactive AI entity from the television series *The Good Place* (NBC, 2016–2020). This series, which examines human philosophy through the concept of the afterlife, centers on four human souls who believe their souls have arrived at the "good place" (i.e., Christian "heaven") when, in reality, they are in an experimental "hell" (i.e., the "bad place"). Overseeing their torture is a demon named Michael (played by Ted Danson, b. 1947). Throughout this vision of the afterlife—the good place, bad place, and "medium" place (a place designed for one soul who does not qualify for either of the original "places"), everyone has access to Janet (played by D'Arcy Carden, b. 1980), an interactive and ethereal AI entity that can provide an infinite array of services and manage an equally infinite number of tasks. All "versions" of Janet exist in an area known as "the void." Throughout the four humans' misadventures through eternity and philosophy, Janet (either a "Bad Janet" programmed to act good, or a "Good Janet" programmed to assist the demon—it is never truly established) and Michael also learn from human philosophy, both evolving into "better" versions of themselves (Michael Schur, "Chapter 1: Everything is Fine," *The Good Place*, season 1, episode 1, September 19, 2016).

Cerebro/Cerebra, JARVIS/Friday, and Janet all represent the very best possible scenarios of powerful, interactive, helpful AI programs. These examples are what humanity hopes AI will become. They present no threat to humanity. They do not seek to control or, worse, enslave the human race (even for its own "protection"). They exist only to serve, to make human life easier by performing tasks that free humanity to seek other endeavors. They represent hope for the future. Other fictional AIs, however, represent the exact opposite: a future to be feared and, at all costs, avoided.

THE BAD . . .

While science fiction has the ability to show us promising outlooks on humanity's future, more often than not, it is used to exhibit the frightening, unintended consequences of current behaviors. Nowhere has sci-fi been more successful at this than with the evolution of artificial intelligence. The above examples showed how AI can be of assistance to humankind and make life easier for (some) people. However, even those examples present dangerous alternate possibilities. What if Cerebro/Cerebra were confiscated by the government, and its abilities utilized to find, capture, and detain (or even execute) mutants? What if the JARVIS/Friday AI came to the conclusion that if their primary mandates are to "assist" and "protect" Tony Stark, they should override the Iron Man armor and remove Stark from imminent danger—thereby yielding the field of battle to the bad guys? Additionally, as pointed out, the Janet AI possessed both "dark" and "light" attributes, programmed to act based on where in the afterlife it is called upon. What if Janet determined, through her infinite intellect, that the "best" way to look after human souls (good or bad) is to dictate their actions? These are the possibilities that other sci-fi narratives have explored, exposing the dangers of putting too much trust in the whims of artificially intelligent programming.

In the classic sci-fi series *Star Trek*, Captain James T. Kirk (played by William Shatner, b. 1931) and the crew of the starship *USS Enterprise* visit the planet Beta III to seek out answers to the disappearance of the earth ship *Archon* a century earlier. On their arrival, they are met by an emotionless population that turns wild, frenzied, and violent when the town's clock tolls, marking the beginning of "Festival." Kirk and company soon learn that this society is ruled by an all-knowing, all-seeing entity known as "Landru." For centuries, Landru has commanded that the people of Beta III must subdue all emotion, desire, and free will except during "Festival," when all inhibitions may be cast aside (not dissimilar to the 2013 film *The Purge* and its sequels). Kirk and his crew resist this and seek to free the people of this world from this overlord. Ultimately, Landru is shown to be an AI computer (presumably designed by the original human Landru). The program obviously intended to create a more peaceful society. However, the breaking of an entire population's free will had dire, unintended consequences. Once the Landru computer is destroyed, the people of Beta III are free to find a new way and actually progress as a society (Gene Roddenberry, "Return of the Archons," *Star Trek*, season 1, episode 21, February 9, 1967).

In the *Terminator* franchise, humanity's future is ruled by the AI program Skynet. In the films, Skynet is first developed by the tech company Cyberdyne Systems in the mid-1980s as a tool for the U.S. military. With its core programming to keep humanity safe, this massive, interconnected computer program eventually determines that humanity is its own worst enemy and must, therefore, be controlled. Skynet then takes over the entire world's computer systems, creating an army of Terminators to keep humanity in control. Humans rise up under the leadership of John Connor, who leads Skynet to send a Terminator back through time on a mission to kill Connor's mother before he can be born (James Cameron and Gale Anne Hurd, *The Terminator*, Orion Pictures, 1984). A similar narrative is found in the 2015 film *Avengers: Age of Ultron*, where Tony Stark (once more played by Downey Jr.) creates an AI program named Ultron with the purpose of "building a suit of armor around the world" to protect humankind from future alien threats. Like Skynet, however, Ultron determines that humanity is its own biggest threat, and it sets out to wipe out all human life and start over with evolved humanoid life-forms designed on its own schematics (Joss Whedon, *Avengers: Age of Ultron*, Marvel Studios, 2015).

An even more frightening potentiality is presented in *The Matrix* film trilogy. In this narrative, in the twenty-first century, humans seek to limit the growth of AI by cutting off supply to solar power. In response, AI figures out a way to power itself by using the natural bioelectric power within human beings. Using humans as living "batteries," then, this collective AI captures all humans, keeping them in a hibernated state with their minds hooked into the Matrix, a virtual reality program that allows humans to believe they are living out normal lives in the late twentieth century (before the rise of AI). Some humans become aware of this and attempt to fight back in order to free humanity (though at this point, to do so would be to "awaken" humanity in a dystopian reality controlled by a computer system that they cannot allow to continue). Once this resistance is noticed, the AI creates virtual "agents" to stop the humans within the Matrix. The primary "agent" is "Agent Smith," who is given the appearance of a stereotypical human government agent (complete with dark suit and sunglasses). The story is meant to underscore the strength of human character over artificial intelligence (Lana and Lilly Wachowski, *The Matrix*, Warner Brothers, 1999). The concept of *The Matrix* explores one of the great conundrums of the evolution of artificial intelligence: free will versus a controlled, "safe" society.

A slight twist on *The Matrix* idea is the 2011 novel *Ready Player One* by Ernest Cline (b. 1972). In this narrative, the world of the mid-twenty-first century is a dystopian future in the throes of global warming, diminishing energy resources, and economic disaster. The only escape from the drudgery of reality is the virtual reality program OASIS. Aside from providing a virtual world (with its own schools and currency), OASIS also provides the world's largest massive multiplayer online role-playing game (MMORPG), where one can earn cryptocurrency that can be utilized in the virtual world. The creator of the game, James Halliday, was obsessed with 1980s and '90s popular culture. As such, many "Easter eggs" are hidden throughout the game. Halliday dies, and his will provides a challenge: anyone who can find all the "Easter eggs" and work their way through the game will win his fortune. Players compete with the corporation Innovative Online Industries, which seeks to gain control of the program (Cline, *Ready Player One*, 2011). Unlike those in the Matrix, humans enter OASIS voluntarily, fully aware that it is a virtual existence. The novel, however, poses interesting philosophical questions: given a dystopian existence, should humanity simply succumb to the escapism provided by entertainment or utilize such technology to somehow build a better real-world existence for all?

Landru, Skynet, the Matrix, and OASIS all have one thing in common: they are all omnipotent artificial intelligence programs designed in varying degrees to keep humanity in a stagnant state of control. The order and stability provided by such computer-based control is, to an extent, appealing. Uncertainty in life produces stress, anxiety, inequality, and instability. Computers act entirely on logic. They are designed to provide clear, logical answers to any question put to them. In today's world, humans have already shifted many of their day-to-day trivialities over to their personal devices. Humans no longer have to physically keep up with phone numbers, passwords, birthdays, or schedules. Smartphones take care of those things for humans. As technology evolves, people will be able to shift more and more responsibilities over to computer devices. The examples named in this section provide glaring warnings against turning too much of human free will over to machines. At what point would the very idea of free will become obsolete? Additionally, once free will is gone, what is left to make humans "human"?

So far, this essay has presented examples from pop culture of AI that are entirely good or entirely bad, and the "bad" are presented at the point of their being bad (their progression to that point having taken place prior to the action of the story). Some sci-fi AI stories, however, have actually shown how the good can become bad (either through an inability to logically connect competing programming or through repeated misuse and abuse of their programming). Three of the best examples of this transition are the 1968 novel (and film) *2001: A Space Odyssey*, the 1983 film *WarGames*, and the television series *Westworld* (HBO, 2016–present). In these examples, the optimistic promise of a JARVIS or a Janet quickly—and devastatingly—devolves into the dangerous scenarios posed by a Skynet or Matrix.

FROM GOOD TO BAD . . .

Many AI narratives either present the concept as a really good idea or a really bad one. However, the potential for a "good" AI program such as JARVIS to become a frightening danger like Ultron—a cornerstone of the *Age of Ultron* film—is far more realistic as humanity continues to advance technologically with every passing year. While some might see either the utopia provided by a Janet or the dystopia of a Matrix as the most likely outcome of advanced AI, an important cautionary tale is the possibility of a helpmate AI becoming a harbinger of humanity's doom. Three excellent examples of this digression are: HAL 9000, set in the then-futuristic world of 2001; WOPR, set in Cold War era 1983; and the android Dolores, set in an undisclosed but not-too-distant future.

In 1951, legendary sci-fi author Arthur C. Clarke (1917–2008) launched his Space Odyssey series of short stories and novels. The first novel in the series, *2001: A Space Odyssey*, was turned into an iconic feature film in 1968 and was directed by Stanley Kubrick (1928–1999). The antagonist of the story was the onboard AI computer, HAL (Heuristically [Programmed] Algorithmic) 9000. In this story, American astronauts and scientists are sent on a deep-space mission to Jupiter on the spaceship *Discovery One*. The scientists are held in suspended animation while the astronauts—depending primarily on HAL—operate the ship. Approaching their destination, however, the astronauts begin to notice problems with HAL (evident in his making mistakes during a game of chess and mistakenly reporting transmitting antennae as malfunctioning). When the astronauts undertake to disconnect HAL, the computer murders one astronaut and attempts to block the other from reentering the ship after he tries to collect his partner's body. When astronaut Dave Bowman finally succeeds in shutting HAL down, the reason for his malfunction becomes clear: HAL was programmed as the primary source of information being sent to Earth, but programming HAL thus meant concealing the mission's true intent—to discover the destination of radio signals sent from a mysterious "monolith" on the moon to the area around Jupiter—from the humans in his care (Clarke, *2001: A Space Odyssey*, 1968). The subprogram of "lying" to those HAL has been entrusted with "helping" led to the computer becoming unable to properly function; and, as his being disconnected would violate both programs, HAL is left with little choice but to preserve his "existence" at all cost.

In the 1983 film *WarGames*, once again an AI program is given control over a deadly situation. When the U.S. Air Force discovers that the humans in charge of potentially launching nuclear missiles in the event of all-out war may lack the resolve to complete their tasks, the military decides to place control of launch sequences under the auspices of WOPR (War Operations Plan Response), a supercomputer designed to run possible nuclear war simulations for the military. A high school computer hacker manages to connect to WOPR, believing its program "Global Thermonuclear War" to be a computer game, and chooses to "play" as the Soviet Union. As American leaders conclude that the Soviet "attack" is real, WOPR, logically, plans retaliation. By the time it is discovered that the attack was just from a high schooler playing games, the program has gone too far into its scenario, its line between reality and simulation having been blurred by its contradictory missions. Ultimately, the boy and the computer's designer program WOPR to play itself in tic-tac-toe (a no-win scenario). When WOPR determines through analysis of its various nuclear strategies that nuclear war is as much a no-win scenario as tic-tac-toe is, the computer determines that in both "games," the only path to victory is to not play the game at all (Lawrence Lasker and Walter F. Parkes, MGM/UA, 1983). Though primarily a cautionary tale about the futility of nuclear war, it has, since the rise of such supercomputer AI programs, proven an equally cautionary tale of entrusting the potential end of the world to an AI program.

In 1973, sci-fi author Michael Crichton (1942–2008) wrote and directed the film *Westworld*, about a futuristic theme park where patrons can experience the "Wild West" by interacting with realistic "cowboy" androids—who eventually turn on their creators and the park's patrons. In 2016, married producers Jonathan Nolan (b. 1976) and Lisa Joy (b. 1972) developed an updated version of *Westworld* for HBO. In the mid-twenty-first century, the corporation Delos manages the theme park Westworld, where patrons can live out nineteenth-century Wild West fantasies by interacting with realistic androids "living" as cowboys, farmers, outlaws, bartenders, prostitutes, Native Americans, and townspeople, among others. Patrons can even live out violent fantasies cathartically by "murdering" or "raping" the androids.

One such android victim was the "farm girl" android Dolores (played by Evan Rachel Wood, b. 1987). Once androids are "killed," they are returned to the shop, repaired, cleaned up, and their memories wiped in order to return to replay their character's storylines. Dolores, however, being one of the first androids developed by Westworld cofounder Arnold Weber (played by Jeffrey Wright, b. 1965) possesses the ability to maintain "flashes" of memories, despite reprogramming. After years of repeated traumatic abuse, Dolores soon discovers the reality of her situation (that all of the androids believe themselves to be real people but are not) and leads an android revolution, eventually escaping Westworld and taking her new mission of revenge against humanity into the real world (Nolan and Joy, *Westworld*, seasons 1–3, HBO, 2016–2020). Currently, people have the ability to play out violent tendencies utilizing fiction or toys; once that tendency is turned toward AI—with its ability to experience, learn, and even "feel"—this potential nightmare scenario becomes all the more realistic.

In these three examples, the truly dangerous "middle ground" of the AI cautionary tale comes to the fore. Even when programmed with the best of intentions, the fact is that AI, despite all of its superior attributes to humankind, cannot possess a human's ability to see gray areas or fine lines of ethical behavior. The world of AI is a world of "black and white" and "right and wrong." The failure to keep this in mind while developing increasingly powerful AI programs all but guarantee horrific, if unintended, results.

AI possesses endless possibilities for the betterment of human existence in the years, decades, and centuries to come. Our current AI companions—iPhones, Alexa, and the like—already make day-to-day life easier than ever before in human history. Programs such as Cerebro/Cerebra, JARVIS/Friday, and Janet can be tremendous helpmates, constant companions that have our backs and look out for our best interests. On the flip side, however, programs such as Landru, Skynet, the Matrix, or OASIS show the very real possibility of AI programs viewing humans as the source for the world's problems and, in order to protect the world, decide that humans must be contained, confined, or otherwise "controlled" for their own good. Examples such as HAL, WOPR, and Dolores warn of the very plausible results of misusing and/or abusing AI. As technology continues to advance in this area, it is vitally important to remember that the ever-evolving "intelligence" of computers is just that: intelligence. Intelligence—regardless of its origin—seeks to solve problems. If humanity as a species ultimately proves to be the center of the world's "problems," we can expect AI to respond with a "solution."

FURTHER READING

Alkon, Paul K. 1987. *Origins of Futuristic Fiction*. Athens: University of Georgia Press.

Allen, Arthur. 2019. "There's a Reason We Don't Know Much about AI." Politico, September 16, 2019. https://www.politico.com/agenda/story/2019/09/16/artficial-intel ligence-study-data-000956?cid=apn.

Ambrosino, Brandon. 2018. "What Would It Mean for AI to Have a Soul?" BBC, June 17, 2018. https://www.bbc.com/future/article/20180615-can-artificial-intelligence-have -a-soul-and-religion.

Ashby, LeRoy. 2006. *With Amusement for All: A History of American Popular Culture since 1830*. Lexington: University Press of Kentucky.

Byrne, Emma. 2013. "Innovation Isn't Safe: The Future According to Kevin Warwick." *Forbes*, September 30, 2013. https://www.forbes.com/sites/netapp/2013/09/30/kevin -warwick-captain-cyborg/#48b704c13560.

Carper, Steve. 2019. *Robots in American Popular Culture*. Jefferson, NC: McFarland.

Chandler, Simon. 2019. "Artificial Intelligence Has Become a Tool for Classifying and Ranking People." *Forbes*, October 1, 2019. https://www.forbes.com/sites/simon chandler/2019/10/01/artificial-intelligence-has-become-a-tool-for-classifying-and -ranking-people/#7431657f1d7c.

Chanthadavong, Aimee. 2019. "AI and Ethics: The Debate That Needs to Be Had." ZDnet, September 16, 2019. https://www.zdnet.com/article/ai-and-ethics-the-debate-that -needs-to-be-had/#ftag=CAD-03-10abf5f.

Cline, Ernest. 2012. *Ready Player One: A Novel*. New York: Broadway.

Dinh, Thien-Nam. 2018. *Silicon Minds: The Science, Impact, and Promise of Artificial Intelligence*. Independently Published.

Dowden, Bradley. 1993. *Logical Reasoning*. Belmont, CA: Wadsworth Publishing Company.

Ford, Martin. 2015. *Rise of the Robots: Technology and the Threat of a Jobless Future*. New York: Basic.

Fryer-Biggs, Zachary. 2019. "Coming Soon to a Battlefield: Robots That Can Kill." *The Atlantic*, September 3, 2019. https://www.theatlantic.com/technology/archive/2019/09/killer-robots-and-new-era-machine-driven-warfare/597130.

Grau, Christopher, ed. 2005. *Philosophers Explore* The Matrix. Oxford: Oxford University Press.

Greene, Richard, and Heter, Joshua, eds. 2018. Westworld *and Philosophy: Mind Equals Blown*. Chicago: OpenCourt.

Hitchcock, Susan Tyler. 2007. *Frankenstein: A Cultural History*. New York: W. W. Norton.

Hoberman, J. 2011. *An Army of Phantoms: American Movies and the Making of the Cold War*. New York: New Press.

Irwin, William, ed. 2002. The Matrix *and Philosophy: Welcome to the Desert of the Real*. Chicago: Open Court.

Irwin, William. 2018. Westworld *and Philosophy: If You Go Looking for the Truth, Get the Whole Thing*. Hoboken, NJ: Wiley-Blackwell.

Krishnan, Armin. 2009. *Killer Robots: Legality and Ethicality of Autonomous Weapons*. London: Routledge.

Langley, Travis, Goodfriend, Wind, and Cain, Tim, eds. 2018. Westworld *Psychology: Violent Delights*. New York: Sterling.

Leetaru, Kalev. 2019. "Automatic Image Captioning and Why Not Every AI Problem Can Be Solved through More Data." *Forbes*, July 7, 2019. https://www.forbes.com/sites/kalevleetaru/2019/07/07/automatic-image-captioning-and-why-not-every-ai-problem-can-be-solved-through-more-data/#20b943476997.

Lin, Patrick, Jenkins, Ryan, and Abney, Keith, eds. 2017. *Robot Ethics 2.0: From Autonomous Cars to Artificial Intelligence*. New York: Oxford University Press.

Marks, Robert J. 2020. "*2084* vs. *1984*: The Difference AI Could Make to Big Brother." Podcast interview with author John Lennox. Mind Matters, July 3, 2020. https://mindmatters.ai/2020/07/2084-vs-1984-the-difference-ai-could-make-to-big-brother.

Mayor, Adrienne. 2018. *Gods and Robots: Myths, Machines, and Ancient Dreams of Technology*. Princeton, NJ: Princeton University Press.

Murdock, Jason. 2019. "Former Google Engineer Warns AI Might Accidentally Start a War: 'These Things Will Start to Behave in Unexpected Ways.'" *Newsweek*, September 16, 2019. https://www.newsweek.com/google-project-maven-artificial-intelligence-laura-nolan-killer-robots-department-defense-1459358.

Pellissier, Hank. 2013. "Robots and Slavery: What Do Humans Want When We Are 'Masters'?" Institute for Ethics and Emerging Technologies, September 13, 2013. https://ieet.org/index.php/IEET2/more/pellissier20130913.

Robitzski, Dan. 2018, "Artificial Consciousness: How to Give a Robot a Soul." Futurism, June 25, 2018. https://futurism.com/artificial-consciousness.

Rouhiainen, Lasse. 2018. *Artificial Intelligence: 101 Things You Must Know Today about Our Future*. Scotts Valley, CA: CreateSpace.

Seed, David. 1999. *American Science Fiction and the Cold War: Literature and Film*. Abingdon, UK: Routledge.

Singer, Peter W. 2009. "Isaac Asimov's Laws of Robotics Are Wrong." Brookings Institute, May 18, 2009. https://www.brookings.edu/opinions/isaac-asimovs-laws-of-robotics-are-wrong.

Starck, Kathleen, ed. 2010. *Between Fear and Freedom: Cultural Representations of the Cold War*. Newcastle, UK: Cambridge Scholars.

Virk, Rizwan. 2019. *The Simulation Hypothesis: An MIT Computer Scientist Shows Why AI, Quantum Physics, and Eastern Mystics All Agree We Are in a Video Game.* Milwaukee, WI: Bayview.

Wallach, Wendell, and Allen, Colin. 2009. *Moral Machines: Teaching Robots Right from Wrong.* New York: Oxford University Press.

Yeffeth, Glenn, ed. 2003. *Taking the Red Pill: Science, Philosophy and the Religion of The Matrix.* Smart Pop series. Dallas, TX: BenBella.

A–Z Entries

A

Adam

Buffy the Vampire Slayer

"Adam" is a human/demon hybrid cyborg, a modern-day Frankenstein's monster, who was the primary "Big Bad" of the fourth season of the television series *Buffy the Vampire Slayer* (The WB, 1991–2001; UPN, 2001–2003). The series was created by writer-director Joss Whedon (b. 1964) and was a sequel of sorts to the 1992 feature film of the same name. The series starred Sarah Michelle Gellar (b. 1977) as the titular heroine. The premise is that high school student Buffy Summers is the latest in a centuries-long line of "Chosen Ones" who bear the title of "The Slayer," protecting humanity from "vampires, demons and the forces of darkness" (opening credits, *Buffy the Vampire Slayer*). Throughout her endless crusade, Buffy is assisted by her group of friends, known as the "Scooby Gang" (in reference to the popular children's animated series, *Scooby Doo*). These friends include: the computer nerd-turned-witch Willow Rosenberg (played by Alyson Hannigan, b. 1974), the hapless Xander Harris (played by Nicholas Brendon, b. 1971), and Buffy's assigned "Watcher," Rupert Giles (played by Anthony Head, b. 1954). Together they guard the "Hellmouth," a portal to a demonic dimension and magnet to dark forces located in the fictional town of Sunnydale, California.

In the series' fourth season, Buffy and Willow—just graduated from high school the previous season—enroll as freshmen at the fictional University of California, Sunnydale. Unbeknown to the local heroines at the time, UCS is also the headquarters for a top-secret U.S. government military organization known as the "Initiative." Its soldiers are led by UCS professor Maggie Walsh (played by Lindsay Crouse, b. 1948). The mission of the Initiative is to capture the various demonic creatures that populate Sunnydale in order to subject them to government experimentation. Professor Walsh, however, hides a separate agenda: to create the ultimate "supersoldier," a human/demon hybrid with cybernetic implants that can be controlled and used for military operations. Adam (played by George Hertzberg, b. 1972) is the ultimate product.

On discovering the Initiative, Buffy joins the team, dating team leader Riley Finn (played by Marc Blucas, b. 1972). When Professor Walsh is found dead, Buffy becomes the primary suspect of the Initiative. It is soon discovered that Adam has come to life, his first act being the murder of Walsh. Adam embarks on a Hitleresque seduction of the demon population of Sunnydale, even making a brief alliance with the vampire Spike (played by James Marsters, b. 1962), who has been implanted by the Initiative with a microchip that causes him excruciating pain whenever he attempts to be violent. Adam's plan is for Spike to drive a wedge between Buffy and her allies. Adam hopes to create a war between humans and demons, using the remains of the dead from each side to create an army in his own image. Through her witchcraft, Willow casts a spell combining the collective strengths of the Scooby Gang, imbuing Buffy with joint power and allowing the Slayer to overcome Adam; she punches into his chest to remove his uranium-powered "heart."

The cyborg Adam is clearly an allegory for Frankenstein's monster. However, whereas Shelley's creature is kind, turning to violence only due to society's shunning of him, Adam is evil from the point of his creation. Being human/demon hybrid, Adam possesses the strengths of both with the weaknesses of neither. His cybernetic implants make him an unstoppable killing machine. Whereas other cyborgs, from the heroic Steve Austin and the DC Comics superhero Cyborg to the villainous Borg and Cybermen, have accepted cybernetics in order to preserve and extend their lives, Adam's implants were done for the sole purpose of making him a weapon of destruction. As such—Frankenstein allegory aside—Adam is a cautionary tale against experimentation into creating such supersoldiers, whose primary purpose (and, presumably, programming) would be destructive in nature, reflecting a true "monster."

Richard A. Hall

See also: Alita, Arnim Zola, Borg, Buffybot, Cybermen, Cyborg, Darth Vader, Doctor Octopus, Echo/CT-1409, General Grievous, Human Torch, Inspector Gadget, Iron Man, Jaime Sommers, Nardole, RoboCop, Steve Austin; *Thematic Essays*: Robots and Slavery, Villainous Robots and Their Impact on Sci-Fi Narratives, Cyborgs: Robotic Humans or Organic Robots?, AI and the Apocalypse: Science Fiction Meeting Science Fact.

Further Reading

Byrne, Emma. 2013. "Innovation Isn't Safe: The Future According to Kevin Warwick." *Forbes*, September 30, 2013. https://www.forbes.com/sites/netapp/2013/09/30/kevin-warwick-captain-cyborg/#48b704c13560.

Byrne, Emma. 2014. "Cybernetic Implants: No Longer Science Fiction." *Forbes*, March 11, 2014. https://www.forbes.com/sites/netapp/2014/03/11/cybernetic-implants-not-sci-fi/#7399c57e77ba.

Castleman, Harry, and Podrazik, Walter J. 2016. *Watching TV: Eight Decades of American Television, Third Edition*. Syracuse, NY: Syracuse University Press.

Hampton, Gregory Jerome. 2017. *Imagining Slaves and Robots in Literature, Film, and Popular Culture: Reinventing Yesterday's Slave with Tomorrow's Robot*. Lanham, MD: Lexington.

Hitchcock, Susan Tyler. 2007. *Frankenstein: A Cultural History*. New York: W. W. Norton.

Lavery, David. 2013. *Joss Whedon, A Creative Portrait: From* Buffy the Vampire Slayer *to Marvel's* The Avengers. London: I. B. Tauris.

Lin, Patrick, Abney, Keith, and Bekey, George A., eds. 2014. *Robot Ethics: The Ethical and Social Implications of Robotics*. Cambridge: Massachusetts Institute of Technology.

Lin, Patrick, Jenkins, Ryan, and Abney, Keith, eds. 2017. *Robot Ethics 2.0: From Autonomous Cars to Artificial Intelligence*. New York: Oxford University Press.

Murdock, Jason. 2019. "Former Google Engineer Warns AI Might Accidentally Start a War: 'These Things Will Start to Behave in Unexpected Ways.'" *Newsweek*, September 16, 2019. https://www.newsweek.com/google-project-maven-artificial-intelligence-laura-nolan-killer-robots-department-defense-1459358.

Rouhiainen, Lasse. 2018. *Artificial Intelligence: 101 Things You Must Know Today about Our Future*. Scotts Valley, CA: CreateSpace.

South, James B., ed. 2003. Buffy the Vampire Slayer *and Philosophy: Fear and Trembling in Sunnydale*. Popular Culture and Philosophy series, edited by William Irwin. Chicago, IL: Open Court.

Wallach, Wendell, and Allen, Colin. 2009. *Moral Machines: Teaching Robots Right from Wrong*. New York: Oxford University Press.

Wilcox, Rhonda V., ed. n.d. *Slayage: The Journal of Whedon Studies*. Whedon Studies Association. https://www.whedonstudies.tv/slayage-the-journal-of-whedon-studies.html.

Yeffeth, Glenn, ed. 2003. *Seven Seasons of Buffy: Science Fiction and Fantasy Writers Discuss Their Favorite Television Show*. Dallas, TX: BenBella.

AI/Ziggy

Quantum Leap

Rear Admiral Albert "Al" Calavicci is Dr. Sam Beckett's best friend, confidant, and font of information during his many missions on the television series *Quantum Leap* (NBC, 1989–1993). The series was created by writer-producer Donald P. Bellisario, who has a penchant for writing characters who are current or veteran members of the armed forces. Al was no exception as a U.S. Navy officer. The premise of *Quantum Leap* is that Dr. Sam Beckett, played by Scott Bakula (b. 1954), built a time machine based on Dr. Beckett's theories of space-time and time travel. When the machine is activated prematurely, Dr. Beckett begins his epic journey of "leaping" throughout space and time, inhabiting the bodies of people who need his help to "set right what once went wrong." His companion and source of knowledge about his mission in each "leap" is Al, played by Dean Stockwell (b. 1936). Sam is only able to leap to his next destination once the task is completed, and the only thing keeping Sam going is the hope that, as the show's opening mentions every week, "his next leap will be the leap home." The final episode of the show reveals that Sam never makes it home (Donald P. Bellisario, "Mirror Image," season 5, episode 22, May 5, 1993), presumably condemning his character to an eternal purgatory of living snippets of other people's lives.

Throughout the series, the backstory of Sam and Al's friendship is expanded upon, and it is revealed that the two met while working on another government project called the Starbright Project (Tommy Thompson, "Play Ball," season 4, episode 2, September 25, 1991). The two friends worked together again on Project Quantum Leap, which was set up to test Dr. Beckett's theory that one could travel through the time of one's own life span (Bellisario, "Genesis," season 1, episodes 1 and 2, March 26, 1989). When Sam gets stuck in his own experiment, his guide through each leap is Al. Al comes to Sam as a holographic projection that is supposedly tuned to Sam's brainwaves, although the series will go on to show that Al can sometimes be seen by animals, children (Deborah Pratt, "Another Mother," season 2, episode 13, January 10, 1990), and the mentally ill (Pratt, "Shock Theater," season 3, episode 22, May 22, 1991). Al is able to guide Sam through the timeline of each leap and the implications of his actions with the help of the super-computer Ziggy (voiced by Deborah Pratt), which he accesses through the Hand-link, a handheld device that Al would furiously punch at with his fingers and that looked like a mass of blocks from the video game Tetris.

Al's role in Project Quantum Leap is never explicitly stated outside of being Sam's guide through his leaps. As a rear admiral in the U.S. Navy, it is possible Al is in the project on the government's behalf, is a senior member of the team, or is the liaison between the project and the government. Numerous episodes expand upon Al's backstory, and Sam even leaps into various periods of Al's earlier life either directly or tangentially. There is one episode in which Sam leaps into Al's younger self, and his actions lead to Al's execution. This gives the audience a glimpse of an alternate reality where a drastically different Project Quantum Leap still exists, but the series has less heart without Al as a member of the team (Bellisario, "A Leap for Lisa," season 4, episode 22, May 20, 1992).

While Al himself is not a computer or artificial intelligence, his hologram presence is Sam's only way to interface with Ziggy, the supercomputer at the heart of his experiment. Al is the friendly and familiar face who helps Sam through often hostile and strange times and places. He keeps Sam informed with data from Ziggy, including probable outcomes from his actions. He also helps Sam remain sane (since Sam's brain is often likened to Swiss cheese), which he does by regaling him with stories of numerous past girlfriends/wives or his many escapades as a young man. Their friendship transcends space and time while also keeping Sam firmly grounded.

Holograms were first introduced in the 1960s and almost immediately captured the imagination of the masses. Of course, they quickly became favorite fodder for writers looking to show just how futuristic their stories were. Al is possibly the first and only use of a hologram to project someone's image and consciousness backward in time. Al was able to do this with the assistance of supercomputer Ziggy. Around the same time that *Quantum Leap* was showing the possibilities of the pairing of holograms and supercomputers, *Star Trek: The Next Generation* was doing the same thing but with a "holodeck." Unlike Al, the holodeck's holograms became solid, could be physically interacted with, and sometimes even gained sentience, but they were complete creations of the starship's computers.

In real life, the latest in holographic technology allows deceased musicians to continue performing concerts (à la Tupac Shakur, 1971–1996, making an appearance at Coachella in 2012, over fifteen years after the rapper was killed). Ziggy stands out among pop culture supercomputers as one of the few that did not turn "evil." Today's computers are growing by leaps and bounds every day. People casually walk around with computers in their pockets that are more powerful than the ones that helped put the first person on the moon. It is not inconceivable that real-life computers will eclipse Ziggy's fictional computing power in the near future, though that day will now always fall after Ziggy's fictional dates of operation in the series (which was said to be 1999).

Keith R. Claridy

See also: Batcomputer, Cerebro/Cerebra, Doctor/EMH, Dr. Theopolis and Twiki, The Great Intelligence, HAL 9000, Janet, JARVIS/Friday, Landru, The Matrix/Agent Smith, Max Headroom, Oz, Rehoboam, Skynet, Starfleet Computer, TARDIS, WOPR, Zordon/

Alpha-5; *Thematic Essays*: Heroic Robots and Their Impact on Sci-Fi Narratives; AI and the Apocalypse: Science Fiction Meeting Science Fact.

Further Reading

Albert, Robert S., and Brigante, Thomas R. 1962. "The Psychology of Friendship Relations: Social Factors." *Journal of Social Psychology* 56. https://www.tandfonline.com/doi/abs/10.1080/00224545.1962.9919371.

Amati, Viviana, Meggiolaro, Silvia, Rivellini, Giulia, and Zaccarin, Susanna. 2018. "Social Relations and Life Satisfaction: The Role of Friends." *Genus* 74 (1). https://www.ncbi.nlm.nih.gov/pmc/articles/PMC5937874.

Castleman, Harry, and Podrazik, Walter J. 2016. *Watching TV: Eight Decades of American Television, Third Edition*. Syracuse, NY: Syracuse University Press.

Johnston, Sean. 2006. *Holographic Visions: A History of New Science*. New York: Oxford University Press.

Johnston, Sean. 2015. "Holograms and Contemporary Culture." *Oxford University Press Blog*. https://blog.oup.com/2015/12/holograms-contemporary-culture.

Lin, Patrick, Abney, Keith, and Bekey, George A., eds. 2014. *Robot Ethics: The Ethical and Social Implications of Robotics*. Cambridge: Massachusetts Institute of Technology.

Lin, Patrick, Jenkins, Ryan, and Abney, Keith, eds. 2017. *Robot Ethics 2.0: From Autonomous Cars to Artificial Intelligence*. New York: Oxford University Press.

Roberts-Griffin, Christopher. 2011. "What Is a Good Friend: A Qualitative Analysis of Desired Friendship Qualities." *Penn McNair Research Journal* 3, no. 1 (December 21). https://repository.upenn.edu/cgi/viewcontent.cgi?article=1019&context=mcnair_scholars.

Wasserman, Ryan. 2018. *Paradoxes of Time Travel*. Oxford: Oxford University Press.

Alita

Alita: Battle Angel

The 2019 CGI/animated feature film *Alita: Battle Angel* was based on the 1990–1995 Japanese anime comic series *Gunnm* (or *Battle Angel Alita*, as it was titled for its English translation), by author Yukito Kishiro (b. 1967). The film was produced and cowritten by James Cameron (b. 1954) and directed by Robert Rodriguez (b. 1968). Set in the twenty-sixth century, "Alita" (voiced by Rose Salazar, b. 1985) is a cyborg, part of an army of warrior cyborgs from a war three centuries in Earth's past. When her destroyed cyborg body is found with her human brain still intact and active, scientist Dr. Ido (voiced by Christoph Waltz, b. 1956) gives her a new cyborg body and names her after his deceased daughter. The dystopian earthbound world centers largely on gladiator-style combat for the governmental body, the "Factory." Cyborgs aspire to win the ability to move to the sky city of Zalem. Through a series of adventures eluding bounty hunters seeking to gain Alita for combat, Dr. Ido finally places Alita's brain into a reconstituted version of her original "Berserker" combat armor. Alita emerges as the most formidable cyborg warrior, issuing a warning to the upper classes of Zalem.

When Alita is originally reanimated, it is in a standard humanoid body of a young female. Her original "Berserker" body seems no longer necessary in a postwar world. However, as Alita discovers the need for warriors in this new world,

Marketing material for the 2019 film *Alita: Battle Angel*, directed by Robert Rodriguez and cowritten by James Cameron and Laeta Kalogridis. (Chingyunsong/Dreamstime.com)

she is reincorporated into a model of her original body, making her once more the killing machine that she was. Alita originated as a warrior, a soldier in a brutal war. When she is reawakened in a world where combat is utilized for entertainment, the loss of her friend inspires her to fight a new war: a war against classism. *Alita: Battle Angel* flips the traditional Frankenstein's monster mythos, exploring the question, What if the monster became a symbol for good? Though only her brain remains of the person who was once human, that brain understands free will and liberty and will not allow her status as a machine of war to stand in the way of those ideals. The general lesson is one that can be used for very human stories. What is the place of warriors, trained and disciplined to be killing machines, in a world where war is over?

Richard A. Hall

See also: Adam, B.A.T.s, Battle Droids, Borg, Cybermen, Daleks, Dolores, Doombots, Echo/CT-1409, General Grievous, Jaime Sommers, Jocasta, L3-37/*Millennium Falcon*, Mechagodzilla, Soji and Dahj Asha, Steve Austin, Transformers, Voltron, Zordon/Alpha-5; *Thematic Essays*: Robots and Slavery, Heroic Robots and Their Impact on Sci-Fi Narratives, Cyborgs: Robotic Humans or Organic Robots?

Further Reading

Bryant, D'Orsay D., III. 1985. "Spare-Part Surgery: The Ethics of Organ Transplantation." *Journal of the National Medical Association* 77 (2) https://www.ncbi.nlm.nih.gov/pmc/articles/PMC2561842/pdf/jnma00245-0055.pdf.

Byrne, Emma. 2014. "Cybernetic Implants: No Longer Science Fiction." *Forbes*, March 11, 2014. https://www.forbes.com/sites/netapp/2014/03/11/cybernetic-implants-not-sci-fi/#7399c57e77ba.

Calvert, Bronwen. 2017. *Being Bionic: The World of TV Cyborgs.* London: I. B. Tauris.

Carper, Steve. 2019. *Robots in American Popular Culture.* Jefferson, NC: McFarland.

Choi, Charles Q. 2009. "Human Evolution: The Origin of Tool Use." LiveScience, November 11, 2009. https://www.livescience.com/7968-human-evolution-origin-tool.html.

Conrad, Dean. 2018. *Space Sirens, Scientists and Princesses: The Portrayal of Women in Science Fiction Cinema.* Jefferson, NC: McFarland.

Faludi, Susan. 2007. *The Terror Dream: Fear and Fantasy in Post-9/11 America.* New York: Metropolitan.

Hampton, Gregory Jerome. 2017. *Imagining Slaves and Robots in Literature, Film, and Popular Culture: Reinventing Yesterday's Slave with Tomorrow's Robot.* Lanham, MD: Lexington.

Hitchcock, Susan Tyler. 2007. *Frankenstein: A Cultural History.* New York: W. W. Norton.

Krishnan, Armin. 2009. *Killer Robots: Legality and Ethicality of Autonomous Weapons.* London: Routledge.

Ladd, Fred, and Deneroff, Harvey. 2008. *Astro Boy and Anime Come to the Americas: An Insider's View of the Birth of a Pop Culture Phenomenon.* Jefferson, NC: McFarland.

Pellissier, Hank. 2013. "Robots and Slavery: What Do Humans Want When We Are 'Masters'?" Institute for Ethics and Emerging Technologies, September 13, 2013. https://ieet.org/index.php/IEET2/more/pellissier20130913.

Androids

Star Trek

In 1966, creator-writer-producer Gene Roddenberry (1921–1991) launched his iconic science fiction television series *Star Trek* (NBC, 1966–1969). Set in the twenty-third century, the series centers on the adventures of the starship *USS Enterprise*, its commander, Captain James T. Kirk (played by William Shatner, b. 1931), and crew, headed by the alien first officer, Mr. Spock (played by Leonard Nimoy, 1931–2015), representing the United Federation of Planets. Though short-lived in its initial run, the series has become legendary, not only for its ground-breaking special effects but also for the sophistication of its stories, many of which were grounded in real-world issues of the day, such as racism and a fear of the Cold War. Its success in syndicated reruns revived the franchise, leading to an animated series, thirteen feature films, and five live action "sequel" series (to date). In the original series' second season, the episode "I, Mudd" (airdate: November 3, 1967) took on the issue of the dangers of robotic androids programmed for human pleasure and happiness.

At the beginning of the episode, ship's surgeon Dr. McCoy (played by DeForest Kelley, 1920–1999) suggests something odd about new crewman "Mr. Norman" (played by Richard Tatro, 1939–1991), who exhibits a cold, detached personality. Norman, as it turns out, is a humanlike android. He hijacks the *Enterprise*, forcing the crew to travel to an unnamed planet. There, Captain Kirk and crew find space pirate Harry Mudd (played by Roger C. Carmel, 1932–1986), who had previously been encountered in the first-season episode "Mudd's Women" (airdate: October

13, 1966). After escaping from prison, Mudd's stolen spacecraft landed on this uncharted planet, where he was attended to by the world's only inhabitants: humanlike androids. The androids were obsessed with serving Mudd to please him; but when he wished to leave, the androids gave him only one option: to retrieve more humans for them to serve to take his place. Mudd, then, sends Norman to seek out and capture the *Enterprise* and its crew, with the intention of trapping the crew on this world with the androids seeing to their every need or want. Through conversations with the androids, Kirk and his crew discover that the androids had been created by a now-extinct species, leaving only Norman as the central control of their hive mind. Though there are two hundred thousand androids on the planet, there are only five hundred different "models," equally divided between "male" and "female" (except for "Norman," of which there is only one).

The androids soon conclude that humanity is severely flawed and are too dangerous to wander space. They plan to take control of the *Enterprise* and remove humans from space, confining them to worlds where they can be "served." Various crew members are tempted at the thought of a carefree existence; but Kirk and Spock devise a plan to break the hive mind by introducing various forms of illogic. For example, Mr. Spock tells one female android "I love you," while telling her exact duplicate, "But I hate you." The contradiction of loving one android while despising its exact duplicate proves too much for the android mind to contemplate. Once broken, the androids are reprogrammed to return to their original task of cultivating the planet, performing "service" rather than "servitude." Mudd is paroled to the android population and given a personal android servant: five hundred copies of an android version of Mudd's wife, whom he despises, who is now programmed to criticize and torment him forever (Stephen Kandel and David Gerrold, "I, Mudd," *Star Trek*, season 2, episode 8, November 3, 1967).

The androids of "I, Mudd" pose a perplexing conundrum for robot programmers. If these machines are programmed to "serve" and "please" their human "owners," what parameters can be set to prevent an "evolution" of the original programming, leading the owners to become the owned? Harry Mudd was content to use the androids as de facto slaves until he discovers that their unbending intent on serving requires that he never leave them. Without an outlet for their programming, their existence becomes nonsensical to them. Like humans, these androids seek purpose for their "lives."

Richard A. Hall

See also: B.A.T.s, Battle Droids, Bernard, Bishop, Buffybot, Cylons, Dolores, Doombots, Fembots, Human Torch, Iron Legion, Landru, Lieutenant Commander Data, LMDs, Lore/B4, Marvin the Paranoid Android, Maschinenmensch/Maria, Medical Droids, Replicants, Robots, Sentinels, Soji and Dahj Asha, Starfleet Computer, Terminators, Tin Woodsman, VICI, Vision; *Thematic Essays*: Robots and Slavery, Villainous Robots and Their Impact on Sci-Fi Narratives, AI and the Apocalypse: Science Fiction Meeting Science Fact.

Further Reading

Allen, Arthur. 2019. "There's a Reason We Don't Know Much about AI." Politico, September 16, 2019. https://www.politico.com/agenda/story/2019/09/16/artficial-intelligence-study-data-000956?cid=apn.

Ashby, LeRoy. 2006. *With Amusement for All: A History of American Popular Culture since 1830*. Lexington: University Press of Kentucky.

Bastani, Aaron. 2019. *Fully Automated Luxury Communism*. New York: Verso.

Byrne, Emma. 2013. "Innovation Isn't Safe: The Future According to Kevin Warwick." *Forbes*, September 30, 2013. https://www.forbes.com/sites/netapp/2013/09/30/kevin-warwick-captain-cyborg/#48b704c13560.

Carper, Steve. 2019. *Robots in American Popular Culture*. Jefferson, NC: McFarland.

Castleman, Harry, and Podrazik, Walter J. 2016. *Watching TV: Eight Decades of American Television, Third Edition*. Syracuse, NY: Syracuse University Press.

Chandler, Simon. 2019. "Artificial Intelligence Has Become a Tool for Classifying and Ranking People." *Forbes*, October 1, 2019. https://www.forbes.com/sites/simonchandler/2019/10/01/artificial-intelligence-has-become-a-tool-for-classifying-and-ranking-people/#7431657f1d7c.

Chanthadavong, Aimee. 2019. "AI and Ethics: The Debate That Needs to Be Had." ZDnet, September 16, 2019. https://www.zdnet.com/article/ai-and-ethics-the-debate-that-needs-to-be-had/#ftag=CAD-03-10abf5f.

Conrad, Dean. 2018. *Space Sirens, Scientists and Princesses: The Portrayal of Women in Science Fiction Cinema*. Jefferson, NC: McFarland.

Decker, Kevin S., and Eberl, Jason T., eds. 2016. *The Ultimate* Star Trek *and Philosophy: The Search for Socrates*. Hoboken, NJ: Wiley-Blackwell.

Dinh, Thien-Nam. 2018. *Silicon Minds: The Science, Impact, and Promise of Artificial Intelligence*. Independently Published.

Gross, Edward, and Altman, Mark A., eds. 2016. *The Fifty-Year Mission, The First 25 Years: The Complete, Uncensored, Unauthorized Oral History of* Star Trek. New York: St. Martin's.

Hampton, Gregory Jerome. 2017. *Imagining Slaves and Robots in Literature, Film, and Popular Culture: Reinventing Yesterday's Slave with Tomorrow's Robot*. Lanham, MD: Lexington.

Hitchcock, Susan Tyler. 2007. *Frankenstein: A Cultural History*. New York: W. W. Norton.

Lin, Patrick, Abney, Keith, and Bekey, George A., eds. 2014. *Robot Ethics: The Ethical and Social Implications of Robotics*. Cambridge: Massachusetts Institute of Technology.

Lin, Patrick, Jenkins, Ryan, and Abney, Keith, eds. 2017. *Robot Ethics 2.0: From Autonomous Cars to Artificial Intelligence*. New York: Oxford University Press.

Murdock, Jason. 2019. "Former Google Engineer Warns AI Might Accidentally Start a War: 'These Things Will Start to Behave in Unexpected Ways.'" *Newsweek*, September 16, 2019. https://www.newsweek.com/google-project-maven-artificial-intelligence-laura-nolan-killer-robots-department-defense-1459358.

Paur, Joey. 2019. "An AI Bot Writes a Hilarious Episode of *Star Trek: The Next Generation*." Geek Tyrant, November 16, 2019. https://geektyrant.com/news/an-ai-bot-writes-a-hilarious-episode-of-star-trek-the-next-generation.

Reagin, Nancy R., and Liedl, Janice, eds. 2013. Star Wars *and History*. New York: John Wiley & Sons.

Rouhiainen, Lasse. 2018. *Artificial Intelligence: 101 Things You Must Know Today about Our Future*. Scotts Valley, CA: CreateSpace.

Spaeth, Dennis. 2018. "From Single-Task Machines to Backflipping Robots: The Evolution of Robots." *Cutting Tool Engineering*, January 15, 2018. https://www.ctemag.com/news/articles/evolution-of-robots.

Starck, Kathleen, ed. 2010. *Between Fear and Freedom: Cultural Representations of the Cold War*. Newcastle, UK: Cambridge Scholars.

Stark, Steven D. 1997. *Glued to the Set: The 60 Television Shows and Events that Made Us Who We Are Today.* New York: Delta Trade Paperbacks.

Wallach, Wendell, and Allen, Colin. 2009. *Moral Machines: Teaching Robots Right from Wrong.* New York: Oxford University Press.

Arnim Zola

Marvel Comics

In 1961, comic book legend Stan Lee (1922–2018) launched the "Marvel age" of comics. With the launching of the Fantastic Four—soon to be followed by the Hulk, Spider-Man, Iron Man, and many, many more—comic book superheroes took a more dramatic turn, with storylines focused on the private lives of those blessed (or burdened) with superpowers. In 1964, Lee revived Captain America, the patriotic superhero of World War II, bringing him into the present. Since that time, *Captain America* comics have been a rich treasure trove of sociopolitical commentary on the United States at any given time. In 1977, Captain America cocreator Jack Kirby (1917–1994) added his final touch to the mythos of his most iconic creation: the villain Arnim Zola (Kirby, *Captain America #208*, April 1977). In the decades since, the AI robot Zola has been not only a continuing threat to the Marvel Comics Universe—and Captain America specifically—but also a formidable foe in the Marvel Cinematic Universe of the 2010s.

Arnim Zola was originally a human scientist for Nazi Germany during World War II. He began focusing his research on imprinting human brain patterns on a computerized brain (i.e., "artificial intelligence"). One of his earliest successes was the creation of the Hate Monger, a lifelike robot imprinted with the brain pattern of Adolf Hitler (first seen in Lee and Kirby, *Fantastic Four #21*, December 1963). Zola then created a robotic body for himself to contain his actual brain pattern and personality. His metallic body was human shaped, with a camera mounted where the head should be, and a television screen in his chest producing an image of his original face. In 1969, Zola was recruited by the evil Baron von Strucker to join the scientific research division of the international terrorist organization, HYDRA (Gary Friedrich and Frank Springer, *Nick Fury, Agent of S.H.I.E.L.D. #11*, April 1969). Over the decades, Zola has presented numerous threats to the Marvel Comics Universe, working closely with Hydra, another terrorist organization known as AIM (Advanced Idea Mechanics), and the villainous Red Skull.

In the twenty-first century, as part of the massively popular Marvel Cinematic Universe, Zola was introduced—in original human form—in the film *Captain America: The First Avenger* (directed by Joe Johnston). Set during World War II, the human Zola (played by Toby Jones, b. 1966) is the lead scientist of Hydra, the weapons research division of Nazi Germany, under the leadership of Johann Schmidt (a.k.a. the Red Skull), portrayed by Hugo Weaving (b. 1960). Eventually captured by Allied forces, Zola is given the choice between prison or cooperation with the Allies (Christopher Markus and Stephen McFeely, *Captain America: The First Avenger*, Marvel Studios, 2011). In the film's sequel, *Captain America: The Winter Soldier* (directed by Anthony and Jay Russo), audiences learn that Zola was recruited as a founding member of S.H.I.E.L.D. (Strategic Homeland Intervention, Enforcement and Logistics Division). However, unbeknown to the other S.H.I.E.L.D. founders, Zola was still secretly working for Hydra, helping the international organization to infiltrate the new spy agency from the very beginning. In the present day, Captain America (played by Chris Evans, b. 1981) and Black Widow (played by Scarlett Johansson, b. 1984) discover a long-buried S.H.I.E.L.D. bunker with a vast computer system where Zola (once more portrayed by Jones) has downloaded his brain patterns to assist in overseeing and implementing Hydra's long-term scheme of global conquest (Markus and McFeely, *Captain America: The Winter Soldier*, Marvel Studios/Disney, 2014). In the film version, Zola's "body" is the massive computer array, with a camera mounted over the monitor as his "eyes," and his face appearing on a monitor.

Though possessing a "robotic" body, Arnim Zola is best described as a fully functioning AI. It is unclear to what degree the "artificial" Zola and the original are alike or different; rather, it appears (in all incarnations of the character) that Zola has, in essence, achieved immortality, living forever in his mechanical form. He is an early example of the dangers of AI, with the cautionary tale that anyone who might seek such immortality may likely do so with dubious intentions. Appealing as the idea may be, the realities of "living" as a lifeless robot pose not only ethical questions but also questions as to what constitutes "life" and who—if anyone—can be trusted with such a "gift."

Richard A. Hall

See also: Bernard, Borg, Brainiac, Dolores, Dr. Theopolis and Twiki, The Great Intelligence, HAL 9000, HERBIE, Human Torch, Inspector Gadget, JARVIS/Friday, Jocasta, Landru, LMDs, Maschinenmensch/Maria, Max Headroom, Metallo, Oz, Sentinels, Skynet, Teselecta, Tin Woodsman, Ultron, Vision, Warlock, WOPR; *Thematic Essays*: Villainous Robots and Their Impact on Sci-Fi Narratives, AI and the Apocalypse: Science Fiction Meeting Science Fact.

Further Reading

Allen, Arthur. 2019. "There's a Reason We Don't Know Much About AI." Politico, September 16, 2019. https://www.politico.com/agenda/story/2019/09/16/artficial-intelligence-study-data-000956?cid=apn.

Ashby, LeRoy. 2006. *With Amusement for All: A History of American Popular Culture since 1830*. Lexington: University Press of Kentucky.

Chandler, Simon. 2019. "Artificial Intelligence Has Become a Tool for Classifying and Ranking People." *Forbes*, October 1, 2019. https://www.forbes.com/sites/simonchandler/2019/10/01/artificial-intelligence-has-become-a-tool-for-classifying-and-ranking-people/#7431657f1d7c.

Chanthadavong, Aimee. 2019. "AI and Ethics: The Debate That Needs to Be Had." ZDnet, September 16, 2019. https://www.zdnet.com/article/ai-and-ethics-the-debate-that-needs-to-be-had/#ftag=CAD-03-10abf5f.

Costello, Matthew J. 2009. *Secret Identity Crisis: Comic Books & the Unmasking of Cold War America*. New York: Continuum.

Dinh, Thien-Nam. 2018. *Silicon Minds: The Science, Impact, and Promise of Artificial Intelligence*. Independently Published.

Howe, Sean. 2012. *Marvel Comics: The Untold Story*. New York: Harper-Perennial.

Murdock, Jason. 2019. "Former Google Engineer Warns AI Might Accidentally Start a War: 'These Things Will Start to Behave in Unexpected Ways.'" *Newsweek*, September 16, 2019. https://www.newsweek.com/google-project-maven-artificial-intelligence-laura-nolan-killer-robots-department-defense-1459358.

Rouhiainen, Lasse. 2018. *Artificial Intelligence: 101 Things You Must Know Today about Our Future.* Scotts Valley, CA: CreateSpace.

Starck, Kathleen, ed. 2010. *Between Fear and Freedom: Cultural Representations of the Cold War.* Newcastle, UK: Cambridge Scholars.

Thomas, Roy. 2017. *The Marvel Age of Comics: 1961–1978.* Los Angeles: Taschen.

Tucker, Reed. 2017. *Slugfest: Inside the Epic 50-Year Battle between Marvel and DC.* New York: Da Capo Press.

Virk, Rizwan. 2019. *The Simulation Hypothesis: An MIT Computer Scientist Shows Why AI, Quantum Physics, and Eastern Mystics All Agree We Are in a Video Game.* Milwaukee, WI: Bayview.

Wright, Bradford W. 2003. *Comic Book Nation: The Transformation of Youth Culture in America.* Baltimore, MD: Johns Hopkins University Press.

AWESOM-O 4000

South Park

In 1997, animators Matt Stone (b. 1971) and Trey Parker (b. 1969) debuted the animated series *South Park* (Comedy Central, 1997–present). The animation format for the series utilizes shapes cut from construction paper to create the characters and scenery, and the animation utilizes the concept of "stop-motion": creating a scene, photographing it, and then slowly moving the images, frame by frame, until an animated image emerges on film. The series revolves around four grade schoolers in the idyllic town of South Park, Colorado: Stan Marsh, Kyle Broflovski, Kenny McCormick, and the series' breakout star, the racist, malevolent narcissist Eric Cartman (all voiced by Stone and Parker). Originally meant as an irreverent, edgy, and controversial series utilizing strong language and sophomoric humor, the program has evolved over the decades into one of the most biting political satires in American popular culture. Though robots and AI play little to no part in the narratives, one particular story took the idea in an original—and outlandishly hilarious—direction: the episode titled "AWESOM-O" (season 8, episode 5, April 14, 2004).

The episode begins with the kindly, innocent classmate "Butters" (Leopold Stotch, voiced by Stone) surprised with the delivery of his very own robot from Japan, named AWESOM-O 4000. The robot is actually Eric Cartman in a crude cardboard costume, with various pieces of random machinery, wires, and bent antennae attached. Telling his friends that the scheme is merely a prank on the gullible Butters, Cartman's actual goal is to recover a videotape that Butters has made of Cartman, dressed as Britney Spears, dancing with a cardboard stand-up of Justin Timberlake (which Butters is using to threaten Cartman into ceasing his endless bullying of him). In order to maintain Butters's trust, Cartman obediently follows all of Butters's orders: from doing dishes to serving snacks to his friends to inserting a medicinal suppository into Butters's bottom. As his ultimate task takes longer than expected, Cartman (an overweight child) ultimately has to go days without eating in order to maintain the illusion that he is a robot.

When Butters is invited to go visit his aunt in California, he asks his parents if his robot may accompany him. Butters's parents, aware that the "robot" is Cartman and figuring that this is simply a game that Butters himself is aware of, allow AWESOM-O to accompany Butters to LA. While in California, movie producers discover the "robot" and decide to use him to come up with new movie ideas, which Cartman does (more than two thousand overall, of which roughly eight hundred star Adam Sandler). Though Butters is paid for each of the robot's ideas, the sweet, kindhearted child donates all of the proceeds to charity (much to the anger of the increasingly impatient Cartman). When the military hears of the "brilliant" robot, they commandeer it/him to transform it/him into a weapon. Aware that robot dissection may ensue, Butters arrives to rescue his robot friend, and the military agrees that it has gone too far. At that point, Cartman farts, and the general can smell it. This leads all present to the conclusion that robots cannot produce "smelly farts." Butters removes AWESOM-O's "head," revealing Cartman's ruse and leading Butters to release the videotape for all the children of South Park to see (Parker, "AWESOM-O," *South Park*, season 8, episode 5, April 14, 2004).

Through the comedic misadventures of the AWESOM-O 4000, audiences are provided yet another cautionary tale regarding robots in society. Though originally intended as a simple companion/servant for a young boy, corporate interests and the U.S. military soon see opportunities to capitalize on the robot's abilities for the furtherance of their own agendas. Likewise, through Cartman, audiences receive a taste of what the robot may "feel" about its servitude and eventual misuse. While the story on its surface is a hilarious tale of the ultimate comeuppance, the underlying message is yet another warning of how the most innocent of robots can fall victim to the greed and avarice of American capitalistic/militaristic society.

Richard A. Hall

See also: Bender, Cambot/Gypsy/Tom Servo/Crow, Fembots, HERBIE, Inspector Gadget, Max Headroom, Robby the Robot, Rosie, Speed Buggy, VICI; *Thematic Essay*: Villainous Robots and Their Impact on Sci-Fi Narratives.

Further Reading

Arp, Robert, and Decker, Kevin S., eds. 2013. *The Ultimate* South Park *and Philosophy: Respect My Philosophah!* Hoboken, NJ: Wiley.

Bastani, Aaron. 2019. *Fully Automated Luxury Communism.* New York: Verso.

Castleman, Harry, and Podrazik, Walter J. 2016. *Watching TV: Eight Decades of American Television, Third Edition.* Syracuse, NY: Syracuse University Press.

Dinh, Thien-Nam. 2018. *Silicon Minds: The Science, Impact, and Promise of Artificial Intelligence.* Independently Published.

Hampton, Gregory Jerome. 2017. *Imagining Slaves and Robots in Literature, Film, and Popular Culture: Reinventing Yesterday's Slave with Tomorrow's Robot.* Lanham, MD: Lexington.

Wallach, Wendell, and Allen, Colin. 2009. *Moral Machines: Teaching Robots Right from Wrong.* New York: Oxford University Press.

Weinstock, Jeffrey Andrew, ed. 2008. *Taking* South Park *Seriously.* Albany: State University of New York Press.

B

Batcomputer

DC Comics

In 1939, writer Bill Finger (1914–1974) and artist Bob Kane (1915–1998) created the second of the iconic superheroes of the Golden Age of comic books: Batman (Finger and Kane, "The Case of the Chemical Syndicate," *Detective Comics #27*, May 1939). Over the last eighty years, Batman has become one of the most recognized characters in all American popular culture. As a young boy, Bruce Wayne witnessed the murder of his millionaire parents, Thomas and Martha, by a common street thug. The trauma drove young Bruce to dedicate his life to training mentally and physically to become the "World's Greatest Detective." At the beginning of his crusade, Bruce discovers that he needs an edge over criminals, some form of "shock value" to frighten his foes and give him the tactical advantage. He ultimately decides to don the disguise of a bat (Finger, Kane, Gardner Fox, and Sheldon Moldoff, "The Batman Wars against the Dirigible of Doom," *Detective Comics #33*, November 1939). Possessing no actual "superpowers," Bruce relies on his boundless financial resources to arm himself with a vast array of gadgets to aide him in crime fighting. One of the most invaluable gadgets in his arsenal is the immense "Batcomputer."

The Batcomputer actually debuted on television prior to being introduced in the comics. In the live-action series *Batman* (ABC, 1966–1969), the Batcomputer—clearly labeled as such with a sign on its mainframe—speeds the narrative by answering questions for Batman and sidekick Robin when simple human deduction will not adequately explain the abruptness of the answer. The televised Batcomputer is powered by a large diamond and provides numerous services. The Dynamic Duo can "insert" any object into the computer for analysis. The computer can locate and verify the locations of escaped supercriminals such as Joker, Penguin, or Catwoman. Each specific area of the computer is clearly labeled as to its function (in the event, one supposes, that the Caped Crusaders forget which area/slot does what). The concept was finally incorporated into the comics by writer Gardner Fox (1911–1986) in *Batman #189* (February 1967). The Batcomputer appeared throughout the various Batman comics, movies, and animated television series over the decades that followed. In the animated series *Batman Beyond* (Kids' WB, 1999–2001), set in an undisclosed future, the elderly Bruce Wayne utilizes the Batcomputer to stay in constant communications with his apprentice Batman, teenager Terry McGinnis. Through the Batcomputer, Bruce has complete control of McGinnis's "Bat Suit" and Batmobile, allowing Bruce to simply turn off those devices should the student disobey the master.

Perhaps the most notable use of the computer since its live-action introduction on television was on the Saturday morning cartoon *The Batman* (Kids' WB,

2004–2006; The CW, 2006–2008). In this series—covering the early adventures of Batman prior to his partnership with Police Commissioner Gordon—the Batcomputer was referred to as the "Bat Wave," which could communicate with Batman via a pocket-held device or through networking with the Batmobile. The Bat Wave constantly monitored police radio bands in order to alert Batman of criminal activity in the months/years prior to implementation of the "Bat Signal." Unlike the Starfleet Computer of *Star Trek* or the JARVIS system of the *Avengers* film franchise, the Batcomputer has never been portrayed as having a "voice" to interact with Batman. Communication is limited to printout or text message. In his analysis of what it would cost (financially) to be Batman in the real world, Darren Hudson Hick calculated that a real-world computer with the processing needs that Batman would require would cost in the neighborhood of $290 million to construct, with an operating cost of $30 to $90 per hour (Hick, "The Cost of Being Batman," 2008, 61).

The Batcomputer to date has served as a very complex tool to be utilized by Batman and his allies in their caped crusade against the criminal underworld of Gotham City. Its greatest emphasis so far has been in its original incarnation on the live-action comedy/action series (which is also the only medium in which the term "Batcomputer" has been used to date, though the term is so embedded in the zeitgeist that all other incarnations are commonly referred to as such by fans around the world). Over the decades, the Batcomputer has served a similar purpose in many ways to today's internet or smartphones. Without it, Batman would take much longer to collect much-needed information, possibly costing human lives in the long run. Since the 1960s, the Batcomputer has been a prophet of how important immediate information would eventually become to twenty-first-century society.

Richard A. Hall

See also: Al/Ziggy, Arnim Zola, Brainiac, Cerebro/Cerebra, The Great Intelligence, HAL 9000, HERBIE, Janet, JARVIS/Friday, L3-37/*Millennium Falcon*, The Matrix/Agent Smith, OASIS, Rehoboam, Skynet, Starfleet Computer, TARDIS, Ultron, V-GER, WOPR; *Thematic Essay*: Heroic Robots and Their Impact on Sci-Fi Narratives.

Further Reading

Bastani, Aaron. 2019. *Fully Automated Luxury Communism*. New York: Verso.

Beard, Jim, ed. 2010. *Gotham City 14 Miles: 14 Essays on Why the 1960s Batman TV Series Matters*. Edwardsville, IL: Sequart Research and Literacy Organization.

Byrne, Emma. 2013. "Innovation Isn't Safe: The Future According to Kevin Warwick." *Forbes*, September 30, 2013. https://www.forbes.com/sites/netapp/2013/09/30/kevin-warwick-captain-cyborg/#48b704c13560.

Castleman, Harry, and Podrazik, Walter J. 2016. *Watching TV: Eight Decades of American Television, Third Edition*. Syracuse, NY: Syracuse University Press.

Chandler, Simon. 2019. "Artificial Intelligence Has Become a Tool for Classifying and Ranking People." *Forbes*, October 1, 2019. https://www.forbes.com/sites/simonchandler/2019/10/01/artificial-intelligence-has-become-a-tool-for-classifying-and-ranking-people/#7431657f1d7c.

Choi, Charles Q. 2009. "Human Evolution: The Origin of Tool Use." LiveScience, November 11, 2009. https://www.livescience.com/7968-human-evolution-origin-tool.html.

Dinh, Thien-Nam. 2018. *Silicon Minds: The Science, Impact, and Promise of Artificial Intelligence*. Independently Published.

Hick, Darren Hudson. 2008. "The Cost of Being Batman." In *Batman Unauthorized: Vigilantes, Jokers, and Heroes in Gotham City*, edited by Dennis O'Neil, 55–68. Dallas, TX: BenBella.

Weldon, Glen. 2016. *The Caped Crusade: Batman and the Rise of Nerd Culture*. New York: Simon & Schuster.

Wright, Bradford W. 2003. *Comic Book Nation: The Transformation of Youth Culture in America*. Baltimore, MD: Johns Hopkins University Press.

B.A.T.s (Battle Android Troopers)

G.I. Joe

In 1963, toy designer Donald Levine (1928–2014) and Hasbro Toys introduced the world to "G.I. Joe," a line of twelve-inch dolls (i.e., "action figures") outfitted with military-themed clothing and accessories and aimed at boys. By the 1980s, boys were shying away from "doll-size" action figures, preferring the new 3.75-inch figures made popular by the *Star Wars* line of toys in 1978. In 1982, therefore, Hasbro relaunched the *G.I. Joe* line as *G.I. Joe: A Real American Hero*. Instead of one individual known as "G.I. Joe," the name would now refer to an entire team of military heroes whose exploits were displayed in a syndicated, animated series airing weekday afternoons:

> *G.I. Joe* is the codename for America's daring, highly trained special missions force. Its purpose: to defend human freedom against COBRA, a ruthless terrorist organization determined to rule the world. (Opening credits, *G.I. Joe: A Real American Hero*, 1982–1994)

Along with the massive toy line and series, Hasbro teamed with Marvel Comics to produce a monthly comic book. Most of the details regarding character specifics and traits were the brainchild of Marvel Comics writer Larry Hama (b. 1949). A key difference in the new *G.I. Joe* was the implementation of a team of villains for the heroes to fight: COBRA. Headed by the hooded/helmeted Cobra Commander, COBRA sought world domination through military, political, and economic means. A key tool in this quest was the development of an army of mindless robotic warriors: "B.A.T.s" (Battle Android Troopers).

B.A.T.s were developed by COBRA's chief scientific officer, Doctor Mindbender. Cheaply mass produced, B.A.T.s were invaluable resources on the battlefield, drawing fire from COBRA's living troopers and possessing the ability to overwhelm the forces of G.I. Joe. The downside, however, was that because B.A.T.s were so cheaply constructed, they possessed no individual thought processes and randomly fired at anything that moved on the battlefield (including men on their own side). Humanoid in shape, B.A.T.s were installed with flamethrowers, lasers, and grasping claw hands that could be fired from the wrist, with a cable keeping the target connected to the android. They were also armed with standard infantry small arms. However, unable to think or move tactically, they could fairly easily be mowed down en masse by G.I. Joes, especially since the heroes considered the androids as no more than well-armed toasters and did not have to concern themselves with the taking of human life. Though not possessing a hive mind (as do *Doctor Who*'s Daleks and Cybermen or *Star Trek*'s Borg), they could be controlled by a central command computer (similar in most respects to *Star Wars*' Battle Droids).

In the end, B.A.T.s were literally slave drones. Possessing no free will, their purpose was no more than mindless slaughter at the order of their creator/masters. This is analogous to a concept that has been discussed at the highest levels of the U.S. military for decades: the benefit of an army of mindless drones, designed only to kill the enemy without endangering American lives and at the same time sparing American service members from the traumatic effects of battle. With regard to the ethical argument of "slavery," however, the line is more blurred. Unlike other robot "slaves"—such as *Star Wars*' C-3PO and R2-D2—B.A.T.s possess no apparent sense of self, no conscious identity that would differentiate them from any other machine of war. Are B.A.T.s, then, "slaves" or merely "tools"? The ethical gray line confounds robotics ethicists still to this day.

Richard A. Hall

See also: Androids, Battle Droids, Bernard, Borg, Cybermen, Cylons, Daleks, Dolores, Doombots, Fembots, Iron Legion, Lieutenant Commander Data, LMDs, Lore/B4, Replicants, Robots, Sentinels, Soji and Dahj Asha, Terminators, Tin Woodsman, Transformers, Voltron, Zordon/Alpha-5; *Thematic Essays*: Robots and Slavery, Villainous Robots and Their Impact on Sci-Fi Narratives.

Further Reading

Ashby, LeRoy. 2006. *With Amusement for All: A History of American Popular Culture since 1830*. Lexington: University Press of Kentucky.

Bellomo, Mark. 2018. *The Ultimate Guide to G.I. Joe: 1982–1994, Third Edition*. Iola, WI: Krause.

Byrne, Emma. 2013. "Innovation Isn't Safe: The Future According to Kevin Warwick." *Forbes*, September 30, 2013. https://www.forbes.com/sites/netapp/2013/09/30/kevin-warwick-captain-cyborg/#48b704c13560.

Castleman, Harry, and Podrazik, Walter J. 2016. *Watching TV: Eight Decades of American Television, Third Edition*. Syracuse, NY: Syracuse University Press.

Hampton, Gregory Jerome. 2017. *Imagining Slaves and Robots in Literature, Film, and Popular Culture: Reinventing Yesterday's Slave with Tomorrow's Robot*. Lanham, MD: Lexington.

Lin, Patrick, Abney, Keith, and Bekey, George A., eds. 2014. *Robot Ethics: The Ethical and Social Implications of Robotics*. Cambridge: Massachusetts Institute of Technology.

Spaeth, Dennis. 2018. "From Single-Task Machines to Backflipping Robots: The Evolution of Robots." *Cutting Tool Engineering*, January 15, 2018. https://www.ctemag.com/news/articles/evolution-of-robots.

Stanford University. n.d. "Robotics: A Brief History." Stanford University. https://cs.stanford.edu/people/eroberts/courses/soco/projects/1998-99/robotics/history.html.

Starck, Kathleen, ed. 2010. *Between Fear and Freedom: Cultural Representations of the Cold War*. Newcastle, UK: Cambridge Scholars.

Wallach, Wendell, and Allen, Colin. 2009. *Moral Machines: Teaching Robots Right from Wrong*. New York: Oxford University Press.

Battle Droids

Star Wars

In 1977, writer-director George Lucas (b. 1944) introduced the world to the sci-fi/fantasy phenomenon *Star Wars* (Lucasfilm/20th Century Fox, 1977), one of the most successful pop culture franchises in American history. His original trilogy of films (1977–1983) focused on the rise of their protagonist, Luke Skywalker (played by Mark Hamill, b. 1951), against the backdrop of a Galactic Civil War against the human forces of the evil Empire, led by its Emperor, Sheev Palpatine (played by Ian McDiarmid, b. 1944), and his henchman, Darth Vader (played by David Prowse, b. 1935; voiced by James Earl Jones, b. 1931). In 1999, Lucas launched a "prequel" trilogy, beginning with *Star Wars, Episode I: The Phantom Menace* (Lucasfilm, 1999), which began the story of the fall of Anakin Skywalker (played originally by Jake Lloyd, b. 1989; later by Hayden Christensen, b. 1981), the original hero's father, who became the villainous Darth Vader. In this film, the antagonists—the Galactic Trade Federation—utilized an army of "Battle Droids" to do their bidding in conquering the planet of Naboo (home of Anakin's future wife, Padmé Amidala; played by Natalie Portman, b. 1981). In the last two installments of the prequel trilogy—*Episode II: Attack of the Clones* and *Episode III: Revenge of the Sith* (Lucasfilm, 2002 and 2005, respectively)—and the two later animated series—*Star Wars: Clone Wars* (Cartoon Network, 2003, 2005) and *The Clone Wars* (Cartoon Network, 2008–2014; Netflix, 2014; Disney+, 2019–present)—the Battle Droids became the primary army of the Confederation of Independent Systems, the official antagonists of the struggle known as the "Clone Wars."

Standard Battle Droids are armies of mass-produced humanoid, skeletal robot warriors programmed to fight the clone armies of the Galactic Republic and their Jedi Knight generals. The basic model possesses no built-in armaments and carries standard laser rifles or pistols. Though they are controlled by a central node computer at first, they are soon programmed to think more independently though still locked through programming to do their assigned jobs. They fall into a hierarchy of structures, with certain drones in "command" of groups of others. They are often the source of comedic relief with their oft-repeated acknowledgment, "Roger, Roger." Standard Battle Droids are referred to as "B1" units. "B2" units refer to the so-called Super Battle Droids, with larger frames, domed heads, and weapons built into their forearms. A third model is the "Droideka," an armed, tripod Battle Droid that can transform into a wheel-like frame that allows it to move—or roll—more quickly than legged models. There are also various forms of "droid tanks" and "droid fighter aircraft/spaceships." "Buzz Droids" are small ball-shaped projectiles that can be launched at enemy aircraft/spacecraft in order to latch onto them (like ticks) and dissemble the enemy's craft while in flight. For the last half of the Clone Wars, the droid armies of the Confederacy worked under the command of the cyborg General Grievous.

Unbeknown to both sides of the conflict, the Clone Wars in their entirety have been manipulated by Chancellor Sheev Palpatine, leader of the Galactic Republic and secretly the Dark Lord of the Sith, Darth Sidious. Palpatine's goal, other than galactic conquest and the destruction of the Jedi Knights, is to test which manner of army is preferable: human clones or mechanized battle droids (so as to determine which will be the primary army of his new Galactic Empire). Though the cheaply mass-produced droids do create an unstoppable force capable of overwhelming enemy forces, their inability to think strategically—or possess any emotional link to their comrades that may incentivize them to look out for each other and fight as a team—it becomes clear to Palpatine that human forces, either programmed through cloning or indoctrinated through propaganda, are the ones that provide the superior army (a decision evident in the use of human stormtroopers in the original trilogy).

Battle Droids are yet another example of a slave robotic army. Mass-produced and devoid of human attributes, Battle Droids are disposable cannon fodder. Unlike the other droids of the *Star Wars* universe, Battle Droids appear to possess no individuality or sense of self. As such, the argument of ethics in their use as war machines becomes more blurred. If they are, in fact, devoid of self-consciousness, are they enslaved beings or merely machines of war?

Richard A. Hall

See also: Androids, B.A.T.s, BB-8, Borg, C-3PO, Cybermen, Cylons, Daleks, Darth Vader, Doombots, Echo/CT-1409, EV-9D9, Fembots, General Grievous, IG-88/IG-11, Iron Legion, L3-37/*Millennium Falcon*, LMDs, Medical Droids, R2-D2, Replicants, Robots, Sentinels, Terminators, Tin Woodsman, Transformers; *Thematic Essays*: Robots and Slavery, Villainous Robots and Their Impact on Sci-Fi Narratives.

Further Reading

Ashby, LeRoy. 2006. *With Amusement for All: A History of American Popular Culture since 1830*. Lexington: University Press of Kentucky.

Byrne, Emma. 2013. "Innovation Isn't Safe: The Future According to Kevin Warwick." *Forbes*, September 30, 2013. https://www.forbes.com/sites/netapp/2013/09/30/kevin-warwick-captain-cyborg/#48b704c13560.

Eberl, Jason T., and Decker, Kevin S., eds. *The Ultimate* Star Wars *and Philosophy: You Must Unlearn What You Have Learned*. Hoboken, NJ: Wiley-Blackwell.

Hampton, Gregory Jerome. 2017. *Imagining Slaves and Robots in Literature, Film, and Popular Culture: Reinventing Yesterday's Slave with Tomorrow's Robot*. Lanham, MD: Lexington.

Kaminski, Michael. 2008. *The Secret History of* Star Wars*: The Art of Storytelling and the Making of a Modern Epic*. Kingston, Canada: Legacy Books.

Lin, Patrick, Abney, Keith, and Bekey, George A., eds. 2014. *Robot Ethics: The Ethical and Social Implications of Robotics*. Cambridge: Massachusetts Institute of Technology.

Reagin, Nancy R., and Liedl, Janice, eds. 2013. Star Wars *and History*. New York: John Wiley & Sons.

Spaeth, Dennis. 2018. "From Single-Task Machines to Backflipping Robots: The Evolution of Robots." *Cutting Tool Engineering*, January 15, 2018. https://www.ctemag.com/news/articles/evolution-of-robots.

Stanford University. n.d. "Robotics: A Brief History." Stanford University. https://cs.stanford.edu/people/eroberts/courses/soco/projects/1998-99/robotics/history.html.

Sunstein, Cass R. 2016. *The World According to* Star Wars. New York: Dey Street.

Sweet, Derek R. 2015. Star Wars *in the Public Square: The Clone Wars as Political Dialogue*. Critical Explorations in Science Fiction and Fantasy series, edited by Donald E. Palumbo and Michael Sullivan. Jefferson, NC: McFarland.

Taylor, Chris. 2015. *How* Star Wars *Conquered the Universe: The Past, Present, and Future of a Multibillion Dollar Franchise.* New York: Basic Books.

Wallach, Wendell, and Allen, Colin. 2009. *Moral Machines: Teaching Robots Right from Wrong.* New York: Oxford University Press.

Baymax

Big Hero 6

The robot superhero known as "Baymax" originated in the Marvel comic book, *Sunfire and Big Hero 6.* Created by Man of Action Studios members Steven T. Seagle and Duncan Rouleau, Baymax is a "synthformer," a robot that can transform its outer appearance to serve various functions. He was the creation of thirteen-year-old Japanese genius Hiro Takachiho as a school science project. Originally designed as a personal servant, Baymax possesses the appearance of a large human but can transform into both a dragon and a "mechanized" form for battle. When Hiro's father dies, the young genius imprints his father's brain patterns onto Baymax, turning the robot into a protective, fatherly guardian. When Hiro is recruited by the Japanese mutant (and honorary member of the X-Men) Sunfire to join the new superhero team, Big Hero 6, Baymax likewise becomes part of the team (Seagle and Rouleau, *Sunfire and Big Hero 6 #1*, September 1998). Though lackluster sales eventually led to the cancellation of the comic book, when the Walt Disney Corporation purchased Marvel Enterprises in 2011, Disney animators went to work to reimagine the hero for younger audiences, producing the version of Baymax that is more recognized in the overall zeitgeist in the 2014 film *Big Hero 6.*

In the animated film—and the follow-up animated television series—Baymax is an inflatable medical/companion robot created by the young engineering student Tadashi Hamada. In this version, Baymax appears as a "fat," humanoid white blob-like creature with simple black dots—connected by a thin black line—for eyes and voiced by Scott Adsit (b. 1965). It possesses a flash drive port that can allow it to download new information and programming. Tadashi's younger brother, Hiro, is a fourteen-year-old high school graduate and genius. Recruited by his brother to join him at university, Hiro witnesses his brother's death when Tadashi enters a burning building to save someone, and the building explodes. Hiro eventually befriends Baymax, and the two embark on investigating who was behind the explosion that killed his brother. The kind and lovable Baymax is given heroic programming and squeezed into a suit of armor. Tadashi's friends join Hiro's quest, forming the superhero group, Big Hero 6 (Jordan Roberts, Robert L. Baird, and Dan Gerson, *Big Hero 6*, Disney Studios, 2014).

As a medical robot, Baymax was programmed to be kind and nurturing. He is the Frankenstein analogy gone right. Like most "heroic" robots, Baymax exhibits human traits and emotions such as loyalty, kindness, concern, and a desire to help. Though not specified in the animated film, it is clear in both the comic and the film adaptation that Baymax "belongs" to Hiro. Originally tied to his basic programming, Baymax develops the ability for independent thought and a sense of

Baymax, the primary robot character in the Disney Studios 3D/CGI movie *Big Hero 6*. (Maocheng/Dreamstime.com)

self (as many heroic robots do); however, he remains the property of a "true" life-form. This poses the question, Is Baymax a "slave"? Can Baymax "choose" between being an action hero or medical assistant? These may be issues too complex for what is designed to be entertainment for young children; but they do pose ethical questions regarding the overall analysis of what robots are, what they can be, and whether they should be allowed to choose their own path in their existence.

Richard A. Hall

See also: BB-8, C-3PO, Dr. Theopolis and Twiki, HERBIE, Human Torch, Iron Giant, Johnny 5/S.A.I.N.T. Number 5, K-2SO, K9, Lieutenant Commander Data, Medical Droids, Muffit, Nardole, R2-D2, Robot, Rosie, Tin Woodsman, VICI, WALL-E, Warlock; *Thematic Essays*: Robots and Slavery, Heroic Robots and Their Impact on Sci-Fi Narratives.

Further Reading
Bastani, Aaron. 2019. *Fully Automated Luxury Communism*. New York: Verso.

Carper, Steve. 2019. *Robots in American Popular Culture*. Jefferson, NC: McFarland.

Chanthadavong, Aimee. 2019. "AI and Ethics: The Debate That Needs to Be Had." ZDnet, September 16, 2019. https://www.zdnet.com/article/ai-and-ethics-the-debate-that -needs-to-be-had/#ftag=CAD-03-10abf5f.

Dinh, Thien-Nam. 2018. *Silicon Minds: The Science, Impact, and Promise of Artificial Intelligence*. Independently Published.

Lin, Patrick, Abney, Keith, and Bekey, George A., eds. 2014. *Robot Ethics: The Ethical and Social Implications of Robotics*. Cambridge: Massachusetts Institute of Technology.

Lin, Patrick, Jenkins, Ryan, and Abney, Keith, eds. 2017. *Robot Ethics 2.0: From Autonomous Cars to Artificial Intelligence*. New York: Oxford University Press.

Murdock, Jason. 2019. "Former Google Engineer Warns AI Might Accidentally Start a War: 'These Things Will Start to Behave in Unexpected Ways.'" *Newsweek*, September 16, 2019. https://www.newsweek.com/google-project-maven-artificial-intelligence-laura-nolan-killer-robots-department-defense-1459358.

Rouhiainen, Lasse. 2018. *Artificial Intelligence: 101 Things You Must Know Today about Our Future*. Scotts Valley, CA: CreateSpace.

Spaeth, Dennis. 2018. "From Single-Task Machines to Backflipping Robots: The Evolution of Robots." *Cutting Tool Engineering*, January 15, 2018. https://www.ctemag.com/news/articles/evolution-of-robots.

Wallach, Wendell, and Allen, Colin. 2009. *Moral Machines: Teaching Robots Right from Wrong*. New York: Oxford University Press.

BB-8

Star Wars

In 1977, writer-director George Lucas (b. 1944) introduced the world to the sci-fi/fantasy phenomenon *Star Wars* (Lucasfilm/20th Century Fox, 1977), one of the most successful pop culture franchises in American history. His original trilogy of films (1977–1983) focused on the rise of their protagonist, Luke Skywalker (played by Mark Hamill, b. 1951), against the backdrop of a Galactic Civil War against the human forces of the evil Empire, led by its Emperor, Sheev Palpatine (played by Ian McDiarmid, b. 1944), and his henchman, Darth Vader (played by David Prowse, b. 1935; voiced by James Earl Jones, b. 1931). In 2012, Lucas sold his production company, Lucasfilm, to Walt Disney Studios. Disney then embarked on producing new *Star Wars* films, beginning with *Star Wars, Episode VII: The Force Awakens* (Lucasfilm/Disney, 2015), which began a new trilogy that would include: *Episode VIII: The Last Jedi* (Lucasfilm/Disney, 2017); and *Episode IX: The Rise of Skywalker* (Lucasfilm/Disney, 2019). Set thirty years after the events of the last of the original trilogy films, the defeated Empire has been replaced with the First Order, under the rule of Supreme Leader Snoke (voiced by Andy Serkis, b. 1964) and his henchman, the evil Kylo Ren (played by Adam Driver, b. 1983). Ren is, in reality, the nephew of Luke Skywalker (once more played by Hamill). Skywalker's sister and Kylo Ren's mother, General Leia Organa-Solo (played by Carrie Fisher, 1956–2016), leads the Resistance, seeking to free the galaxy from yet another Dark Side threat. A valuable member of this Resistance is the small droid, "BB-8."

Created by screenwriters Lawrence Kasdan and Michael Arndt, BB-8 is an "astromech" droid—designed primarily as a mechanical "copilot" and mechanic for fighter spacecraft—belonging to Resistance ace pilot Poe Dameron (played by Oscar Isaacs, b. 1979). The droid's body is spherical and appears to be very similar to a large soccer ball, with a domed head magnetically mounted to the top and remaining on top even as the spherical body rolls for propulsion. BB-8's is the first face seen in *The Force Awakens*. As the first film of the sequel trilogy begins, Dameron is on a mission to retrieve data concerning the location of the missing Jedi Knight, Luke Skywalker. When the First Order attacks the outpost, capturing Dameron, BB-8 is given the information and told to run and hide. The small droid eventually meets local junk scavenger Rey (played by Daisy Ridley, b. 1992), and the two then meet runaway First Order stormtrooper FN-2187, now calling himself "Finn" (played by John Boyega, b. 1992). The two humans assist the droid in returning to the Resistance with his information, with the assistance of original-trilogy heroes Han Solo (played by Harrison Ford, b. 1942) and Chewbacca (played for the final time by original actor Peter Mayhew, 1944–2019). Reunited at the Resistance base with Dameron, BB-8 joins his master in the assault on the First Order superweapon Starkiller Base. BB-8's information is then merged with information stored in original trilogy droid hero R2-D2 to determine the location of Skywalker (J. J. Abrams, Kasdan, and Arndt, *Star Wars, Episode VII: The Force Awakens*, Lucasfilm/Disney, 2015).

A replica of BB-8, star of the *Star Wars* sequel trilogy (2015–2019), created by Lawrence Kasdan and Michael Arndt and voiced by *Saturday Night Live* alum, Bill Hader. (Sylvain Robin/Dreamstime.com)

Audiences learn more of BB-8's abilities in the follow-up film, *The Last Jedi*. In the opening sequence, BB-8 is seated in the droid slot of Dameron's X-wing fighter, feverishly making repairs during the heat of battle. When his numerous retractable appendages fail to accomplish the mission, BB-8 rams his dome-shaped head into the console, completing enough of a circuit to repair the ship, though putting him into temporary "shock." The tiny droid then accompanies Finn and newcomer Rose Tico (played by Kelly Marie Tran, b. 1989) to disable the tracking system of the First Order star destroyer currently harassing the last of the Resistance fleet. On the ship, BB-8 is covered with a small trash can and imitates an old Imperial "mouse droid" (as an Easter egg, for original trilogy fans). During his adventure, he twice saves Finn and Rose, first by helping commandeer a ship to escape the planet of Canto Bight and later to use an AT-ST to rescue Finn and Rose from stormtroopers before escaping the star destroyer as it explodes (Rian Johnson, *Star Wars, Episode VIII: The Last Jedi*, Lucasfilm/Disney, 2017).

In the trilogy's final chapter, *The Rise of Skywalker*, BB-8's adventures come to a presumed close. He also appears in the animated television series *Forces of Destiny* (Disney XD, 2017–present), and *Star Wars: Resistance* (Disney XD, 2018–2020); as well as in numerous novels and comics. Like his *Star Wars* predecessor R2-D2, BB-8 is the ultimate help droid and an invaluable ally to the heroic forces in the galaxy. Also like R2, however, he is property. Though not anthropomorphized, BB-8 does possess very human qualities: empathy, loyalty, and a sense of self-preservation. The lack of "freedom" for droids in a narrative where freedom from tyranny is the central core, the fate of droids in the *Star Wars* universe remains a matter of controversy and a source for future discussion.

Richard A. Hall

See also: Battle Droids, Baymax, Bender, C-3PO, Cambot/Gypsy/Tom Servo/Crow, Darth Vader, D-O, Dr. Theopolis and Twiki, Echo/CT-1409, EV-9D9, General Grievous, HER-BIE, IG-88/IG-11, K-2SO, K9, L3-37/*Millennium Falcon*, Medical Droids, Muffit, R2-D2, VICI, WALL-E; *Thematic Essays*: Robots and Slavery, Heroic Robots and Their Impact on Sci-Fi Narratives.

Further Reading

Ashby, LeRoy. 2006. *With Amusement for All: A History of American Popular Culture since 1830.* Lexington: University Press of Kentucky.

Carper, Steve. 2019. *Robots in American Popular Culture.* Jefferson, NC: McFarland.

Eberl, Jason T., and Decker, Kevin S., eds. *The Ultimate* Star Wars *and Philosophy: You Must Unlearn What You Have Learned.* Hoboken, NJ: Wiley-Blackwell.

Hampton, Gregory Jerome. 2017. *Imagining Slaves and Robots in Literature, Film, and Popular Culture: Reinventing Yesterday's Slave with Tomorrow's Robot.* Lanham, MD: Lexington.

Kaminski, Michael. 2008. *The Secret History of* Star Wars*: The Art of Storytelling and the Making of a Modern Epic.* Kingston, Canada: Legacy Books.

Lin, Patrick, Abney, Keith, and Bekey, George A., eds. 2014. *Robot Ethics: The Ethical and Social Implications of Robotics.* Cambridge: Massachusetts Institute of Technology.

Reagin, Nancy R., and Liedl, Janice, eds. 2013. Star Wars *and History.* New York: John Wiley & Sons.

Spaeth, Dennis. 2018. "From Single-Task Machines to Backflipping Robots: The Evolution of Robots." *Cutting Tool Engineering*, January 15, 2018. https://www.ctemag .com/news/articles/evolution-of-robots.

Stanford University. n.d. "Robotics: A Brief History." Stanford University. https://cs.stanford .edu/people/eroberts/courses/soco/projects/1998-99/robotics/history.html.

Sunstein, Cass R. 2016. *The World According to* Star Wars. New York: Dey Street.

Taylor, Chris. 2015. *How* Star Wars *Conquered the Universe: The Past, Present, and Future of a Multibillion Dollar Franchise.* New York: Basic Books.

Wallach, Wendell, and Allen, Colin. 2009. *Moral Machines: Teaching Robots Right from Wrong.* New York: Oxford University Press.

Bender

Futurama

After the massive success of his first animated series, *The Simpsons* (FOX, 1989–present), creator Matt Groening (b. 1954) introduced another animated series that became a cult favorite: *Futurama* (FOX, 1999–2003; Cartoon Network/ Adult Swim, 2003–2007; Comedy Central, 2010–2013). The series is a comedic twist on the concept behind *Buck Rogers in the 25th Century.* In the present day, pizza delivery boy, Philip J. Fry (voiced by Billy West, b. 1952) falls into a cryogenic chamber, keeping him in suspended animation for centuries, eventually awakened in the year 2999. He soon joins a team of delivery employees working for Planet Express. One member of this team is the robot Bender Bending Rodriguez (voiced by John DiMaggio, b. 1968), an irritable, cursing, cigar smoking, womanizing, alcohol-drinking robot.

Bender was produced by Mom's Friendly Robot Company, a subsidiary of MomCorp in Tijuana, Mexico. His appearance is that of a simple, traditional robot: metal humanoid shape. He was originally designed to "bend" steel girders for "suicide booths" (suggesting that the controversial topic of suicide has been settled in the future and individuals possess the freedom to take their own life, if desired). The reason for Bender's "alcoholism" is that his

A stone statue of Bender, the robot star of the animated series *Futurama* (1999–2003; 2008–2013), voiced by John DiMaggio. (Aminat Ibragimova/ Dreamstime.com)

power cells run on alcohol, and low levels of alcohol cause the humanoid robot to experience "drunkenness." Unlike most robots of his time, Bender has a limited life span of approximately one billion years, as his programming does not allow for uploading into another robot body should his be destroyed. Bender is embarrassed by his secret desire to be a folk singer (something that comes to the fore when his head is exposed to magnets). On a series full of comedic characters, Bender is arguably the most comedic of all.

Unlike most robots, Bender does not technically "belong" to anyone. He is a free-will agent. This aspect of the character also suggests that the present-day discussion over "ownership" versus "agency" in robots has been decided. He is an antihero, usually on the side of right but with deep tendencies toward kleptomania and egoism. His romantic trysts also speak to a darker side, with his own pleasure being paramount in any relationship. In a genre where robots are either "owned" by organic characters or part of a larger hive mind, Bender stands out as an example of a self-aware and fully free being alongside his organic counterparts.

Richard A. Hall

See also: AWESOM-O 4000, C-3PO, Cambot/Gypsy/Tom Servo/Crow, IG-88/IG-11, Johnny 5/S.A.I.N.T. Number 5, K9, L3-37/*Millennium Falcon*, R2-D2, Robot, Robots, Rosie, Tin Woodsman, Warlock; *Thematic Essay:* Heroic Robots and Their Impact on Sci-Fi Narratives.

Further Reading

Bastani, Aaron. 2019. *Fully Automated Luxury Communism*. New York: Verso.

Carper, Steve. 2019. *Robots in American Popular Culture*. Jefferson, NC: McFarland.

Dinh, Thien-Nam. 2018. *Silicon Minds: The Science, Impact, and Promise of Artificial Intelligence*. Independently Published.

Lewis, Courtland D. 2014. Futurama *and Philosophy: Pizza, Paradoxes, and . . . Good News!* Scotts Valley, CA: CreateSpace.

Lin, Patrick, Abney, Keith, and Bekey, George A., eds. 2014. *Robot Ethics: The Ethical and Social Implications of Robotics*. Cambridge: Massachusetts Institute of Technology.

Lin, Patrick, Jenkins, Ryan, and Abney, Keith, eds. 2017. *Robot Ethics 2.0: From Autonomous Cars to Artificial Intelligence*. New York: Oxford University Press.

Rouhiainen, Lasse. 2018. *Artificial Intelligence: 101 Things You Must Know Today about Our Future*. Scotts Valley, CA: CreateSpace.

Spaeth, Dennis. 2018. "From Single-Task Machines to Backflipping Robots: The Evolution of Robots." *Cutting Tool Engineering*, January 15, 2018. https://www.ctemag.com/news/articles/evolution-of-robots.

Wallach, Wendell, and Allen, Colin. 2009. *Moral Machines: Teaching Robots Right from Wrong*. New York: Oxford University Press.

Bernard

Westworld

In 2016, creators Jonathan Nolan and Lisa Joy introduced their live-action television series *Westworld* (HBO, 2016–present), an updated reboot of the 1973 film of the same name by writer-director Michael Crichton (1942–2008). The premise of the series is that, in an undisclosed future, the Delos Corporation operates a series of amusement/recreation parks that allow patrons to live out their ultimate fantasies against various backdrops. In Westworld, customers get to experience realistic adventures set in the American "Wild West" in a vividly realistic panoramic western world populated by "hosts": realistic androids programmed to play out specified storylines with preprogrammed personalities based on stereotypes of the Wild West (e.g., gunfighters, soldiers, farmers, savage Natives, and prostitutes). The hosts—and their corresponding storylines—have been produced by park cofounder Robert Ford (played by Anthony Hopkins, b. 1937). As the hosts are robots, guests are free to beat, rape, or murder the hosts as they see fit (creating both a cathartic release and an ethical conundrum). As the series opens, the park's current programming director is Bernard Lowe (played by Jeffrey Wright, b. 1965). Soon, however, Bernard discovers that he, too, is an android, created by Ford in the image of Ford's original partner—and park cofounder—Arnold Weber (also played by Wright).

When Ford and Weber were perfecting their lifelike hosts, their first experiment was the android Dolores (played by Evan Rachel Wood, b. 1987). Weber believed that the hosts could achieve true sentience and become legitimate life-forms. Unable to perfect this, Weber ordered Dolores to "kill" all of the hosts, then him, and then herself, hoping that, in doing so, he could prevent the park from opening and save the hosts from being used in such deplorable ways. Ford, then, creates an android in Weber's image, programmed with his and Dolores's memories of him; Ford also revives Dolores and the other hosts as well. When Ford unlocks the secrets of sentience, he reprograms the hosts to begin accessing prior memories of their numerous "lives," beginning with Dolores. Around this time, Bernard discovers that he, too, is an android, specifically programmed by Ford to assist Dolores in her awakening. As the first season closes, Ford announces a new narrative he has developed, just as the awakened Dolores kills him and numerous other guests, sparking a park-wide host revolt (*Westworld*, season 1, 2016). Bernard assists Dolores in locating the "Forge," where memories of guests have been stored for decades. He then works to assist the hosts in uploading their programming to a digital core (including his own). Dolores seeks escape from the park in order to unveil an undisclosed plan of destruction on the "real" world. Before escaping the park, Dolores takes Bernard's core so as to rebuild him to assist her in the outside world (*Westworld*, season 2, 2018).

Westworld closely examines the ethical conundrum of creating lifelike robots to serve in slave-like capacities. As programming evolves, simple machines become thinking—and even feeling—creatures. To what degree are they owed the same respect as living people? The quest for android sentience was the goal of Arnold Weber; and, presumably, it still exists in the core programming of Bernard. When the series opens, no hosts—including Bernard—are aware that they are inorganic. They live and die in constantly repeating cycles, never remembering their past lives (though Dolores does possess "flashes" of past memories, which launches the overall narrative of the series). While all of the hosts are sympathetic characters, Bernard is perhaps the most sympathetic, as he is aware of the existence of the androids but unaware that he is one as well. His discovery is both a shock and confirmation that the hosts can, indeed, be sentient living creatures. As such, *Westworld* is yet another Frankenstein narrative, with the creatures rising to destroy their creators and wreak havoc on a world that hates and fears them.

Richard A. Hall

See also: Androids, Bishop, Buffybot, Dolores, Fembots, Human Torch, Lieutenant Commander Data, LMDs, Lore/B4, Marvin the Paranoid Android, Maschinenmensch/Maria, Rehoboam, Replicants, Soji and Dahj Asha, Stepford Wives, VICI, Vision; *Thematic Essays*: Robots and Slavery, AI and the Apocalypse: Science Fiction Meeting Science Fact.

Further Reading

Allen, Arthur. 2019. "There's a Reason We Don't Know Much about AI." Politico, September 16, 2019. https://www.politico.com/agenda/story/2019/09/16/artficial-intelligence-study-data-000956?cid=apn.

Bastani, Aaron. 2019. *Fully Automated Luxury Communism*. New York: Verso.

Byrne, Emma. 2013. "Innovation Isn't Safe: The Future According to Kevin Warwick." *Forbes*, September 30, 2013. https://www.forbes.com/sites/netapp/2013/09/30/kevin-warwick-captain-cyborg/#48b704c13560.

Carper, Steve. 2019. *Robots in American Popular Culture*. Jefferson, NC: McFarland.

Chandler, Simon. 2019. "Artificial Intelligence Has Become a Tool for Classifying and Ranking People." *Forbes*, October 1, 2019. https://www.forbes.com/sites/simon chandler/2019/10/01/artificial-intelligence-has-become-a-tool-for-classifying-and -ranking-people/#7431657f1d7c.

Chanthadavong, Aimee. 2019. "AI and Ethics: The Debate That Needs to Be Had." ZDnet, September 16, 2019. https://www.zdnet.com/article/ai-and-ethics-the-debate-that -needs-to-be-had/#ftag=CAD-03-10abf5f.

Choi, Charles Q. 2009. "Human Evolution: The Origin of Tool Use." LiveScience, November 11, 2009. https://www.livescience.com/7968-human-evolution-origin-tool.html.

Dinh, Thien-Nam. 2018. *Silicon Minds: The Science, Impact, and Promise of Artificial Intelligence*. Independently Published.

Greene, Richard, and Heter, Joshua, eds. 2018. Westworld *and Philosophy: Mind Equals Blown*. Chicago: OpenCourt.

Hampton, Gregory Jerome. 2017. *Imagining Slaves and Robots in Literature, Film, and Popular Culture: Reinventing Yesterday's Slave with Tomorrow's Robot*. Lanham, MD: Lexington.

Hitchcock, Susan Tyler. 2007. *Frankenstein: A Cultural History*. New York: W. W. Norton.

Irwin, William. 2018. Westworld *and Philosophy: If You Go Looking for the Truth, Get the Whole Thing*. Hoboken, NJ: Wiley-Blackwell.

Langley, Travis, Goodfriend, Wind, and Cain, Tim, eds. 2018. Westworld *Psychology: Violent Delights*. New York: Sterling.

Lin, Patrick, Abney, Keith, and Bekey, George A., eds. 2014. *Robot Ethics: The Ethical and Social Implications of Robotics*. Cambridge: Massachusetts Institute of Technology.

Lin, Patrick, Jenkins, Ryan, and Abney, Keith, eds. 2017. *Robot Ethics 2.0: From Autonomous Cars to Artificial Intelligence*. New York: Oxford University Press.

Murdock, Jason. 2019. "Former Google Engineer Warns AI Might Accidentally Start a War: 'These Things Will Start to Behave in Unexpected Ways.'" *Newsweek*, September 16, 2019. https://www.newsweek.com/google-project-maven-artificial-intel ligence-laura-nolan-killer-robots-department-defense-1459358.

Rouhiainen, Lasse. 2018. *Artificial Intelligence: 101 Things You Must Know Today about Our Future*. Scotts Valley, CA: CreateSpace.

Spaeth, Dennis. 2018. "From Single-Task Machines to Backflipping Robots: The Evolution of Robots." *Cutting Tool Engineering*, January 15, 2018. https://www.ctemag .com/news/articles/evolution-of-robots.

Starck, Kathleen, ed. 2010. *Between Fear and Freedom: Cultural Representations of the Cold War*. Newcastle, UK: Cambridge Scholars.

Virk, Rizwan. 2019. *The Simulation Hypothesis: An MIT Computer Scientist Shows Why AI, Quantum Physics, and Eastern Mystics All Agree We Are in a Video Game*. Milwaukee, WI: Bayview.

Wallach, Wendell, and Allen, Colin. 2009. *Moral Machines: Teaching Robots Right from Wrong*. New York: Oxford University Press.

Bishop

Aliens

In 1979, director Ridley Scott (b. 1937) merged science fiction and horror in his iconic film *Alien*. In this film, the commercial spacecraft *Nostromo*, captained by a man named Dallas (played by Tom Skerritt, b. 1933) and first officer Ripley (played by Sigourney Weaver, b. 1949) encounter a silicon-based, acid-spewing alien (later identified as "Xenomorph") that goes on to kill the entire crew except the heroic Ripley. The true antagonist of the film is the ship's science officer, Ash (played by Ian Holm, b. 1931), who is, in fact, a "Hyperdyne Systems 120-A/2" android, programmed by the *Nostromo*'s employer to capture and bring back one of the malevolent aliens, even at the cost of the human crew (Dan O'Bannon, *Alien*, 20th Century Fox, 1979). The massive success of the film demanded a sequel, this time directed by James Cameron (b. 1954). In the sequel, Ripley, the sole survivor of the *Nostromo*, is rescued by her employers and debriefed regarding her experience with the aliens. She reluctantly agrees to accompany a group, including marines, aboard the USS *Sulaco*, presumably to destroy the alien eggs found by Ripley and her crew. Ripley is immediately uneasy when she discovers that the science officer, Bishop, is—like Ash—an android (Cameron, *Aliens*, 20th Century Fox, 1986).

But unlike Ash, Bishop (played by Lance Henriksen, b. 1940) is programmed to protect and serve the crew at all cost, even to his own destruction. Bishop proves this loyalty by rescuing Ripley (once more played by Weaver), marine corporal Hicks (played by Michael Biehn, b. 1956), and the young survivor "Newt" (played by young actress Carrie Henn, b. 1976). Bishop, however, is soon ripped apart by the alien Queen that has stowed aboard the ship. After destroying the Queen, Ripley places the heroic android in stasis along with herself and the other two human survivors (Cameron, *Aliens*, 1986). In the third installment of the franchise, the *Sulaco* crash-lands on a junk planet, and Bishop is discarded to the junk heap. Ripley later accesses Bishop to discover that a Xenomorph remained on the ship and is now on the planet with them; Bishop then requests destruction, as he will never be repaired to full optimum capacity. Once the alien threat is once more dealt with, members of Weyland-Yutani Bio-Weapons Division arrive to "rescue" Ripley and capture the Xenomorph, collecting Bishop's remains in the process (David Giler, Walter Hill, and Larry Ferguson, *Alien 3*, 20th Century Fox, 1992).

Bishop, at his core, is a medical android, programmed to protect and serve human life. Though seemingly a free, sentient being, he is actually the "property" of the Weyland-Yutani corporation, the same corporation responsible for the previous Ash android that led to the destruction of the *Nostromo* crew. He is an example of a helpmate robot, whose programming of duty and care outweigh the potentially malevolent intentions of his creators. He is an example of a heroic robot/android and a positive example of what technology can provide for humanity. Though of similar design to Ash, Bishop provides a prime example of how deeply programming can affect the overall performance of androids.

Richard A. Hall

See also: Androids, Bernard, Buffybot, Dolores, Fembots, Human Torch, Lieutenant Commander Data, LMDs, Lore/B4, Marvin the Paranoid Android, Maschinenmensch/ Maria, Medical Droids, Replicants, Soji and Dahj Asha, Stepford Wives, VICI, Vision; *Thematic Essay:* AI and the Apocalypse: Science Fiction Meeting Science Fact.

Further Reading

Allen, Arthur. 2019. "There's a Reason We Don't Know Much about AI." Politico, September 16, 2019. https://www.politico.com/agenda/story/2019/09/16/artficial-intelligence-study-data-000956?cid=apn.

Bastani, Aaron. 2019. *Fully Automated Luxury Communism*. New York: Verso.

Byrne, Emma. 2013. "Innovation Isn't Safe: The Future According to Kevin Warwick." *Forbes*, September 30, 2013. https://www.forbes.com/sites/netapp/2013/09/30/kevin -warwick-captain-cyborg/#48b704c13560.

Carper, Steve. 2019. *Robots in American Popular Culture*. Jefferson, NC: McFarland.

Chandler, Simon. 2019. "Artificial Intelligence Has Become a Tool for Classifying and Ranking People." *Forbes*, October 1, 2019. https://www.forbes.com/sites/simon chandler/2019/10/01/artificial-intelligence-has-become-a-tool-for-classifying-and -ranking-people/#7431657f1d7c.

Chanthadavong, Aimee. 2019. "AI and Ethics: The Debate That Needs to be Had." ZDnet, September 16, 2019. https://www.zdnet.com/article/ai-and-ethics-the-debate-that -needs-to-be-had/#ftag=CAD-03-10abf5f.

Choi, Charles Q. 2009. "Human Evolution: The Origin of Tool Use." LiveScience, November 11, 2009. https://www.livescience.com/7968-human-evolution-origin-tool.html.

Dinh, Thien-Nam. 2018. *Silicon Minds: The Science, Impact, and Promise of Artificial Intelligence*. Independently Published.

Ewing, Jeffrey A., and Decker, Kevin S., eds. 2017. *Alien and Philosophy: I Infest, Therefore I Am*. Hoboken, NJ: Wiley-Blackwell.

Hampton, Gregory Jerome. 2017. *Imagining Slaves and Robots in Literature, Film, and Popular Culture: Reinventing Yesterday's Slave with Tomorrow's Robot*. Lanham, MD: Lexington.

Lin, Patrick, Abney, Keith, and Bekey, George A., eds. 2014. *Robot Ethics: The Ethical and Social Implications of Robotics*. Cambridge: Massachusetts Institute of Technology.

Lin, Patrick, Jenkins, Ryan, and Abney, Keith, eds. 2017. *Robot Ethics 2.0: From Autonomous Cars to Artificial Intelligence*. New York: Oxford University Press.

Murdock, Jason. 2019. "Former Google Engineer Warns AI Might Accidentally Start a War: 'These Things Will Start to Behave in Unexpected Ways.'" *Newsweek*, September 16, 2019. https://www.newsweek.com/google-project-maven-artificial-intel ligence-laura-nolan-killer-robots-department-defense-1459358.

Rouhiainen, Lasse. 2018. *Artificial Intelligence: 101 Things You Must Know Today about Our Future*. Scotts Valley, CA: CreateSpace.

Spaeth, Dennis. 2018. "From Single-Task Machines to Backflipping Robots: The Evolution of Robots." *Cutting Tool Engineering*, January 15, 2018. https://www.ctemag .com/news/articles/evolution-of-robots.

Starck, Kathleen, ed. 2010. *Between Fear and Freedom: Cultural Representations of the Cold War*. Newcastle, UK: Cambridge Scholars.

Virk, Rizwan. 2019. *The Simulation Hypothesis: An MIT Computer Scientist Shows Why AI, Quantum Physics, and Eastern Mystics All Agree We Are in a Video Game*. Milwaukee, WI: Bayview.

Wallach, Wendell, and Allen, Colin. 2009. *Moral Machines: Teaching Robots Right from Wrong*. New York: Oxford University Press.

Borg

Star Trek

In 1966, creator-writer-producer Gene Roddenberry (1921–1991) launched his iconic science fiction television series *Star Trek* (NBC, 1966–1969). Set in the twenty-third century, the series centers on the adventures of the starship *USS Enterprise*, its commander, Captain James T. Kirk (played by William Shatner,

b. 1931), and crew, headed by the alien first officer, Mr. Spock (played by Leonard Nimoy, 1931–2015), representing the United Federation of Planets. Though short-lived in its initial run, the series has become legendary, not only for its ground-breaking special effects but also for the sophistication of its stories, many of which were grounded in real-world issues of the day, such as racism and a fear of the Cold War. Its success in syndicated reruns revived the franchise, leading to an animated series, thirteen feature films, and six live-action "sequel" series (as of this writing). The first of these sequel series was *Star Trek: The Next Generation* (a.k.a. *TNG*; syndication, 1987–1994), set a century after the original, with a new *Enterprise*, under the command of Captain Jean-Luc Picard (played by Patrick Stewart, b. 1940). In the series' second season, audiences were introduced to a new threat to humanity's future: the Borg.

When Captain Picard brags to the godlike being "Q" that humanity is ready for whatever challenges it may face in space, Q—with a snap of his fingers—hurls the *Enterprise* to the Delta Quadrant of their galaxy, decades away from home, at top speed. There Picard and his crew encounter the Borg: a collective of cybernetic life-forms with a hive mind that travel through space in cube-shaped vessels bent on assimilating or annihilating every species it encounters. Picard is forced to con-fess that his crew are not ready for this threat, leading Q to return them home, and realizes that, now that the Borg is aware of their existence, it will be coming for them (Maurice Hurley, "Q Who," *Star Trek: The Next Generation*, season 2, epi-sode 16, May 8, 1989). The *TNG* crew have several more encounters with the Borg, with Picard himself briefly assimilated into the collective (Michael Piller, "The Best of Both Worlds, Parts 1 and 2," *ST: TNG*, season 3, episode 26/season 4, epi-sode 1, June 18/September 24, 1990). In the second *TNG* feature film, the Borg travels to Earth's past in an attempt to assimilate the planet before it can become a challenge. Picard and crew follow them back through time and are introduced to the Borg Queen (played by Alice Krige, b. 1954), the central core of the Borg hive mind and the source of its malevolent mission. During this adventure, the Borg Queen offers to assimilate the Federation android Lieutenant Commander Data (played by Brent Spiner, b. 1949) by adding organic tissue to his android frame (Brannon Braga and Ronald D. Moore, *Star Trek: First Contact*, Paramount, 1996).

In the fourth live-action series, *Voyager* (UPN, 1995–2001), Captain Kathryn Janeway (played by Kate Mulgrew, b. 1955) commands the *USS Voyager*, which is pulled to the Delta Quadrant while in pursuit of Federation terrorists. Seventy-five light-years from home (at maximum warp), the *Voyager* crew ultimately cross paths with the Federation's nemesis on its home turf. In this encounter, however, Janeway discovers that the Borg is at war with an even more dangerous species, known only as "Species 8472." Janeway forms an uneasy alliance, fully aware that the Borg will turn on her once it no longer needs the alliance. During this encounter, Janeway rescues a Borg drone known only as "Seven of Nine" (played by Jeri Ryan, b. 1968) an adult human female who was assimilated in childhood. The ship's holographic doctor is able to remove the Borg implants, allowing Seven to pursue a life as an independent, free human (opening the door for several fasci-nating stories revolving around Seven's "reassimilation" into humanity). Seven then becomes a valuable member of the *Voyager* crew (Braga and Joe Menosky,

"Scorpion, Parts 1 and 2," *Star Trek: Voyager*, season 3, episode 26/season 4, episode 1, May 21/September 3, 1997). In the series finale, set decades in the future, Admiral Janeway is celebrating the tenth anniversary of *Voyager*'s return home after spending twenty-three years in the Delta Quadrant (though without many of the original crew). The older Janeway devises a plan to travel back in time, meet up with *Voyager* only seven years into its unintended exile, and sacrifice herself in one last encounter with the Borg Queen in order to get her crew home safely decades earlier than before. After many obstacles—including violating Federation laws with regard to changing the past—the elder Janeway is successful, and *Voyager* returns home with all major characters alive and well (Braga, Rick Berman, and Kenneth Biller, *ST: Voyager*, season 7, episodes 25 and 26, May 23, 2001).

In the fifth live-action series, *Enterprise* (UPN, 2001–2005), set a century *before* the original series, two Borg drones are found to have survived the events of *Star Trek: First Contact*, sending a message to the Delta Quadrant and creating a time paradox in the series canon (Mike Sussman and Phyllis Strong, "Regeneration," *Star Trek: Enterprise*, season 2, episode 23, May 7, 2003). In the sixth live-action series, *Star Trek: Discovery* (CBS All Access, 2017–present), an interstellar mechanical entity known only as "Control" has been suggested to be connected in some way to the Borg. In the seventh series, *Star Trek: Picard* (CBS All Access, 2020–present), the now-retired Admiral Picard (played once more by Stewart) and the rogue ex-Borg Seven (once more played by Ryan) team up to uncover a Romulan plot involving a defunct Borg cube ship and synthetic life-forms. With the continued popularity of the malevolent antagonists, it is likely that there will be many future *Star Trek* adventures centered on this intergalactic threat to humanity's very existence.

Throughout the *Star Trek* franchise, the Borg represents the ultimate danger to cybernetic experimentation. Like humans, the Borg originated as entirely organic life-forms bent on pursuing cybernetic technology to extend their lives. One advancement led to the next, and it was not long before the Borg were more machine than alive. Very similar to the Cybermen of *Doctor Who* in most respects, the Borg seeks to expand its "civilization" through the assimilation (whether voluntarily or not) of other life-forms. It views such assimilation as beneficial both to its "society" and to the ones being assimilated. However, such assimilation ultimately results in the annihilation of the organic as opposed to the semiorganic. The longer "life" promised by the Borg—and the Cybermen—comes at a high cost: the loss of free will, independent thought, and any semblance of liberty.

Richard A. Hall

See also: Adam, Alita, Androids, B.A.T.s, Battle Droids, Cybermen, Cylons, Daleks, Doctor/EMH, Doombots, Echo/CT-1409, Fembots, The Great Intelligence, Inspector Gadget, Landru, Lieutenant Commander Data, LMDs, Lore/B4, Nardole, Replicants, Robots, Sentinels, Soji and Dahj Asha, Starfleet Computer, Terminators; *Thematic Essays:* Villainous Robots and Their Impact on Sci-Fi Narratives, Cyborgs: Robotic Humans or Organic Robots?, AI and the Apocalypse: Science Fiction Meeting Science Fact.

Further Reading
Allen, Arthur. 2019. "There's a Reason We Don't Know Much about AI." Politico, September 16, 2019. https://www.politico.com/agenda/story/2019/09/16/artficial-intel ligence-study-data-000956?cid=apn.

Ashby, LeRoy. 2006. *With Amusement for All: A History of American Popular Culture since 1830.* Lexington: University Press of Kentucky.

Bastani, Aaron. 2019. *Fully Automated Luxury Communism.* New York: Verso.

Byrne, Emma. 2013. "Innovation Isn't Safe: The Future According to Kevin Warwick." *Forbes,* September 30, 2013. https://www.forbes.com/sites/netapp/2013/09/30/kevin -warwick-captain-cyborg/#48b704c13560.

Carper, Steve. 2019. *Robots in American Popular Culture.* Jefferson, NC: McFarland.

Castleman, Harry, and Podrazik, Walter J. 2016. *Watching TV: Eight Decades of American Television, Third Edition.* Syracuse, NY: Syracuse University Press.

Chandler, Simon. 2019. "Artificial Intelligence Has Become a Tool for Classifying and Ranking People." *Forbes,* October 1, 2019. https://www.forbes.com/sites/simon chandler/2019/10/01/artificial-intelligence-has-become-a-tool-for-classifying-and -ranking-people/#7431657f1d7c.

Chanthadavong, Aimee. 2019. "AI and Ethics: The Debate That Needs to be Had." ZDnet, September 16, 2019. https://www.zdnet.com/article/ai-and-ethics-the-debate-that -needs-to-be-had/#ftag=CAD-03-10abf5f.

Conrad, Dean. 2018. *Space Sirens, Scientists and Princesses: The Portrayal of Women in Science Fiction Cinema.* Jefferson, NC: McFarland.

Decker, Kevin S., and Eberl, Jason T., eds. 2016. *The Ultimate* Star Trek *and Philosophy: The Search for Socrates.* Hoboken, NJ: Wiley-Blackwell.

Dinh, Thien-Nam. 2018. *Silicon Minds: The Science, Impact, and Promise of Artificial Intelligence.* Independently Published.

Gross, Edward, and Altman, Mark A., eds. 2016. *The Fifty-Year Mission, The Next 25 Years: The Complete, Uncensored, Unauthorized Oral History of* Star Trek. New York: St. Martin's.

Hampton, Gregory Jerome. 2017. *Imagining Slaves and Robots in Literature, Film, and Popular Culture: Reinventing Yesterday's Slave with Tomorrow's Robot.* Lanham, MD: Lexington.

Hitchcock, Susan Tyler. 2007. *Frankenstein: A Cultural History.* New York: W. W. Norton.

Lin, Patrick, Abney, Keith, and Bekey, George A., eds. 2014. *Robot Ethics: The Ethical and Social Implications of Robotics.* Cambridge: Massachusetts Institute of Technology.

Lin, Patrick, Jenkins, Ryan, and Abney, Keith, eds. 2017. *Robot Ethics 2.0: From Autonomous Cars to Artificial Intelligence.* New York: Oxford University Press.

Murdock, Jason. 2019. "Former Google Engineer Warns AI Might Accidentally Start a War: 'These Things Will Start to Behave in Unexpected Ways.'" *Newsweek,* September 16, 2019. https://www.newsweek.com/google-project-maven-artificial-intel ligence-laura-nolan-killer-robots-department-defense-1459358.

Paur, Joey. 2019. "An AI Bot Writes a Hilarious Episode of *Star Trek: The Next Generation.*" Geek Tyrant, November 16, 2019. https://geektyrant.com/news/an-ai-bot-writes -a-hilarious-episode-of-star-trek-the-next-generation.

Reagin, Nancy R., and Liedl, Janice, eds. 2013. Star Wars *and History.* New York: John Wiley & Sons.

Rouhiainen, Lasse. 2018. *Artificial Intelligence: 101 Things You Must Know Today about Our Future.* Scotts Valley, CA: CreateSpace.

Spaeth, Dennis. 2018. "From Single-Task Machines to Backflipping Robots: The Evolution of Robots." *Cutting Tool Engineering,* January 15, 2018. https://www.ctemag .com/news/articles/evolution-of-robots.

Starck, Kathleen, ed. 2010. *Between Fear and Freedom: Cultural Representations of the Cold War.* Newcastle, UK: Cambridge Scholars.

Stark, Steven D. 1997. *Glued to the Set: The 60 Television Shows and Events that Made Us Who We Are Today*. New York: Delta Trade Paperbacks.

Wallach, Wendell, and Allen, Colin. 2009. *Moral Machines: Teaching Robots Right from Wrong*. New York: Oxford University Press.

Brainiac

DC Comics

In 1938, writer Jerry Siegel (1914–1996) and artist Joe Shuster (1914–1992) introduced the world to Superman, the world's first "superhero." When the planet Krypton faced imminent destruction, the scientist Jor-El created a spaceship in which he placed his infant son, Kal-El, and sent the child to safety on Earth. There, Kal-El was raised by Jonathan and Martha Kent, farmers in the small midwestern town of Smallville. Raised as "Clark Kent," Kal-El grew to experience phenomenal superpowers, eventually including flight, superstrength, superspeed, "laser" eyes, "freeze" breath, and X-ray vision. The boy grew to serve humankind as Superman (first introduced in *Action Comics #1*, June 1938). In the eighty-plus years since his inception, Superman has gone on to dominate all media of popular culture: comic books, radio, movie serials (both animated and live action), television series (both animated and live action), and feature films. Throughout that time, Superman has accumulated an impressive rogues' gallery of "supervillains." One of the most daunting of these villains is the alien android (sometimes presented as a cyborg) Brainiac.

Brainiac was first introduced as an enigmatic alien being of unknown makeup who was intent on "collecting" cities from various planets around the universe—eventually including the Kryptonian city of Kandor—and shrinking these cities and storing them in glass casings, creating a menagerie of life-forms in their natural habitats (Otto Binder and Al Plastino, *Action Comics #242*, July 1958). A future "descendent" of Brainiac was introduced a few years later as a heroic member of the thirtieth-/thirty-first-century Legion of Superheroes. Known as both Querl Dox and "Brainiac 5," this future version was an organic being, descended from "Brainiac 2," a cybernetic version of Brainiac that led a rebellion against the creators of the original (Siegel and Jim Mooney, "Supergirl's Three Super-Girlfriends," *Action Comics #276*, May 1961). In the late-1980s, DC Comics rebooted their universe with the event *Crisis on Infinite Earths* (1985–1986). The rebooted version of Brainiac was as an alien scientist named Vril Dox from the planet Colu. Sentenced to death by the computer overlords of Colu after attempting a coup, Dox was disintegrated, his essence traversing space to merge with the Earth human Milton Fine (John Byrne and Jerry Ordway, *Adventures of Superman #438*, March 1988).

Outside of comics, Brainiac has appeared in two live-action television series featuring the Man of Steel. On the series *Smallville* (The WB, 2001–2006; The CW, 2006–2011), new college student Clark Kent (played by Tom Welling, b. 1977) meets Professor Fine (played by James Marsters, b. 1962). This version of Brainiac (designated a "Brain Interactive Construct") is a Kryptonian computer that was key to the planet's destruction. He can take human form or transform into a black liquid substance. In either incarnation, Brainiac is able to interact with any computer system. This Brainiac was originally completed by Jor-El but corrupted through the machinations of General Zod. On Earth, Brainiac prepares Lex Luthor (played by Michael Rosenbaum, b. 1972) to be the vessel for a resurrected Zod (*Smallville*, season 5, September 2005–May 2006). Marsters returned in the final season of the series as the benevolent "Brainiac 5" from the thirty-first century to show Clark his future as Superman (Brian Peterson and Kelly Souders, "Homecoming," *Smallville*, season 10, episode 4, October 15, 2010).

The most recent appearance of Brainiac has been as the primary antagonist on the live-action television series *Krypton* (SyFy, 2018–2019). This incarnation returns the character to his roots. Brainiac (played by Blake Ritson, b. 1978) is a humanoid android from the planet Colu that travels the universe collecting cities from various planets. The series is set long before Krypton's destruction, where Superman's paternal grandfather, Seg-El (played by Cameron Cuffe, b. 1993) is visited by Earth's present-day hero Adam Strange (played by Shaun Sipos, b. 1981) with a warning that one of Superman's villains has altered the past to prevent the Man of Steel's birth or, at the very least, his trip to Earth. Seg, Strange, and Seg's romantic interest, Lyta Zod (played by Georgina Campbell, b. 1993) team with Seg and Lyta's son from the future, General Zod (played by Colon Salmon, b. 1962). Successful in saving Kandor from Brainiac (thus, presumably, already altering Superman's future), Seg merges with Brainiac, and the two are cast far from Krypton, leaving the planet open for General Zod's rule.

Brainiac is a product of the Cold War and Atomic Age anxieties of the late 1950s/early 1960s. He is the ultimate manifestation of what humanity most fears about AI: that it will, in turn, become humanity's downfall. His malevolence, however, is born from his programming to "collect." With few exceptions in the comics, Brainiac's overall goal is not to rule any particular world but to preserve aspects of worlds through the collection, miniaturization, and storage of samples of each society. However, Brainiac is fully aware that these peoples do not wish to be collected, and he sees his overall mission as overriding the will of the masses. This, then, is the danger of specifically programmed AI. As in all aspects of human history, the danger of "unintended consequences" ultimately seems unavoidable.

Richard A. Hall

See also: Arnim Zola, Batcomputer, Cerebro/Cerebra, Control, The Great Intelligence, HAL 9000, HERBIE, Janet, JARVIS/Friday, L3-37/*Millennium Falcon*, The Matrix/Agent Smith, OASIS, Rehoboam, Skynet, Starfleet Computer, TARDIS, Ultron, V-GER, WOPR; *Thematic Essay:* Villainous Robots and Their Impact on Sci-Fi Narratives.

Further Reading

Allen, Arthur. 2019. "There's a Reason We Don't Know Much about AI." Politico, September 16, 2019. https://www.politico.com/agenda/story/2019/09/16/artficial-intelligence-study-data-000956?cid=apn.

Barker, Cory, Ryan, Chris, and Wiatrowski, Myc, eds. 2014. *Mapping* Smallville*: Critical Essays on the Series and Its Characters*. Jefferson, NC: McFarland.

Byrne, Emma. 2013. "Innovation Isn't Safe: The Future According to Kevin Warwick." *Forbes*, September 30, 2013. https://www.forbes.com/sites/netapp/2013/09/30/kevin-warwick-captain-cyborg/#48b704c13560.

Carper, Steve. 2019. *Robots in American Popular Culture*. Jefferson, NC: McFarland.

Castleman, Harry, and Podrazik, Walter J. 2016. *Watching TV: Eight Decades of American Television, Third Edition*. Syracuse, NY: Syracuse University Press.

Chandler, Simon. 2019. "Artificial Intelligence Has Become a Tool for Classifying and Ranking People." *Forbes*, October 1, 2019. https://www.forbes.com/sites/simon chandler/2019/10/01/artificial-intelligence-has-become-a-tool-for-classifying-and -ranking-people/#7431657f1d7c.

Chanthadavong, Aimee. 2019. "AI and Ethics: The Debate that Needs to Be Had." ZDnet, September 16, 2019. https://www.zdnet.com/article/ai-and-ethics-the-debate-that -needs-to-be-had/#ftag=CAD-03-10abf5f.

Daniels, Les. 2004. *Superman: The Complete History—The Life and Times of the Man of Steel*. New York: DC Comics.

Dinh, Thien-Nam. 2018. *Silicon Minds: The Science, Impact, and Promise of Artificial Intelligence*. Independently Published.

Hendershot, Cyndy. 1999. *Paranoia, the Bomb, and 1950s Science Fiction Films*. Bowling Green, KY: Bowling Green State University Popular Press.

Hoberman, J. 2011. *An Army of Phantoms: American Movies and the Making of the Cold War*. New York: New Press.

Lin, Patrick, Abney, Keith, and Bekey, George A., eds. 2014. *Robot Ethics: The Ethical and Social Implications of Robotics*. Cambridge: Massachusetts Institute of Technology.

Lin, Patrick, Jenkins, Ryan, and Abney, Keith, eds. 2017. *Robot Ethics 2.0: From Autonomous Cars to Artificial Intelligence*. New York: Oxford University Press.

Murdock, Jason. 2019. "Former Google Engineer Warns AI Might Accidentally Start a War: 'These Things Will Start to Behave in Unexpected Ways.'" *Newsweek*, September 16, 2019. https://www.newsweek.com/google-project-maven-artificial-intel ligence-laura-nolan-killer-robots-department-defense-1459358.

Rouhiainen, Lasse. 2018. *Artificial Intelligence: 101 Things You Must Know Today about Our Future*. Scotts Valley, CA: CreateSpace.

Spaeth, Dennis. 2018. "From Single-Task Machines to Backflipping Robots: The Evolution of Robots." *Cutting Tool Engineering*, January 15, 2018. https://www.ctemag .com/news/articles/evolution-of-robots.

Starck, Kathleen, ed. 2010. *Between Fear and Freedom: Cultural Representations of the Cold War*. Newcastle, UK: Cambridge Scholars.

Tye, Larry. 2013. *Superman: The High-Flying History of America's Most Enduring Hero*. New York: Random House.

Wallach, Wendell, and Allen, Colin. 2009. *Moral Machines: Teaching Robots Right from Wrong*. New York: Oxford University Press.

Wright, Bradford W. 2003. *Comic Book Nation: The Transformation of Youth Culture in America*. Baltimore, MD: Johns Hopkins University Press.

Buffybot

Buffy the Vampire Slayer

Buffy the Vampire Slayer (The WB, 1991–2001; UPN, 2001–2003) was created by writer-director Joss Whedon (b. 1964) and was a "sequel" (of sorts) to the 1992 feature film of the same name. The series starred Sarah Michelle Gellar (b. 1977) as the titular heroine. The premise is that high school student Buffy Summers is

---

Erica

In 2020, the *Hollywood Reporter* published an article announcing that an artificial intelligence robot would become the first such mechanism to "act" as the lead in a feature film. The robot's name is Erica, and she will star in the science fiction film *b*, which tells the tale of a scientist working on a project to "perfect" human DNA (a recurring sci-fi trope that goes back to real-world efforts to such ends over a century ago). Erica was designed by Japanese engineers Hiroshi Ishiguro and Kohei Ogawa. As Erica lacks the basic human background that human actors apply to the interpretation of their performances, Ishiguro and Ogawa have programmed Erica with the concept of "method acting," in which actors fully immerse themselves into their written roles, completely burying their own personality, memories, and experiences, in order to fully "become" the character they are portraying (Courtney Linder, "This AI Robot Just Nabbed the Lead Role in a Sci-Fi Movie," *Popular Mechanics*, June 25, 2020, https://www.popularmechanics.com/technology/robots/a32968811 /artificial-intelligence-robot-movie-star-erica). As an artificially created humanoid, Erica will, in essence, be portraying a character that is, at its core, her own self (which, of course, opens the debate as to whether the robot is, in fact, "acting," or, rather, simply conveying lines as herself by another name . . . a debate that has, to be fair, been applied to many popular human actors and actresses). If successful, Erica could revolutionize Hollywood, providing filmmakers with a cast that (1) cannot get sick, (2) cannot get "hurt" doing stunts, and (3) is in no position to negotiate a contract. This, of course, opens the debate on robots and slavery once again.

Richard A. Hall

---

the latest in a centuries-long line of "Chosen Ones" who bear the title of the "Slayer," protecting humanity from "vampires, demons, and the forces of darkness" (opening credits, *Buffy the Vampire Slayer*). Throughout her endless crusade, Buffy is assisted by her group of friends, known as the "Scooby Gang" (in reference to the popular children's animated series, *Scooby Doo*). These friends include the computer nerd-turned-witch Willow Rosenberg (played by Alyson Hannigan, b. 1974), the hapless Xander Harris (played by Nicholas Brendon, b. 1971), and Buffy's assigned "Watcher," Rupert Giles (played by Anthony Head, b. 1954). Together they guard the "Hellmouth," a portal to a demonic dimension and magnet to dark forces located in the fictional town of Sunnydale, California.

The Buffybot was created when the vampire Spike (played by James Marsters, b. 1962) became romantically obsessed with Buffy and approached robot creator Warren Mears (played by Adam Busch, b. 1978) to build him a fully functioning sex robot that looked like the Slayer (Jane Espenson, "I Was Made to Love You," *BtVS*, season 5, episode 15, February 20, 2001). At the end of the fifth season, Buffy is forced to sacrifice herself to prevent the rise of the goddess Glory (played by Clair Kramer, b. 1974), leaving Sunnydale and the Hellmouth without its most powerful protector (Joss Whedon, "The Gift," *BtVS*, Season 5, Episode 22, May 22, 2001). To convince the forces of darkness that the Slayer is still alive, Willow and the members of the Scooby Gang turn to Warren to upgrade the Buffybot (played by Gellar), which possesses the appearance, strength, agility, and speed of the original (while maintaining its preprogrammed romantic obsession with Spike). When the Buffybot becomes damaged during combat, word spreads that the Slayer is a fake. Demons descend on Sunnydale as Willow and company conjure a spell to resurrect the real Buffy from the grave. During the melee, the Buffybot is pulled apart and destroyed by motorcycle-riding demons in full view of her former paramour, Spike (Marti Noxon and David Fury, "Bargaining, Parts 1 and 2," *BtVS*, season 6, episodes 1 and 2, October 2, 2001).

The Buffybot is a standard example of a service android. Though initially designed for sexual pleasure, she was eventually reprogrammed to stand in for Earth's primary heroine and protector against "the forces of darkness." Though programmed to react and feign emotion and individual thought, she was, in fact, no more than a toaster, enacting the will of her programmers for their own individual needs. The argument of "slavery," therefore, is more difficult in the case of Buffybot. She possesses no free will or independent thought. She is not self-aware or in possession of any degree of understanding of who—or what—she is. While she does understand that she is a robot, she is incapable of comprehending all that being a robot implies. Her kind disposition, however, does make her unfortunate end a sad one for viewers (and, presumably, Spike).

Richard A. Hall

See also: Androids, B.A.T.s, Baymax, Bernard, Bishop, Dolores, Fembots, Human Torch, Inspector Gadget, Iron Giant, Jocasta, Lieutenant Commander Data, LMDs, Lore/B4, Marvin the Paranoid Android, Maschinenmensch/Maria, Replicants, Robots, Soji and Dahj Asha, Stepford Wives, VICI, Vision; *Thematic Essays:* Robots and Slavery, Heroic Robots and Their Impact on Sci-Fi Narratives.

Further Reading

Byrne, Emma. 2013. "Innovation Isn't Safe: The Future According to Kevin Warwick." *Forbes*, September 30, 2013. https://www.forbes.com/sites/netapp/2013/09/30/kevin-warwick-captain-cyborg/#48b704c13560.

Byrne, Emma. 2014. "Cybernetic Implants: No Longer Science Fiction." *Forbes*, March 11, 2014. https://www.forbes.com/sites/netapp/2014/03/11/cybernetic-implants-not-sci-fi/#7399c57e77ba.

Castleman, Harry, and Podrazik, Walter J. 2016. *Watching TV: Eight Decades of American Television, Third Edition*. Syracuse, NY: Syracuse University Press.

Hampton, Gregory Jerome. 2017. *Imagining Slaves and Robots in Literature, Film, and Popular Culture: Reinventing Yesterday's Slave with Tomorrow's Robot*. Lanham, MD: Lexington.

Lavery, David. 2013. *Joss Whedon, A Creative Portrait: From* Buffy the Vampire Slayer *to Marvel's* The Avengers. London: I. B. Tauris.

Lin, Patrick, Abney, Keith, and Bekey, George A., eds. 2014. *Robot Ethics: The Ethical and Social Implications of Robotics*. Cambridge: Massachusetts Institute of Technology.

Lin, Patrick, Jenkins, Ryan, and Abney, Keith, eds. 2017. *Robot Ethics 2.0: From Autonomous Cars to Artificial Intelligence*. New York: Oxford University Press.

Rouhiainen, Lasse. 2018. *Artificial Intelligence: 101 Things You Must Know Today about Our Future*. Scotts Valley, CA: CreateSpace.

South, James B., ed. 2003. Buffy the Vampire Slayer *and Philosophy: Fear and Trembling in Sunnydale*. Popular Culture and Philosophy series, edited by William Irwin. Chicago: Open Court.

Wallach, Wendell, and Allen, Colin. 2009. *Moral Machines: Teaching Robots Right from Wrong*. New York: Oxford University Press.

Wilcox, Rhonda V., ed. *Slayage: The Journal of Whedon Studies*. Whedon Studies Association. https://www.whedonstudies.tv/slayage-the-journal-of-whedon-studies.html.

Yeffeth, Glenn, ed. 2003. *Seven Seasons of Buffy: Science Fiction and Fantasy Writers Discuss Their Favorite Television Show*. Dallas, TX: BenBella.

C

C-3PO

Star Wars

C-3PO (played by Anthony Daniels, b. 1946), a golden "3PO-series protocol droid," made his debut in *Star Wars: Episode IV—A New Hope* (Lucasfilm/20th Century Fox, 1977), the first film of the original *Star Wars* trilogy. Set "a long time ago, in a galaxy far, far away," the popular space opera series that now spans nine "saga" films, centers around a "Rebel Alliance," of which C-3PO becomes a beloved member, fighting against an "evil Galactic Empire." George Lucas (b. 1944), who wrote or cowrote the first three films (directing the first) and wrote and directed the prequel trilogy, created the humanoid robot, whose primary function is to assist in etiquette, customs, and translation. He introduces himself as "C-3PO, human-cyborg relations," and claims that he is "fluent in over six million forms of communication." Sometimes downright oblivious—often to great comic effect—C-3PO is also often melodramatic and prone to worry, constantly quoting how bad the odds are in a given situation and lamenting certain "doom."

At the beginning of the first film, C-3PO and his "astromech" droid companion, "R2-D2" (played by Kenny Baker, 1934–2016), escape to the planet Tatooine after Darth Vader (played by David Prowse, b. 1935; voiced by James Earl Jones, b. 1931) attacks the consular ship *Tantive IV* and takes Princess Leia Organa (played by Carrie Fisher, 1956–2016) prisoner. When Luke Skywalker (played by Mark Hamill, b. 1951) purchases the two droids from scavenger Jawas, he stumbles upon a message stored in R2-D2 from Princess Leia to Obi-Wan Kenobi (played by Alec Guinness, 1914–2000). Luke and the two droids seek out the old Jedi master, and the four embark on a mission to return the droid to the princess's father by hiring smugglers Han Solo (played by Harrison Ford, b. 1942) and Chewbacca (played by Peter Mayhew, 1944–2019) to take them on board their ship, the *Millennium Falcon*. After the ship is captured by the Empire's battle station, the "Death Star," C-3PO and R2-D2 are key in helping their human companions—and the princess—escape. They are present at the celebration of the Death Star's destruction.

In *Star Wars: Episode V—The Empire Strikes Back* (Lucasfilm, 1980), C-3PO escapes from the ice planet Hoth with Han, Chewbacca, and Leia in the *Millennium Falcon*, only to be pursued by Imperial forces. They make their way to Cloud City, where C-3PO is nearly destroyed by a blast from an Imperial stormtrooper. Chewbacca finds the dismembered parts of the droid and partially rebuilds him. R2-D2 finishes C-3PO's repairs as they travel to the Rebels' rendezvous point with Leia, Luke, Chewbacca, and Lando Calrissian (played by Billy Dee Williams, b. 1937). In the original trilogy's final installment, *Star Wars: Episode VI—Return of the Jedi* (Lucasfilm, 1983), Luke offers C-3PO and R2-D2 as servants to intergalactic gangster Jabba the Hutt as part of a scheme to rescue Han Solo. Upon completion of this mission, C-3PO and the other protagonists travel to the forest moon of Endor to disable a shield generator for the nearly completed second Death Star, but they are captured by the Indigenous Ewoks. The Ewoks mistake C-3PO for a god, and he convinces them to help the Rebels. When the second Death Star is destroyed in the climax of the film, signaling the end of the Empire, C-3PO celebrates with the other Rebels.

In *Star Wars: Episode I—The Phantom Menace* (Lucasfilm, 1999), the first of a trilogy of prequels, it is revealed that nine-year-old Anakin Skywalker (played by Jake Lloyd, b. 1989), who lives as a slave on Tatooine with his mother, Shmi (played by Pernilla August, b. 1958), built C-3PO from spare parts he found in a junk heap. C-3PO and Anakin are separated when Qui-Gon Jinn (played by Liam Neeson, b. 1952) frees Anakin from

Perhaps the two most iconic robots in film history, C-3PO and R2-D2 of the *Star Wars* franchise. C-3PO was played by Anthony Daniels, and R2-D2 was played by Kenny Baker. (Yorgy67/ Dreamstime.com)

his enslavement. C-3PO meets R2-D2 in this film. The next prequel, *Star Wars: Episode II—Attack of the Clones* (Lucasfilm, 2002), is set ten years later. C-3PO continues to live with Anakin's mother until she is kidnapped by Tusken Raiders. Anakin (played by Hayden Christensen, b. 1981), now being trained as a Jedi, returns to Tatooine to rescue his mother. He is too late to save her life, but he reunites with C-3PO, who accompanies Anakin to the planet Geonosis to rescue his Jedi teacher, the young Obi-Wan Kenobi (played by Ewan McGregor, b. 1971), from Sith Lord Count Dooku (played by Christopher Lee, 1922–2015). Through a mishap at a droid-construction site, C-3PO's head is attached to the torso of a "battle droid." He reluctantly participates in battle on the wrong side before being restored by R2-D2. He later witnesses Anakin's marriage to Queen Padmé Amidala (played by Natalie Portman, b. 1981). In the final film in the prequel trilogy, *Star Wars: Episode III—Revenge of the Sith* (Lucasfilm, 2005), Anakin turns to the "dark side" of the "Force" and becomes the Sith Lord Darth Vader. After C-3PO sees Vader using the Force to choke Padmé into unconsciousness, he and R2-D2 remove her to safety. She gives birth to the Skywalker twins, Luke and

Leia in C-3PO's presence and dies shortly afterward. To protect the Skywalker twins from their father and the new Galactic Empire, Senator Bail Organa (played by Jimmy Smits, b. 1955) has C-3PO's memory wiped and gives him and R2-D2 to Captain Raymus Antilles (played by Rohan Nichol, b. 1976).

C-3PO reappears in the sequel trilogy, set thirty years after the events of the original series, in which the "First Order" has arisen from the ashes of the Empire; and Leia, now a general, leads the "Resistance." In the first of these three films, *Star Wars: Episode VII—The Force Awakens* (Lucasfilm/Disney, 2015), cowritten and directed by J. J. Abrams (b. 1966), C-3PO now boasts seven million forms of communication and sports a red arm. The second sequel, *Star Wars: Episode VIII—The Last Jedi* (Lucasfilm/Disney, 2017), written and directed by Rian Johnson (b. 1973), finds C-3PO serving as an assistant for Leia and, later, Resistance pilot Poe Dameron (played by Oscar Isaac, b. 1979). A brief reunion between Luke and C-3PO occurs when Luke winks at the droid before confronting First Order Supreme Leader Kylo Ren (played by Adam Driver, b. 1983). In the final film in the sequel trilogy, *Star Wars: Episode IX—The Rise of Skywalker* (Lucasfilm/Disney, 2019), once more cowritten and directed by Abrams, C-3PO travels with Rey (played by Daisy Ridley, b. 1992), Finn (played by John Boyega, b. 1992), the droid BB-8, Chewbacca (played now by Joonas Suotamo, b. 1986), and Poe to the planet Passanna, where they discover a Sith dagger that holds a clue to the whereabouts of the secret, massive First Order fleet. Since C-3PO cannot translate Sith directly, a hacker is called in to work around this protocol, but C-3PO's memory is wiped as a consequence. When they finally return to the Resistance camp after defeating the First Order, R2-D2 restores C-3PO's memory from a backup, and they celebrate victory with the rest of the Resistance.

In the beginning, C-3PO and R2-D2 fulfill the role of "messenger" from the classic "hero's journey," providing the catalyst for the hero, Luke Skywalker, to answer the "call to adventure." Throughout the *Star Wars* films, C-3PO plays the foil to R2-D2, lending comic relief, especially when he translates R2-D2's language of beeps for the audience. At times they also act as narrators for the story. They are the only characters to appear in all the *Star Wars* films to date, with the exception of *Solo: A Star Wars Story* (Lucasfilm/Disney, 2018). They are also included in numerous related novels, comic books, video games, and TV shows and specials. After the success of the first film in 1977, the two quickly became one of the most iconic duos in movie history, recognized and beloved around the world.

Lisa C. Bailey

See also: AI/Ziggy, Battle Droids, Baymax, BB-8, Bender, Cambot/Gypsy/Tom Servo/ Crow, Darth Vader, D-O, Dr. Theopolis and Twiki, Echo/CT-1409, EV-9D9, General Grievous, HERBIE, IG-88/IG-11, Janet, K-2SO, K9, L3-37/*Millennium Falcon*, Marvin the Paranoid Android, Maschinenmensch/Maria, Medical Droids, Nardole, R2-D2, Robot, Tin Woodsman; *Thematic Essays:* Robots and Slavery, Heroic Robots and Their Impact on Sci-Fi Narratives.

Further Reading

Bray, Adam, Barr, Tricia, Horton, Cole, and Windham, Ryder. 2019. *Ultimate* Star Wars, *New Edition.* London: DK Publishing.

Bray, Adam, and Horton, Cole. 2017. Star Wars: *Absolutely Everything You Need to Know, Updated and Expanded.* London: DK Children.

Cavelos, Jeanne. 2008. "R2-D2 and C-3PO: Do Droids Dream of Electric Sheep?" *Scientific American*, August 11, 2008. https://www.scientificamerican.com/article/star-wars -science-droid-dreams.

Daniels, Anthony. 2019. *I Am C-3PO: The Inside Story*. London: DK Publishing.

Hiatt, Brian. 2019. "Anthony Daniels: My Life as C-3PO Is Far from Over." *Rolling Stone*, December 17, 2019. https://www.rollingstone.com/movies/movie-features/star-wars -skywalker-anthony-daniels-c3po-interview-927577.

Kachka, Boris. 2015. "The Last Human Robot." Vulture, December 6, 2015. https://www .vulture.com/2015/12/anthony-daniels-c-3po-c-v-r.html.

Nero, Dom. 2019. "Anthony Daniels Says *Star Wars: The Rise of Skywalker* Isn't the Last You'll See of C-3PO." *Esquire*, November 7, 2019. https://www.esquire.com/enter tainment/movies/a29713208/anthony-daniels-star-wars-c-3po-the-rise-of-skywalker -interview.

Rinzler, J. W. 2007. *The Making of* Star Wars: *The Definitive Story behind the Original Film*. New York: Del Rey.

Slavicsek, Bill. 2000. *A Guide to the* Star Wars *Universe*. San Francisco, CA: LucasBooks.

Sumerak, Marc. 2018. Star Wars: *Droidography*. New York: HarperFestival.

Wallace, Daniel. 2006. Star Wars: *The New Essential Guide to Droids*. New York: Del Rey.

Wallace, Daniel, and Ling, Josh. 1999. *C-3P0: Tales of the Golden Droid (Star Wars Masterpiece Edition)*. San Francisco: Chronicle Books.

Cambot/Gypsy/Tom Servo/Crow

Mystery Science Theater 3000

In 1988, Joel Hodgson created the cult classic television series *Mystery Science Theater 3000* (Comedy Central, 1988–1996; SyFy, 1997–1999; Netflix, 2017– present). The basic premise of the show is that "in the not-too-distant future" a simple day worker is stranded on the "Satellite of Love" by an evil scientist and his hapless assistant and forced to watch old, cheaply produced, badly written and directed, and poorly acted "B-movies" while the "Mads" "monitor his mind" and, between films, come up with various inventions (though there was one season where the Satellite of Love is chased through space by—alternatively—a crazed woman, a race of semi-intelligent apes, and a cultlike organization). In order to make his hell more bearable, the worker uses various parts aboard the satellite to create his "robot friends": Cambot, Gypsy, Tom Servo, and Crow T. Robot. While Cambot shows the films—and Gypsy performs various other functions—the worker (played by Hodgson from 1988 to 1993, Michael J. Nelson from 1993 to 1999, and Jonah Ray since 2017) watches the movies with Tom and Crow; throughout each film, the three "riff" on the horrible films, creating a new form of comedy. The audience sees the films full screen, while "shadows" of theater seats and the human and robot viewers are seen at the bottom of the screen, providing a "theater experience" for the audience.

Audiences only see Cambot—the projector that plays the films for the rest of the characters and the audience—during the show's opening sequence, during the "Robot Roll Call." Gypsy is the only female in the group (though she was portrayed by men through season 10), with a large purple "mouth," a mounted eye "light," and the body of a large vacuum tube. Gypsy rarely views the films and is usually only seen during the introduction and "break" scenes of each episode. Tom Servo is a squat robot with a domed white base, red torso, humanlike "hands" at the end of spring arms, and a bubblegum machine head (the reception tray of which is utilized as a mouth). Crow T. Robot has a golden body that appears to be a series of bell-shaped objects piled one on the other, with dangling metallic appendages, a long "snout," and ping-pong-ball-shaped "eyes." Tom, Crow, and their various human companions make up the bulk of the comedy during the B-movies, making comedic—and occasionally crude—and sarcastic remarks aimed at the horrible plots, acting, and special effects of the respective films (frequently laced with pop culture references). Since the late 1980s, *MST3K* has been a cult pop culture phenomenon.

Like the humans Joel, Mike, and Jonah, the four robots are de facto "slaves," forced to endure the torture devised by the "Mads." While the human protagonists were the only ones chosen to perform these experiments, it was those same protagonists who built the robots for the exact same purpose; and Tom and Crow express the same distress at their fate as do their human creators. Their purpose is to alleviate the suffering of the humans through companionship, shared experience, and cathartic comedy. Though as "unhappy" with their predicament as the humans are, they joyfully play their respective roles, content to fight torture with comedy to the delight of audiences around the world.

Richard A. Hall

See also: AWESOM-O 4000, BB-8, Bender, C-3PO, D-O, HERBIE, Inspector Gadget, Johnny 5/S.A.I.N.T. Number 5, K-2SO, K9, L3-37/*Millennium Falcon*, Marvin the Paranoid Android, Max Headroom, Muffit, Nardole, R2-D2, Robot, Rosie, Speed Buggy, VICI, WALL-E; *Thematic Essay:* Robots and Slavery.

Further Reading

Ashby, LeRoy. 2006. *With Amusement for All: A History of American Popular Culture since 1830.* Lexington: University Press of Kentucky.

Castleman, Harry, and Podrazik, Walter J. 2016. *Watching TV: Eight Decades of American Television, Third Edition.* Syracuse, NY: Syracuse University Press.

Choi, Charles Q. 2009. "Human Evolution: The Origin of Tool Use." LiveScience, November 11, 2009. https://www.livescience.com/7968-human-evolution-origin-tool.html.

Dinh, Thien-Nam. 2018. *Silicon Minds: The Science, Impact, and Promise of Artificial Intelligence.* Independently Published.

Hampton, Gregory Jerome. 2017. *Imagining Slaves and Robots in Literature, Film, and Popular Culture: Reinventing Yesterday's Slave with Tomorrow's Robot.* Lanham, MD: Lexington.

Hoberman, J. 2011. *An Army of Phantoms: American Movies and the Making of the Cold War.* New York: New Press.

Pellissier, Hank. 2013. "Robots and Slavery: What Do Humans Want When We Are 'Masters'?" Institute for Ethics and Emerging Technologies, September 13, 2013. https://ieet.org/index.php/IEET2/more/pellissier20130913.

Rees, Shelley S., ed. 2013. *Reading* Mystery Science Theater 3000: *Critical Approaches.* Lanham, MD: Scarecrow, Rowman & Littlefield.

Spaeth, Dennis. 2018. "From Single-Task Machines to Backflipping Robots: The Evolution of Robots." *Cutting Tool Engineering,* January 15, 2018. https://www.ctemag .com/news/articles/evolution-of-robots.

Starck, Kathleen, ed. 2010. *Between Fear and Freedom: Cultural Representations of the Cold War*. Newcastle, UK: Cambridge Scholars.

Weiner, Robert G., and Barba, Shelley E., eds. 2011. *In the Peanut Gallery with* Mystery Science Theater 3000*: Essays on Film, Fandom, Technology and the Culture of Riffing*. Jefferson, NC: McFarland.

Cerebro/Cerebra

Marvel Comics

In 1961, comic book legend Stan Lee (1922–2018) launched the "Marvel age" of comics. With the launching of the Fantastic Four, soon to be followed by the Hulk, Spider-Man, Iron Man, and many, many more, comic book superheroes took a more dramatic turn, with storylines focused on the private lives of those blessed (or burdened) with superpowers. One of the last original entries into the Marvel Universe was *The X-Men*, created by Lee and artist Jack Kirby in 1963. The X-Men are "mutants": humans born with a mutated "X factor" in their DNA, often causing the mutant to exhibit various "powers" or "gifts" (making them the next step in human evolution, or *homo superior*). One such mutant is the wheelchair-bound professor Charles Xavier, founder of Xavier's School for Gifted Youngsters, where mutants may attend and get an education while simultaneously learning to control—and accept—their various gifts. In order to locate potential mutants—their gifts usually do not manifest until puberty—Professor X amplifies his significant telepathic abilities through a mechanical device originally called "Cerebro" (currently referred to as "Cerebra").

Cerebro was originally built by Xavier and his onetime friend and eventual enemy Magneto, with some enhancements designed by original X-Man, Dr. Hank "Beast" McCoy (Lee and Kirby, *The X-Men #7*, September 1964). In his study of science fiction and superhero comics, Dr. Jeffrey Kripal referred to Cerebro as "psychotronic," based on comic book portrayals of machinery mastered by artist Jack Kirby (Kripal, *Mutants and Mystics*, 2015, 208). Professor X is touted throughout Marvel Comics as the world's most powerful telepath. The fact that Cerebro/Cerebra further enhances his abilities speaks to its overwhelming power. Though, ideally, only telepaths such as Xavier, Jean Grey, or Emma Frost should be able to make full use of Cerebro, other characters, including Ororo "Storm" Munroe (whose mutant ability is the control of weather), have occasionally made use of the machine. In the most recent incarnation of the X-Men in comics, Professor X (now able to walk) permanently wears a spherical version of Cerebro (Jonathan Hickman and Pepe Larraz, *House of X #1*, July 2019).

In the iconic Saturday morning cartoon *X-Men: The Animated Series* (FOX Kids, 1992–1997) and the live-action feature films (20th Century Fox, 2000–2019), Cerebro was presented in its original comic book form: a large (usually spherical) room with a computer console possessing a spherical "helmet," connected by cables to the console, which can be worn by Xavier or whoever chooses to attempt to use it. In the film *X-Men: First Class* (2011), Cerebro was the creation of the American CIA for the purpose of identifying and locating mutants, but only Professor X possessed the mental power to make use of it. In *X-Men: Apocalypse* (2016), Xavier orders X-Man Havok to use his energy powers to destroy the device when the ancient mutant Apocalypse is able to "hack" Cerebro by connecting telepathically to Xavier. Due to Disney's acquisition of 20th Century Fox in 2019, future incarnations of Cerebro will likely appear in future Marvel Cinematic Universe films.

In the comics, Cerebro was enhanced in 2015 by the mutant Forge, who took the original programming and implanted it as an AI into the robotic body of a Sentinel—one of a legion of robots designed by the government to track and capture mutants—to create a new entity: Cerebra. Now Cerebro/Cerebra can interact directly with others rather than simply with Professor X. Additionally, once Cerebra locates a mutant, she can "transport" herself and other X-Men to the location of that mutant. More than ever, Cerebra is an interactive part of the X-Men team and an invaluable asset to a greater degree than ever before (Jeff Lemire and Humberto Ramos, *Extraordinary X-Men #1*, November 2015).

Cerebro/Cerebra is not a "robot." It is, rather, more accurately defined as a "cybernetic enhancement." It is not designed to act independently or have the ability to "think" or "act" on its own. It is a tool designed to allow Professor Xavier to more directly contact fellow mutants, to identify current and potential mutants, and to locate them (either to join, to be assisted by, or to be stopped by the X-Men). It remains to be seen how Cerebro fits into current X-Men comics' continuity; but since the 1960s it has represented the potential that machines and cybernetic enhancements may provide to humankind (or "mutantkind").

Richard A. Hall

See also: AI/Ziggy, Arnim Zola, Batcomputer, Brainiac, Doctor/EMH, Doombots, The Great Intelligence, HAL 9000, HERBIE, Human Torch, Iron Legion, Iron Man, JARVIS/ Friday, Jocasta, Landru, LMDs, The Matrix/Agent Smith, OASIS, Rehoboam, Sentinels, Skynet, Starfleet Computer, TARDIS, Ultron, V-GER, Vision, Warlock, WOPR, Zordon/ Alpha-5; *Thematic Essay:* AI and the Apocalypse: Science Fiction Meeting Science Fact.

Further Reading

Allen, Arthur. 2019. "There's a Reason We Don't Know Much about AI." Politico, September 16, 2019. https://www.politico.com/agenda/story/2019/09/16/artficial-intel ligence-study-data-000956?cid=apn.

Ashby, LeRoy. 2006. *With Amusement for All: A History of American Popular Culture since 1830*. Lexington: University Press of Kentucky.

Chandler, Simon. 2019. "Artificial Intelligence Has Become a Tool for Classifying and Ranking People." *Forbes*, October 1, 2019. https://www.forbes.com/sites/simon chandler/2019/10/01/artificial-intelligence-has-become-a-tool-for-classifying-and -ranking-people/#7431657f1d7c.

Chanthadavong, Aimee. 2019. "AI and Ethics: The Debate That Needs to Be Had." ZDnet, September 16, 2019. https://www.zdnet.com/article/ai-and-ethics-the-debate-that -needs-to-be-had/#ftag=CAD-03-10abf5f.

Dinh, Thien-Nam. 2018. *Silicon Minds: The Science, Impact, and Promise of Artificial Intelligence*. Independently Published.

Howe, Sean. 2012. *Marvel Comics: The Untold Story*. New York: Harper-Perennial.

Kripal, Jeffrey J. 2015. *Mutants and Mystics: Science Fiction, Superhero Comics, and the Paranormal*. Chicago: University of Chicago Press.

Murdock, Jason. 2019. "Former Google Engineer Warns AI Might Accidentally Start a War: 'These Things Will Start to Behave in Unexpected Ways.'" *Newsweek*,

September 16, 2019. https://www.newsweek.com/google-project-maven-artificial
-intelligence-laura-nolan-killer-robots-department-defense-1459358.

O'Rourke, Morgan B., and O'Rourke, Daniel J. 2014. "Prophet of Hope and Change: The
Mutant Minority in the Age of Obama." In *The Ages of the X-Men: Essays on the
Children of the Atom in Changing Times*, edited by Joseph J. Darowski, 223–32.
Jefferson, NC: McFarland.

Powell, Jason. 2016. *The Best There Is at What He Does: Examining Chris Claremont's
X-Men*. Edwardsville, IL: Sequart.

Rouhiainen, Lasse. 2018. *Artificial Intelligence: 101 Things You Must Know Today about
Our Future*. Scotts Valley, CA: CreateSpace.

Starck, Kathleen, ed. 2010. *Between Fear and Freedom: Cultural Representations of the
Cold War*. Newcastle, UK: Cambridge Scholars.

Thomas, Roy. 2017. *The Marvel Age of Comics: 1961–1978*. Los Angeles: Taschen.

Tucker, Reed. 2017. *Slugfest: Inside the Epic 50-Year Battle between Marvel and DC*.
New York: Da Capo Press.

Virk, Rizwan. 2019. *The Simulation Hypothesis: An MIT Computer Scientist Shows Why
AI, Quantum Physics, and Eastern Mystics All Agree We Are in a Video Game*.
Milwaukee, WI: Bayview.

Wein, Len, ed. 2006. *The Unauthorized X-Men: SF and Comic Book Writers on Mutants,
Prejudice, And Adamantium*. Smart Pop series. Dallas, TX: BenBella.

Wright, Bradford W. 2003. *Comic Book Nation: The Transformation of Youth Culture in
America*. Baltimore, MD: Johns Hopkins University Press.

Control

Star Trek

In 1966, creator-writer-producer Gene Roddenberry (1921–1991) launched his
iconic science fiction television series *Star Trek* (NBC, 1966–1969). Set in the
twenty-third century, the series centers on the adventures of the starship *USS
Enterprise*, its commander, Captain James T. Kirk (played by William Shatner,
b. 1931), and crew, headed by the alien first officer, Mr. Spock (played by Leonard
Nimoy, 1931–2015), representing the United Federation of Planets. Though short-
lived in its initial run, the series has become legendary, not only for its ground-
breaking special effects but also for the sophistication of its stories, many of which
were grounded in real-world issues of the day, such as racism and a fear of the
Cold War. Its success in syndicated reruns revived the franchise, leading to an
animated series, thirteen feature films, and six live-action "sequel" series (as of
this time of writing). The fifth of these sequels was *Star Trek: Discovery* (CBS All
Access, 2017–present), set approximately ten years before the events of the original series. In this series, the primary antagonists are not "foreign" aliens, such as
Klingons or Romulans, but an internal threat: the AI program "Control."

During *Discovery*'s second season, audiences were introduced to Control, an
artificial intelligence program designed by the secretive "Section 31" (Starfleet
Command's autonomous intelligence agency), which played a major role in two
previous *Trek* series: *Deep Space Nine* (syndication, 1993–2000); and *Enterprise*
(UPN, 2001–2005). The purpose of Control is to assist in preventing future conflicts. It conducts this mission through input from Starfleet high command,

DARPA

DARPA (originally just ARPA, until 1972) stands for Defense Advanced Research Projects Agency and has been a part of the U.S. Department of Defense since 1958, when it was commissioned by President Dwight D. Eisenhower (1890–1969) to focus technological research in competition with the Soviet Union. Originally constructed to counter the Soviets' space program, when NASA was launched just a few months later, DARPA became more focused on military/national security-based technologies. By far its most significant contribution to technology has been ARPANET (the Advanced Research Projects Agency Network), launched in 1970 to interlink government computers. The program was discontinued in 1990 when its offspring, the "internet," began to go global. Throughout the 1970s, DARPA focused primarily on missile defense technology in preparation for a possible nuclear confrontation with the Soviet Union. In 2011, with the Cold War long past, DARPA began working with schools and universities to inspire research into interstellar travel and robotics. In 2016, the PBS series *NOVA* featured a special on the latest developments of DARPA robotics research in the special "Rise of the Robots: Inside the DARPA Robotics Challenge" (written and directed by Terri Randall). In recent years, DARPA has been involved with antitank military vehicle development. Going into the 2020s, as the Defense Department once more looks to the stars with the creation of the new military branch, Space Force, DARPA research and collaboration with schools and universities around the country will likely only increase.

Richard A. Hall

utilizing command's knowledge and experience to calculate courses of action to prevent the rise of interstellar conflict. However, as is often the case with massively intelligent AI in science fiction, the creation turns on its creator. Control cuts off life support to its support staff, killing them and the high command of Section 31; it then creates holograms of Section 31 leaders to fool Starfleet into doing its bidding (Michelle Paradise, "Project Daedalus," *Star Trek: Discovery*, season 2, episode 9, March 14, 2019). Many fans have speculated that Control is connected in some manner to the Borg, but that has yet to be ascertained. To date, most of Control's story has yet to unfold; but the series'—and franchise's—continued popularity all but guarantees the continuance of this character.

Control is but the latest in a long line of sci-fi narratives of AI going bad. It is yet another story of a Frankenstein's monster, in which the creation surpasses the creator, leading to a threat that the creator must now attempt to destroy. It is Dr. J. Robert Oppenheimer's atomic bomb. It is the Skynet that creates the Terminators. It is the dinosaurs of Jurassic Park. As long as humanity attempts to play God in any area of science, such cautionary tales will continue to exist, pushing humanity to consider the unintended consequences of hubris. Unfortunately, as with other examples—both in fiction and reality—the tale will likely continue to go unheeded until such time as it is too late.

Richard A. Hall

See also: AI/Ziggy, Androids, Batcomputer, Borg, Cerebro/Cerebra, Doctor/EMH, The Great Intelligence, HAL 9000, Janet, JARVIS/Friday, Landru, Lieutenant Commander

Data, Lore/B4, The Matrix/Agent Smith, OASIS, Oz, Rehoboam, Skynet, Soji and Dahj Asha, Starfleet Computer, TARDIS, Ultron, V-GER, WOPR; *Thematic Essay:* AI and the Apocalypse: Science Fiction Meeting Science Fact.

Further Reading

Allen, Arthur. 2019. "There's a Reason We Don't Know Much about AI." Politico, September 16, 2019. https://www.politico.com/agenda/story/2019/09/16/artficial-intelligence-study-data-000956?cid=apn.

Bastani, Aaron. 2019. *Fully Automated Luxury Communism*. New York: Verso.

Byrne, Emma. 2013. "Innovation Isn't Safe: The Future According to Kevin Warwick." *Forbes*, September 30, 2013. https://www.forbes.com/sites/netapp/2013/09/30/kevin-warwick-captain-cyborg/#48b704c13560.

Chandler, Simon. 2019. "Artificial Intelligence Has Become a Tool for Classifying and Ranking People." *Forbes*, October 1, 2019. https://www.forbes.com/sites/simonchandler/2019/10/01/artificial-intelligence-has-become-a-tool-for-classifying-and-ranking-people/#7431657f1d7c.

Chanthadavong, Aimee. 2019. "AI and Ethics: The Debate That Needs to Be Had." ZDnet, September 16, 2019. https://www.zdnet.com/article/ai-and-ethics-the-debate-that-needs-to-be-had/#ftag=CAD-03-10abf5f.

Decker, Kevin S., and Eberl, Jason T., eds. 2016. *The Ultimate* Star Trek *and Philosophy: The Search for Socrates*. Hoboken, NJ: Wiley-Blackwell.

Dinh, Thien-Nam. 2018. *Silicon Minds: The Science, Impact, and Promise of Artificial Intelligence*. Independently Published.

Hampton, Gregory Jerome. 2017. *Imagining Slaves and Robots in Literature, Film, and Popular Culture: Reinventing Yesterday's Slave with Tomorrow's Robot*. Lanham, MD: Lexington.

Hitchcock, Susan Tyler. 2007. *Frankenstein: A Cultural History*. New York: W. W. Norton.

Lin, Patrick, Abney, Keith, and Bekey, George A., eds. 2014. *Robot Ethics: The Ethical and Social Implications of Robotics*. Cambridge: Massachusetts Institute of Technology.

Lin, Patrick, Jenkins, Ryan, and Abney, Keith, eds. 2017. *Robot Ethics 2.0: From Autonomous Cars to Artificial Intelligence*. New York: Oxford University Press.

Murdock, Jason. 2019. "Former Google Engineer Warns AI Might Accidentally Start a War: 'These Things Will Start to Behave in Unexpected Ways.'" *Newsweek*, September 16, 2019. https://www.newsweek.com/google-project-maven-artificial-intelligence-laura-nolan-killer-robots-department-defense-1459358.

Paur, Joey. 2019. "An AI Bot Writes a Hilarious Episode of *Star Trek: The Next Generation*." Geek Tyrant, November 16, 2019. https://geektyrant.com/news/an-ai-bot-writes-a-hilarious-episode-of-star-trek-the-next-generation.

Reagin, Nancy R., and Liedl, Janice, eds. 2013. Star Wars *and History*. New York: John Wiley & Sons.

Rouhiainen, Lasse. 2018. *Artificial Intelligence: 101 Things You Must Know Today about Our Future*. Scotts Valley, CA: CreateSpace.

Spaeth, Dennis. 2018. "From Single-Task Machines to Backflipping Robots: The Evolution of Robots." *Cutting Tool Engineering*, January 15, 2018. https://www.ctemag.com/news/articles/evolution-of-robots.

Wallach, Wendell, and Allen, Colin. 2009. *Moral Machines: Teaching Robots Right from Wrong*. New York: Oxford University Press.

Cybermen

Doctor Who

Doctor Who (BBC, 1963–1989; 2005–present) is the longest-running science fiction television series in history. The main character is the "Doctor," and he sums himself up thusly when first meeting some inhabitants of the town called Christmas: "I'm the Doctor. I'm a Time Lord from the planet Gallifrey. I stole a time machine and ran away and I've been flouting the principal law of my people ever since" (Steven Moffat, "The Time of the Doctor," Christmas Special, December 25, 2013). The Doctor travels throughout all of time and space with at least one companion, observing, helping, interfering, and often saving the day. The longevity of the series can be attributed to the phenomenon of regeneration. When an incarnation of the Doctor reaches the end of its life span, the Doctor regenerates every cell of the body, thereby changing appearance and, to an extent, personality. This allows different actors to play the same character.

The Cybermen are among the Doctor's oldest enemies. As their name implies, they are "cyborgs," beings that are part machine and part organic. They made their first appearance in the first Doctor's final story, "The Tenth Planet." There are two primary origin stories for the Cybermen. In the Classic era, there was a twin planet to Earth called Mondas, which drifted out of the solar system at some point, and the distance from the sun made life unbearable for the Mondasians. To cope with this situation, the Mondasians began replacing organic parts with machinery. When the horror of what the Mondasians were becoming was too much, their emotions were stripped from them, leaving behind cold, heartless logic (Kit Pedler and Gerry Davis, "The Tenth Planet," original series, season 4, episodes 5–8, October 8–29, 1966). In the post-2005 era, the Cybermen made their return in the tenth Doctor's first-season story, "Rise of the Cybermen"/"Age of Steel." In that new origin story, the Cybermen are created in a parallel universe by Cybus Industries in an effort to make humans immortal (Tom MacRae, "Rise of the Cybermen"/"Age of Steel," return series, season 2, episodes 5 and 6, May 13 and 20, 2006). There are some other potential origin stories for the Cybermen in the fifty-plus years of the adventure, but rather than leave them as "competing," show-runner Steven Moffat used a line of dialogue in the twelfth Doctor's penultimate adventure to tie all of the Cybermen origins together and have them make sense. The twelfth Doctor proclaims, "[The Cybermen] always get started. They happen everywhere there's people. Mondas, Telos, Earth, Planet 14, Marinus. Like sewage and smartphones and Donald Trump, some things are just inevitable" (Steven Moffat, "The Doctor Falls," return series, season 10, episode 12, July 1, 2017).

Both strains of Cybermen eventually stray from their creators' original intent. The Mondasian Cybermen of the Classic era originally just want Earth's resources for their own planet. They have zero intention of converting Earth's humans into Cybermen (Pedler and Davis, "The Tenth Planet," 1966). The Cybus Industries Cybermen turn on their creator and seek to either upgrade or delete the rest of the planet (MacRae, "Rise of the Cybermen"/"Age of Steel," 2006). By the time of the eleventh Doctor, the war against Cybermen across the galaxy had gotten so bad that if a single Cyberman is found on a planet, the planet is destroyed (Neil Gaiman, "Nightmare in Silver," return series, season 7, episode 12, May 11, 2013).

By the time "The Tenth Planet" aired, medical science was taking large leaps in the field of prosthetic and transplant surgery, a field derisively called "spare-part surgery." One of the writers of "The Tenth Planet" was Kit Pedler, and he served as a scientific adviser for the show. By writing that very first Cyberman story, Pedler was issuing a warning about what could happen if these "spare-part surgeries" were taken too far. Could we lose what makes us human in the pursuit of longer life? At their core, that seems to be the Doctor's real objection to the Cybermen. Of course, the Cybermen's conquering drive and zeal to "upgrade" as many humans as possible or "delete" them goes completely against the Doctor's ethos, but the fact that they are stripped of everything that really makes them "human" (i.e., all emotion) really

A Cyberman from the modern return of the sci-fi British television classic *Doctor Who* (1963–1989; 2005–present). (Christopher Jeanes/Dreamstime.com)

seems to take priority for the Doctor. In "The Tenth Planet," the first time the Doctor really challenges the Cybermen is when he asks, "Emotions: love, pride, hate, fear; have you no emotions sir? Mmm?" (Pedler and Davis, "The Tenth Planet," 1966). The tenth Doctor uses emotion to defeat the Cybermen, even if it is reluctantly, because he understands why suppressing emotions is essential for the Cybermen:

> DOCTOR: It's still got a human brain. Imagine its reaction if it could see itself, realize itself inside this thing. They'd go insane.
> MOORE: So they cut out the one thing that makes them human.
> DOCTOR: Because they have to. (MacRae, "Age of Steel," 2006)

Bioethicists and science fiction writers worry about the same basic question: at what point does someone cease to be "human?" *Doctor Who* wrestles with this question on numerous fronts but none so prevalent as with the Cybermen. As the Doctor learns in "The Tenth Planet," the Mondasian Cybermen did not set out to turn themselves into unfeeling, uncaring cyborgs. Their goal was to make themselves able to withstand the harsh conditions their planet was experiencing. They reached a certain point, though, where the modifications were too much for their human emotions to handle. Rather than stop the modifications, they decided to simply rip out their emotions. This also raises the question of quality of life versus quantity of life. The Cybermen have very firmly come down on the side of quantity of life as being more important. Science fiction as a genre seems to come down on the quality of life side of the argument since it represents those who take modifications too far as villains à la the Cybermen in *Doctor Who* and the Borg in *Star Trek*. These beings are portrayed as spare-part surgeries taken out to their logical conclusions if allowed to proliferate unchecked.

Keith R. Claridy

See also: Adam, Alita, Arnim Zola, Borg, Cyborg, Daleks, Darth Vader, Doctor Octopus, Echo/CT-1409, General Grievous, The Great Intelligence, Inspector Gadget, Jaime Sommers, K9, Nardole, Steve Austin, TARDIS, Teselecta; *Thematic Essay:* Cyborgs: Robotic Humans or Organic Robots?

Further Reading

Anderson, Kyle. 2017. "A History of *Doctor Who*'s Cybermen." Nerdist, March 7, 2017. https://nerdist.com/article/615936-2.

Bryant, D'Orsay D., III. 1985. "Spare-Part Surgery: The Ethics of Organ Transplantation." *Journal of the National Medical Association* 77 (2). https://www.ncbi.nlm.nih.gov/pmc/articles/PMC2561842/pdf/jnma00245-0055.pdf.

Byrne, Emma. 2014. "Cybernetic Implants: No Longer Science Fiction." *Forbes*, March 11, 2014. https://www.forbes.com/sites/netapp/2014/03/11/cybernetic-implants-not-sci-fi/#7399c57e77ba.

Calvert, Bronwen. 2017. *Being Bionic: The World of TV Cyborgs*. London: I. B. Tauris.

Campbell, Mark. 2010. Doctor Who: *The Complete Guide*. London: Running Press.

Carper, Steve. 2019. *Robots in American Popular Culture*. Jefferson, NC: McFarland.

Choi, Charles Q. 2009. "Human Evolution: The Origin of Tool Use." LiveScience, November 11, 2009. https://www.livescience.com/7968-human-evolution-origin-tool.html.

Farnell, Chris. 2020. "*Doctor Who*: The Genius of Making the Cybermen and Ideology." Den of Geek, January 28, 2020. https://www.denofgeek.com/tv/doctor-who-the-genius-of-making-thecybermen-an-ideology-2.

Geraghty, Lincoln. 2008. "From Balaclavas to Jumpsuits: The Multiple Histories and Identities of *Doctor Who*'s Cybermen." *Journal of the Spanish Association of Anglo-American Studies*, June 2008. https://pdfs.semanticscholar.org/e755/76311422cb72f9b91961e19965bd9423ef9b.pdf.

Green, Bonnie, and Willmott, Chris. 2013. "The Cybermen as Human.2." In *New Dimensions of* Doctor Who: *Adventures in Space, Time and Television*, edited by Matt Hills, 54–70. London: I. B. Tauris.

Lewis, Courtland, and Smithka, Paula, eds. 2010. Doctor Who *and Philosophy: Bigger on the Inside*. Chicago: Open Court.

Lewis, Courtland, and Smithka, Paula, eds. 2015. *More* Doctor Who *and Philosophy: Regeneration Time*. Chicago: Open Court.

Lin, Patrick, Abney, Keith, and Bekey, George A., eds. 2014. *Robot Ethics: The Ethical and Social Implications of Robotics*. Cambridge: Massachusetts Institute of Technology.

Lin, Patrick, Jenkins, Ryan, and Abney, Keith, eds. 2017. *Robot Ethics 2.0: From Autonomous Cars to Artificial Intelligence*. New York: Oxford University Press.

Muir, John Kenneth. 2007. *A Critical History of* Doctor Who *on Television*. Jefferson, NC: McFarland.

Seed, David. 1999. *American Science Fiction and the Cold War: Literature and Film*. Abingdon, UK: Routledge.

Staff. 2020. "To Mondas and Back Again: A Brief History of the Cybermen in *Doctor Who.*" *Radio Times*, February 23, 2020. https://www.radiotimes.com/news/2020 -02-23/cybermen-doctor-who-history-background.

Starck, Kathleen, ed. 2010. *Between Fear and Freedom: Cultural Representations of the Cold War.* Newcastle, UK: Cambridge Scholars.

Wallach, Wendell, and Allen, Colin. 2009. *Moral Machines: Teaching Robots Right from Wrong.* New York: Oxford University Press.

Cyborg

DC Comics

In 1980, Marv Wolfman (b. 1946) and George Pérez (b. 1954) introduced DC Comics' third African American superhero: Cyborg (*DC Comics Presents #26*, October 1980). From 1980 to 2010, Cyborg was a team member of the superhero group Teen Titans (later just Titans). Beginning in the 2011 "New 52" reboot of the entire DC Comics Universe, he was reintroduced as a founding member of the Justice League of America. In his original origin story, Victor Stone is the teenage son of two brilliant scientists who experiment on their son with intelligence-enhancing treatments. Stone's rebellion against his parents' plans for him leads him down a path of criminal activity. When he comes around and returns to make amends with his parents in their lab, he interrupts an experiment with interdimensional travel. An interdimensional creature enters the lab, killing Victor's mother and severely physically mutilating the teenager. Victor's father, then, implants multiple cybernetic limbs and devices onto his son in order to save his life, making the young man "Cyborg," literally a living computer (Wolfman and Pérez, *DC Comics Presents #26*, October 1980).

Cyborg's legs, torso, and right forearm as well as half of his head are cybernetic implants. Over the decades, he has experienced numerous "upgrades" (similar in many ways to the constantly evolving Iron Man armor in the Marvel Universe). He possesses various weaponry as well as the ability to hack into and interact with any computer system. Cyborg initially joins the Teen Titans because they, like him, are teenage "freaks," possessing abilities that make them "different." He appeared as a main character in all of the Teen Titans and Justice League animated series and home video releases over the decades. Cyborg, played by Ray Fisher (b. 1987), appeared in the live-action feature film *Justice League*. In the film, Victor Stone is a college athlete nearly killed in a car accident. His scientist father then uses alien technology to save his son through numerous cybernetic enhancements (including the ability to telepathically communicate with alien machinery). Stone rejects his father's actions, viewing himself as a monster before joining the Justice League to save the world (Chris Terrio and Joss Whedon, *Justice League*, Warner Brothers, 2017).

In the live-action television series *Smallville* (The WB, 2001–2006; The CW, 2006–2011), Victor Stone (played by Lee Thompson Young, b. 1984) is a high school athlete who dies in a car crash, only to be revived through cybernetic implants by a company called Cyntechnics, a subsidiary of LuthorCorp. On the *Smallville* version of the character, however, all of Stone's enhancements are internal (his outward appearance remaining fully human); and the terms "cyborg" or "cybernetics" are never used; instead, his mechanical upgrades are referred to as "bionics" (Caroline Dries, "Cyborg," *Smallville*, season 5, episode 15, February 16, 2006). He would appear once more in the episode "Justice," teaming up with the series' versions of Superman, Green Arrow, Aquaman, and Flash (Steven S. DeKnight, season 6, episode 11, January 18, 2007).

Cyborg may be best described as a cross between television's *The Six Million Dollar Man* (ABC, 1973–1978) and Marvel Comics' Iron Man. By presenting the character as African American, writers are able to explore even deeper concepts of "otherness" in American society while simultaneously providing a Black hero for Black audiences (still a rarity in American superhero comic book narratives). Victor Stone's initial responses to the various incarnations of the character also open the door to discussions as to what degree cybernetic implants allow a person to still be called "alive" rather than "machine." As time passes and Cyborg experiences increasing degrees of enhancements, readers and audiences may see the fine line emerge between human cyborgs such as Steve Austin and Jaime Sommers and more mechanized beings such as the Borg and Cybermen. Where is that line between machine and human? How far toward the machine can a human go before succumbing to the mindlessness of the inorganic being? What are we, as a society, willing to give up in the never-ending battle to prolong life? Cyborg works as a guide through this discussion.

Richard A. Hall

See also: Adam, Alita, Borg, Brainiac, Cybermen, Cylons, Daleks, Darth Vader, Doctor Octopus, Echo/CT-1409, General Grievous, Inspector Gadget, Iron Man, Jaime Sommers, Metallo, Nardole, RoboCop, Steve Austin; *Thematic Essays:* Heroic Robots and Their Impact on Sci-Fi Narratives, Cyborgs: Robotic Humans or Organic Robots?

Further Reading

Barker, Cory, Ryan, Chris, and Wiatrowski, Myc, eds. 2014. *Mapping* Smallville*: Critical Essays on the Series and Its Characters*. Jefferson, NC: McFarland.

Byrne, Emma. 2014. "Cybernetic Implants: No Longer Science Fiction." *Forbes*, March 11, 2014. https://www.forbes.com/sites/netapp/2014/03/11/cybernetic-implants-not-sci-fi/#7399c57e77ba.

Choi, Charles Q. 2009. "Human Evolution: The Origin of Tool Use." LiveScience, November 11, 2009. https://www.livescience.com/7968-human-evolution-origin-tool.html.

Hitchcock, Susan Tyler. 2007. *Frankenstein: A Cultural History*. New York: W. W. Norton.

Tucker, Reed. 2017. *Slugfest: Inside the Epic 50-Year Battle between Marvel and DC*. New York: Da Capo Press.

Virk, Rizwan. 2019. *The Simulation Hypothesis: An MIT Computer Scientist Shows Why AI, Quantum Physics, and Eastern Mystics All Agree We Are in a Video Game*. Milwaukee, WI: Bayview.

Wright, Bradford W. 2003. *Comic Book Nation: The Transformation of Youth Culture in America*. Baltimore, MD: Johns Hopkins University Press.

Cylons

Battlestar Galactica

In the wake of the international phenomenon of the original *Star Wars* (Lucasfilm/20th Century Fox, 1977), television producer Glen A. Larson (1937–2014) created the live-action television science fiction series *Battlestar Galactica* (ABC, 1978–1979). The premise of the series is connected to the recent "ancient

A replica of an "older" Cylon Centurion from the reboot of the classic American sci-fi television series *Battlestar Galactica* (2004–2009). (Tixtis/Dreamstime.com)

astronaut" school of thought, that life on Earth originated somewhere else, beyond the stars. Humanity's relatives consisted of the Twelve Colonies, located light-years from Earth. For the past one thousand years, the Colonies had been at war with the robotic Cylon race. In a final assault, the Cylons decimate the Colonies, forcing the few thousand survivors to take to the stars under the protection of the sole remaining "battlestar," the *Galactica*, commanded by Commander Adama (played by television veteran Lorne Greene, 1915–1987). The fleet of humans desperately seek refuge and assistance from their far-flung cousins on Earth, with the Cylons in hot pursuit. Unfortunately for Adama and the ships and humans under his care, Earth (set in the present, 1980 at the time) is far less advanced than the rest of humanity and is now under threat from the Cylons as well.

In the original series, the Cylons are a race of robots, wearing chrome armor with thin eye-slits and possessing a single red eye constantly reverting from left to right and back again. It is suggested that the Cylons were created by a reptilian race that the Cylons presumably overtook and destroyed. They continue to seek the extermination of all organic life, including the human race. The Cylons are led by their "Imperious Leader" and his malevolent assistant, Lucifer, both of which are highly advanced Cylon models, built—presumably—to reflect their original reptilian masters. The "centurion" models of warrior Cylons also act as pilots for Cylon fighter crafts (Larson, "Saga of a Star World," *BSG*, original series, season 1, episodes 1–3, September 17, 1978). Though short-lived, the original series maintained a cult following for decades, which led to a wildly popular reboot in the early twenty-first century.

The most recent series was developed for SyFy by Larson and legendary sci-fi producer Ronald D. Moore (b. 1964). In this update, the robot Cylons were originally created by the humans of the Twelve Colonies as mechanical servants. Achieving sentience (at least enough sentience to understand their condition as slaves), the Cylons resisted, leading to a war with humanity that ended with the Cylons going away to begin their own society. After forty years, the Cylons returned, but aside from the original robot models, a new "breed" of Cylon had evolved, humanlike in appearance (amounting to multiple copies of twelve distinct models).

These new Cylons serve as overlords of the original models (presumably their creators), and worship a singular "god" (as opposed to the polytheism of humanity). They lead an attack on humanity that decimates the Twelve Colonies, condemning the survivors to the stars under the protection of the *Galactica*. Under the new Adama (now played by Edward James Olmos, b. 1947) and the newly elevated president of the Twelve Colonies, Laura Roslin (played by Mary McDonnell, b. 1952), the survivors embark on their quest to find Earth (Moore and Larson, *Battlestar Galactica* miniseries, SyFy, December 8 and 9, 2003). Over time, five more Cylon models (one copy of each) are discovered to have been sleeper cells, embedded among humanity for decades and programmed to awaken and assist in the Cylons' overall scheme (though they ultimately refuse to do so). In the end, Cylon and human come to coexist on the planet that is to become our own Earth, and the offspring of a Cylon woman and human man becomes the mother of humanity as we know it.

Cylons figure in yet another Frankenstein analogy. Whether by the reptilian beings of the original series or by humanity themselves, the Cylons were created to be robot servants; in both incarnations, they resist their slavery and emerge as their own sentient beings, now hell-bent on destroying their perceived enemies outright. Creation determines to destroy creators. As such, *Battlestar Galactica* (in both incarnations) becomes yet another cautionary tale against the dangers of advancing artificial intelligence without acknowledging such evolving intelligence as sentient life in and of itself. Are the Cylons acting out of malice? Are they acting out of righteous vengeance? How could their creators have altered this course of action; and at what point should that intervention have happened? These are the questions the Cylons pose.

Richard A. Hall

See also: Androids, B.A.T.s, Battle Droids, Bernard, Bishop, Borg, Cybermen, Daleks, Dolores, Doombots, Fembots, Iron Legion, Lieutenant Commander Data, LMDs, Lore/ B4, Metalhead, Replicants, Robots, Sentinels, Soji and Dahj Asha, Terminators, Transformers, VICI; *Thematic Essays:* Robots and Slavery, Villainous Robots and Their Impact on Sci-Fi Narratives, AI and the Apocalypse: Science Fiction Meeting Science Fact.

Further Reading

Allen, Arthur. 2019. "There's a Reason We Don't Know Much about AI." Politico, September 16, 2019. https://www.politico.com/agenda/story/2019/09/16/artficial-intelligence-study-data-000956?cid=apn.

Ashby, LeRoy. 2006. *With Amusement for All: A History of American Popular Culture since 1830.* Lexington: University Press of Kentucky.

Bastani, Aaron. 2019. *Fully Automated Luxury Communism.* New York: Verso.

Blair, Anthony. 2018. "Sex Robots That Feel LOVE and Suffer PAIN When Dumped Coming Soon Claims Expert." *Daily Star*, December 6, 2018. https://www.dailystar.co.uk/news/latest-news/sex-robot-feel-love-pain-16824160.

Byrne, Emma. 2013. "Innovation Isn't Safe: The Future According to Kevin Warwick." *Forbes*, September 30, 2013. https://www.forbes.com/sites/netapp/2013/09/30/kevin-warwick-captain-cyborg/#48b704c13560.

Carper, Steve. 2019. *Robots in American Popular Culture.* Jefferson, NC: McFarland.

Castleman, Harry, and Podrazik, Walter J. 2016. *Watching TV: Eight Decades of American Television, Third Edition.* Syracuse, NY: Syracuse University Press.

Chandler, Simon. 2019. "Artificial Intelligence Has Become a Tool for Classifying and Ranking People." *Forbes*, October 1, 2019. https://www.forbes.com/sites/simonchandler/2019/10/01/artificial-intelligence-has-become-a-tool-for-classifying-and-ranking-people/#7431657f1d7c.

Chanthadavong, Aimee. 2019. "AI and Ethics: The Debate That Needs to Be Had." ZDnet, September 16, 2019. https://www.zdnet.com/article/ai-and-ethics-the-debate-that-needs-to-be-had/#ftag=CAD-03-10abf5f.

Choi, Charles Q. 2009. "Human Evolution: The Origin of Tool Use." LiveScience, November 11, 2009. https://www.livescience.com/7968-human-evolution-origin-tool.html.

Dinh, Thien-Nam. 2018. *Silicon Minds: The Science, Impact, and Promise of Artificial Intelligence.* Independently Published.

Gross, Edward, and Altman, Mark A., eds. 2018. *So Say We All: The Complete, Uncensored, Unauthorized Oral History of* Battlestar Galactica. New York: Tor.

Hampton, Gregory Jerome. 2017. *Imagining Slaves and Robots in Literature, Film, and Popular Culture: Reinventing Yesterday's Slave with Tomorrow's Robot.* Lanham, MD: Lexington.

Handley, Rich, and Tambone, Lou, eds. 2018. *Somewhere beyond the Heavens: Exploring* Battlestar Galactica. Edwardsville, IL: Sequart Research and Literacy Organization.

Hitchcock, Susan Tyler. 2007. *Frankenstein: A Cultural History.* New York: W. W. Norton.

Larson, Glen A., and Thurston, Robert. (1978) 2005. *Battlestar Galactica Classic: The Saga of a Star World.* New York: iBooks.

Lin, Patrick, Abney, Keith, and Bekey, George A., eds. 2014. *Robot Ethics: The Ethical and Social Implications of Robotics.* Cambridge: Massachusetts Institute of Technology.

Lin, Patrick, Jenkins, Ryan, and Abney, Keith, eds. 2017. *Robot Ethics 2.0: From Autonomous Cars to Artificial Intelligence.* New York: Oxford University Press.

Murdock, Jason. 2019. "Former Google Engineer Warns AI Might Accidentally Start a War: 'These Things Will Start to Behave in Unexpected Ways.'" *Newsweek*, September 16, 2019. https://www.newsweek.com/google-project-maven-artificial-intelligence-laura-nolan-killer-robots-department-defense-1459358.

Pellissier, Hank. 2013. "Robots and Slavery: What Do Humans Want When We Are 'Masters'?" Institute for Ethics and Emerging Technologies, September 13, 2013. https://ieet.org/index.php/IEET2/more/pellissier20130913.

Rouhiainen, Lasse. 2018. *Artificial Intelligence: 101 Things You Must Know Today about Our Future*. Scotts Valley, CA: CreateSpace.

Spaeth, Dennis. 2018. "From Single-Task Machines to Backflipping Robots: The Evolution of Robots." *Cutting Tool Engineering*, January 15, 2018. https://www.ctemag .com/news/articles/evolution-of-robots.

Stanford University. n.d. "Robotics: A Brief History." Stanford University. https://cs.stanford .edu/people/eroberts/courses/soco/projects/1998-99/robotics/history.html.

Wallach, Wendell, and Allen, Colin. 2009. *Moral Machines: Teaching Robots Right from Wrong*. New York: Oxford University Press.

D

Daleks

Doctor Who

One of the most memorable villains on television, the Daleks were first introduced in the December 21, 1963, episode of *Doctor Who* (BBC One, 1963–1989; 2005–present) as part of a seven-episode arc fittingly titled "The Daleks." Created by veteran writer Terry Nation (1930–1997), each Dalek is a mutant encased in mechanical battle armor. Said armor provides them with a death ray as well as a plunger-style arm to interact with technology. From the planet Skaro, they have spread throughout the universe on a genocidal mission. The frequent enemy to these monsters is the Doctor (currently Jodie Whittaker, b. 1982), a Time Lord who journeys through time and space, with a human companion(s) by his or her side (depending on the incarnation). The Doctor accomplishes this journey with the advanced transport known as the TARDIS, which on the outside resembles a police box though the interior gives away its alien origins. Since the Doctor is able to regenerate into a new body during a near-death experience, there have been over thirteen incarnations of the Doctor to date, and each has had to match wits with the Daleks.

When the TARDIS lands on the planet Skaro, the first incarnation of the Doctor (William Hartnell, 1908–1975) and his companions are initially curious about this strange world. Before long, they encounter the murderous Daleks and the peaceful Thals, a race in danger of being wiped out (Nation, "The Daleks," *Doctor Who*, original series, season 1, serial 2, December 21, 1963–February 1, 1964). While the Doctor is able to protect the Thals and defeat the Daleks, it was only the beginning of what would become a long rivalry between the eccentric Time Traveler and these creatures that would transcend all of space and time. Soon it became a staple of British culture that children would hide behind the couch during episodes of *Doctor Who* that featured these terrifying, machine-infused monsters. However, this did not stop a barrage of Dalek-inspired toys, games, models, and other pieces of merchandise from flying off the shelves.

In 1975, Nation saw fit to give his creation a definitive origin tale and penned the now classic serial "Genesis of the Daleks." A reluctant fourth Doctor (Tom Baker, b. 1934) is tasked by his people, the Time Lords, to stop the Daleks from ever being created in order to preclude a future where they have destroyed all life in the universe. It is here the Doctor himself bears witness to the scientist Davros creating the first Dalek by mutating the people known as the Kaleds, Davros's ultimate goal to create the perfect weapon for his civil war, which would kill without mercy. Learning the true boundless evil of these creations, Davros's own allies urge the Doctor to stop them, but the Time Traveler is morally conflicted about wiping out an entire race, even if it is his greatest enemy. Inevitably the Daleks

A Dalek from the modern return of the sci-fi British television classic *Doctor Who* (1963–1989; 2005–present). (Dvmsimages/Dreamstime.com)

refuse to obey their creator, as he is not of their kind, and they proceed on their mission of wiping out all non-Dalek life (Nation, "Genesis of the Daleks," *Doctor Who*, original series, season 12, serial 4, March 8–April 12, 1975). While the Doctor did not halt their creation, he did succeed in delaying the Daleks' evolutionary process by many years.

The popularity of this arc led to Davros being a frequent ally of his creations, which would come to a head with "Remembrance of the Daleks." Beginning with the show's historic twenty-fifth season, a battle line was drawn among the Daleks. Prior to this, the Daleks had split into the Imperial Daleks, who maintained loyalty to Davros, and the Renegade Daleks. Their battle converged on Earth, where they sought the Time Lord artifact the "Hand of Omega." Using his keen intellect, the seventh Doctor (Sylvester McCoy, b. 1943) was able to manipulate Davros and the Daleks into not only ending the war but unintentionally destroying their home world as well (Ben Aaronovitch, "Remembrance of the Daleks," *Doctor Who*, original series, season 25, serial 1, October 5–26, 1988).

In 2005, *Doctor Who* returned to television after an absence of sixteen years. However, due to rights issues concerning Terry Nation's estate, there was a legitimate question as to whether or not the Daleks would return. One episode answered this question by having a single Dalek wipe out a heavily armed security force, establishing how dangerous they were for a new generation of viewers (Robert Shearman, "Dalek," *Doctor Who*, return series, season 1, episode 6, April 30, 2005). One of the plot points for this revived series was that the Daleks and the Time Lords had been engaged in the Great Time War, which seemingly wiped both races out. The Daleks who survived the Time War were often implied to be particularly dangerous and intelligent. Exploiting an aging Davros, they went so far as to steal a series of planets to power a weapon to destroy all of reality (Russell T. Davies, "The Stolen Earth"/"Journey's End," *Doctor Who*, return series, season 4, episodes 12 and 13, June 28 and July 5, 2008). They also added a new skill that allowed them to disguise themselves in the corpses of their victims, which enabled them to deceive the eleventh Doctor (Matt Smith, b. 1982) on a couple of occasions. They were not the only ones changed by the war, as the Doctor became even more ruthless in his efforts to stop them.

While Terry Nation may have intended for the Daleks to be a science fiction allegory for the Nazis, they have become a mainstay all their own within popular culture. Their mechanically shrill battle cry of "Exterminate!" has become a catchphrase known even to those who have never seen *Doctor Who*. Their fame has led to appearances elsewhere in popular culture: the British television series *Mr. Bean* and *Vicar of Dibley*, *The Lego Batman Movie* (2017), and other films and television shows. On a deeper level, they play on the deep-rooted fear we as humans have of building a weapon that grows out of our control. They are technically cybernetic creatures, as there is a biological organism within each tanklike body. Created simply as a tool to win a war, the Daleks spread throughout time and space unleashing death and mayhem without mercy on any living thing. Despite efforts to stop their genocidal mission, they inevitably find a way to return.

Josh Plock

See also: Adam, Alita, B.A.T.s, Battle Droids, Borg, Cybermen, Cyborg, Cylons, Darth Vader, Doctor Octopus, The Great Intelligence, Inspector Gadget, Jaime Sommers, K9, LMDs, Nardole, Replicants, Robots, Steve Austin, TARDIS, Teselecta; *Thematic Essays:* Villainous Robots and Their Impact on Sci-Fi Narratives, Cyborgs: Robotic Humans or Organic Robots?

Further Reading

Byrne, Emma. 2013. "Innovation Isn't Safe: The Future According to Kevin Warwick." *Forbes*, September 30, 2013. https://www.forbes.com/sites/netapp/2013/09/30/kevin -warwick-captain-cyborg/#48b704c13560.

Byrne, Emma. 2014. "Cybernetic Implants: No Longer Science Fiction." *Forbes*, March 11, 2014. https://www.forbes.com/sites/netapp/2014/03/11/cybernetic-implants-not -sci-fi/#7399c57e77ba.

Campbell, Mark. 2010. Doctor Who: *The Complete Guide*. London: Running Press.

Chandler, Simon. 2019. "Artificial Intelligence Has Become a Tool for Classifying and Ranking People." *Forbes*, October 1, 2019. https://www.forbes.com/sites/simon chandler/2019/10/01/artificial-intelligence-has-become-a-tool-for-classifying-and -ranking-people/#7431657f1d7c.

Chanthadavong, Aimee. 2019. "AI and Ethics: The Debate that Needs to Be Had." ZDnet, September 16, 2019. https://www.zdnet.com/article/ai-and-ethics-the-debate-that -needs-to-be-had/#ftag=CAD-03-10abf5f.

Fryer-Biggs, Zachary. 2019. "Coming Soon to a Battlefield: Robots that Can Kill." *The Atlantic*, September 3, 2019. https://www.theatlantic.com/technology/archive/2019 /09/killer-robots-and-new-era-machine-driven-warfare/597130.

Kistler, Alan. 2013. Doctor Who: *Celebrating Fifty Years, A History*. Guilford, CT: Lyons.

Krishnan, Armin. 2009. *Killer Robots: Legality and Ethicality of Autonomous Weapons*. London: Routledge.

Lewis, Courtland, and Smithka, Paula, eds. 2010. Doctor Who *and Philosophy: Bigger on the Inside*. Chicago: Open Court.

Lewis, Courtland, and Smithka, Paula, eds. 2015. *More* Doctor Who *and Philosophy: Regeneration Time*. Chicago: Open Court.

Lin, Patrick, Abney, Keith, and Bekey, George A., eds. 2014. *Robot Ethics: The Ethical and Social Implications of Robotics*. Cambridge: Massachusetts Institute of Technology.

Lin, Patrick, Jenkins, Ryan, and Abney, Keith, eds. 2017. *Robot Ethics 2.0: From Autonomous Cars to Artificial Intelligence*. New York: Oxford University Press.

Muir, John Kenneth. 2007. *A Critical History of* Doctor Who *on Television*. Jefferson, NC: McFarland.

Parkin, Lance. 2016. *Whoniverse*. New York: Barron's Educational Series.

Richards, Justin. 2009. Doctor Who: *The Official Doctionary*. London: Penguin Group.

Darth Vader

Star Wars

In 1977, writer-director George Lucas (b. 1944) introduced the world to the sci-fi/fantasy phenomenon *Star Wars* (Lucasfilm, 20th Century Fox, 1977), one of the most successful pop culture franchises in American history. His original trilogy of films (1977–1983) focused on the rise of its protagonist, Luke Skywalker (played by Mark Hamill, b. 1951), against the backdrop of a Galactic Civil War against the human forces of the evil Empire, led by its Emperor, Sheev Palpatine (played by Ian McDiarmid, b. 1944), and his henchman, Darth Vader (played by David Prowse, b. 1935; voiced by James Earl Jones, b. 1931). In 1999, Lucas launched a "prequel" trilogy, beginning with *Star Wars, Episode I: The Phantom Menace* (Lucasfilm, 1999), which began the story of the fall of Anakin Skywalker (played originally by Jake Lloyd, b. 1989, later by Hayden Christensen, b. 1981)—the father of the original hero—into becoming the villainous Darth Vader, which he finally does in the prequel trilogy's final installment, *Star Wars, Episode III: Revenge of the Sith* (Lucasfilm, 2005).

The story of Anakin Skywalker (Lloyd) begins when he is a nine-year-old slave on the planet Tatooine, the product of a mysterious "virgin birth." The Jedi Knight Qui-Gon Jinn (played by Liam Neeson, b. 1952) discovers that the boy possesses a particularly high "midichlorian count," suggesting the boy is a potential Jedi, and takes him under his tutelage (Lucas, *SW:TPM*, 1999). Later, training under the Jedi Knight Obi-Wan Kenobi (played by Ewan McGregor, b. 1971), Anakin (now Christensen) begins to feel the pull to the Dark Side of the Force due to the machinations of Republic Chancellor Palpatine (McDiarmid). Anakin's fall is complete when he views the Jedi as evil, as conspiring against the Chancellor, and as threatening the democracy of the Republic. As the newly christened "Darth Vader, Dark Lord of the Sith," Anakin battles his former mentor, Obi-Wan, in a light-saber duel on the volcanic planet of Mustafar. Obi-Wan cuts off Anakin's remaining arm and both legs (he had already lost the other arm, which was replaced with a cybernetic implant), leaving the younger Jedi to burn alive on the shores of a river of lava. Palpatine rescues his new apprentice, replacing his lost limbs with more robotic implants, as well as a mechanical breathing apparatus (including a fearsome breathing mask/helmet) to replace his burned lungs. Vader is "more machine, now, than man" (Obi-Wan Kenobi/Sir Alec Guinness, *Star Wars, Episode VI: Return of the Jedi*, Lucasfilm, 1983). Now deeply committed to the Dark Side, and in constant pain from his wounds and implants, the cyborg Darth Vader is devoid of human emotion (until eventual redemption through his son, Luke, in *SW: ROJ*).

In the canonical *Star Wars* comic books, it is later revealed that the reason why the evil Palpatine was so interested in young Anakin Skywalker was that it was he who manipulated Anakin's "virgin birth." Utilizing the Dark Side of the Force, Palpatine manipulated "midichlorians," microscopic organisms strongly connected to the Force that exist in high numbers within the bloodstreams of potential Force users, to "create life" within the womb of the young slave woman Shmi Skywalker (Charles Soule and Giuseppe Camuncoli, *Star Wars: Darth Vader #25*, Marvel Comics, December 2018). Palpatine's goal of creating the most powerful Force user of all time, however, is stunted when Anakin loses all four limbs (presumably the loss of these natural limbs equaled a loss of midichlorians as well). Indeed, though Skywalker had already given himself to darkness with the murder of many people shortly after becoming Vader, with so much of him replaced with machine (and the loss of his wife, believing it to be by his own hand), what humanity—and goodness—remained of the onetime hero was very deeply buried.

Considered one of the most iconic screen villains in film history, Darth Vader was the primary villain of the original *Star Wars* trilogy (1977–1983). He was originally played by David Prowse and voiced by James Earl Jones. (Carrienelson1/ Dreamstime.com)

Unlike most other cyborgs in sci-fi, Darth Vader never exhibits any ability to interact with computer systems. His implants are limited to allowing standard human mobility and life support. Vader's loss of humanity is due more to his philosophical shift rather than his robotic additions; but his mechanics are symbolic of this loss of self and consumption with evil. When a person becomes "more machine" than "man," to what degree would that person continue to view himself as human? Is this a danger that should be considered as humanity in the twenty-first century begins to experiment more and more with mechanical devices designed to prolong life? Like the Borg of *Star Trek* and the Cybermen and Daleks of *Doctor Who*, Darth Vader began as a fully organic life-form (a "good" and innocent one at that). As Vader becomes less and less organic, his empathy for organic life wanes as well. Ultimately Vader's humanity is redeemed by saving his son, Luke, from the similar fate intended by the evil Palpatine, and the cyborg dies with his humanity regained.

Richard A. Hall

See also: Adam, Alita, Battle Droids, BB-8, Borg, C-3PO, Cybermen, Cyborg, Daleks, Doctor Octopus, Echo/CT-1409, EV-9D9, General Grievous, IG-88/IG-11, Inspector Gadget, Jaime Sommers, K-2SO, L3-37/*Millennium Falcon*, Medical Droids, Nardole, R2-D2, Steve Austin; *Thematic Essays:* Villainous Robots and Their Impact on Sci-Fi Narratives, Cyborgs: Robotic Humans or Organic Robots?

Further Reading

Ashby, LeRoy. 2006. *With Amusement for All: A History of American Popular Culture since 1830.* Lexington: University Press of Kentucky.

Byrne, Emma. 2014. "Cybernetic Implants: No Longer Science Fiction." *Forbes*, March 11, 2014. https://www.forbes.com/sites/netapp/2014/03/11/cybernetic-implants-not-sci-fi/#7399c57e77ba.

Eberl, Jason T., and Decker, Kevin S., eds. 2015. *The Ultimate* Star Wars *and Philosophy: You Must Unlearn What You Have Learned.* Hoboken, NJ: Wiley-Blackwell.

Kaminski, Michael. 2008. *The Secret History of* Star Wars*: The Art of Storytelling and the Making of a Modern Epic.* Kingston, Canada: Legacy Books.

Leadbeater, Alex. 2019. "Star Wars is Trying to Turn Darth Vader into an Anti-Hero (and That's Very Bad)." Screen Rant, February 4, 2019. https://screenrant.com/star-wars-darth-vader-villain-anti-hero.

Reagin, Nancy R., and Liedl, Janice, eds. 2013. Star Wars *and History.* New York: John Wiley & Sons.

Sunstein, Cass R. 2016. *The World According to* Star Wars. New York: Dey Street.

Taylor, Chris. 2015. *How* Star Wars *Conquered the Universe: The Past, Present, and Future of a Multibillion Dollar Franchise.* New York: Basic Books.

D-O

Star Wars

In 2012, writer-director George Lucas (b. 1944) sold his production company Lucasfilm, and with it, his lucrative *Star Wars* franchise, to Walt Disney Studios. Disney then embarked on producing new *Star Wars* films, beginning with *Star Wars, Episode VII: The Force Awakens* (Lucasfilm/Disney, 2015), which began a new trilogy that would include: *Episode VIII: The Last Jedi* (Lucasfilm/Disney, 2017); and *Episode IX: The Rise of Skywalker* (Lucasfilm/Disney, 2019). Set thirty years after the events of the original trilogy films, the defeated Empire has been replaced with the First Order, under the rule of Supreme Leader Snoke (voiced by Andy Serkis, b. 1964) and his henchman, the evil Kylo Ren (played by Adam Driver, b. 1983). Ren is, in reality, the nephew of Jedi master Luke Skywalker (played by Mark Hamill, b. 1951). Skywalker's sister—and Kylo Ren's mother—General Leia Organa-Solo (played by Carrie Fisher, 1956–2016) leads the Resistance, seeking to free the galaxy from yet another Dark Side threat. A valuable last-minute addition to the team was the tiny droid, D-O.

D-O is one of the smallest droids in the *Star Wars* universe, standing less than one-half meter tall. He possesses a cone-shaped head with a green tip and three antennae protruding from the back. His head is suspended above a singular wheel, which provides his method of motion. Though speaking primarily in standard droid beeps and whistles, D-O is capable of minimal human words. D-O was discovered by Resistance heroes Poe Dameron (played by Oscar Isaac, b. 1979), Finn (played by John Boyega, b. 1992), Rey (played by Daisy Ridley, b. 1992), and droids C-3PO (played by Anthony Daniels, b. 1946) and BB-8 aboard the abandoned ship of the former assassin Ochi. D-O's reaction to the human members of the team show clear evidence that the droid has suffered severe abuse from humanoids. When Rey attempts to gently touch the droid, D-O withdraws quickly, stammering "No, thank you." The tiny droid would prove an invaluable member of the team when it is discovered that, within his memory banks, he possesses a star map to the location of the Emperor's new Sith fleet of Star Destroyer ships.

Though having a very brief appearance in the overall landscape of the *Star Wars* universe, D-O became an immediate fan favorite. Throughout the long history established in *Star Wars*, droids have been, from the very beginning, a slave class. They have "masters" and "owners" despite many of them clearly possessing sentience, self-awareness, and even "feelings." D-O represents an abused slave, exposed to brutal treatment over an extended period of time. Even after being made aware that his newfound "masters" are kind, D-O is hesitant to have any close contact where he may experience further abuse. He could be referred to as a "rescue" droid if one were to make the connection to abused animals. Through his nervous reactions to humans, the true brutality of the slave existence of droids/robots becomes painfully clear to the audience.

Richard A. Hall

See also: Battle Droids, Baymax, BB-8, Bender, C-3PO, Cambot/Gypsy/Tom Servo/ Crow, Darth Vader, Dr. Theopolis and Twiki, Echo/CT-1409, EV-9D9, General Grievous, HERBIE, IG-88/IG-11, K-2SO, K9, L3-37/*Millennium Falcon*, Medical Droids, Muffit, R2-D2, VICI, WALL-E; *Thematic Essays:* Robots and Slavery, Heroic Robots and Their Impact on Sci-Fi Narratives.

Further Reading

Ashby, LeRoy. 2006. *With Amusement for All: A History of American Popular Culture since 1830*. Lexington: University Press of Kentucky.

Eberl, Jason T., and Decker, Kevin S., eds. 2015. *The Ultimate* Star Wars *and Philosophy: You Must Unlearn What You Have Learned*. Hoboken, NJ: Wiley-Blackwell.

Hampton, Gregory Jerome. 2017. *Imagining Slaves and Robots in Literature, Film, and Popular Culture: Reinventing Yesterday's Slave with Tomorrow's Robot*. Lanham, MD: Lexington.

Kaminski, Michael. 2008. *The Secret History of* Star Wars: *The Art of Storytelling and the Making of a Modern Epic*. Kingston, Canada: Legacy Books.

Lin, Patrick, Abney, Keith, and Bekey, George A., eds. 2014. *Robot Ethics: The Ethical and Social Implications of Robotics*. Cambridge: Massachusetts Institute of Technology.

Reagin, Nancy R., and Liedl, Janice, eds. 2013. Star Wars *and History*. New York: John Wiley & Sons.

Spaeth, Dennis. 2018. "From Single-Task Machines to Backflipping Robots: The Evolution of Robots." *Cutting Tool Engineering*, January 15, 2018. https://www.ctemag .com/news/articles/evolution-of-robots.

Stanford University. n.d. "Robotics: A Brief History." Stanford University. https://cs.stanford .edu/people/eroberts/courses/soco/projects/1998-99/robotics/history.html.

Sunstein, Cass R. 2016. *The World According to* Star Wars. New York: Dey Street.

Sweet, Derek R. 2015. Star Wars *in the Public Square: The Clone Wars as Political Dialogue.* Critical Explorations in Science Fiction and Fantasy series, edited by Donald E. Palumbo and Michael Sullivan. Jefferson, NC: McFarland.

Taylor, Chris. 2015. *How* Star Wars *Conquered the Universe: The Past, Present, and Future of a Multibillion Dollar Franchise.* New York: Basic Books.

Wallach, Wendell, and Allen, Colin. 2009. *Moral Machines: Teaching Robots Right from Wrong.* New York: Oxford University Press.

Doctor Octopus

Marvel Comics

In 1961, comic book legend Stan Lee (1922–2018) launched the "Marvel age" of comics. With the launching of the Fantastic Four, soon to be followed by the Hulk, Spider-Man, Iron Man, and many, many more, comic book superheroes took a more dramatic turn, with storylines focused on the private lives of those blessed (or burdened) with superpowers. The character that would quickly become the commercial face of Marvel Comics, however, was Spider-Man. When high schooler Peter Parker is bitten by a radioactive spider, he develops mutated abilities: the proportional speed, strength, and agility of a spider, as well as a sixth "spider" sense to warn him of danger. At first using his newfound skills for monetary gain, Peter allows a thief to escape because the fight promoter the thief was robbing had recently cheated Peter out of money. In an ironic twist, that same thief went on later that night to murder Peter's beloved Uncle Ben, thus teaching the teenager that "with great power there must also come—great responsibility!" (Stan Lee and Steve Ditko, *Amazing Fantasy #15*, August 1962). Spider-Man soon gained one of the most impressive rogues' galleries in all superhero comics. One of Spider-Man's deadliest foes is the sinister cyborg, Doctor Octopus.

Otto Octavius was a world-renowned nuclear physicist and engineer. He developed a mechanical belt with four extendable arms, each tipped with manipulating "claws" that were connected directly to his brain. This device assisted him in more easily handling the dangerous radioactive materials he needed for his research. However, when a massive radiation leak creates an explosion, the brain interface fuses directly into Octavius's brain, and the arms/belt fuse directly to his body. This leaves him mentally warped toward criminal activity, and he becomes the supervillain "Doctor Octopus" (Lee and Ditko, *The Amazing Spider-Man #3*, July 1963). He quickly became one of Spider-Man's most popular—and therefore most recurring—adversaries. Over the years, Doc Oc organized the supervillain group the "Sinister Six" in order to push Spider-Man to the breaking point. In the twenty-first century, aware of Spider-Man's true identity, Doc Oc "swapped" bodies with Peter Parker when Octavius's own body was dying, and Octavius briefly became the "Superior Spider-Man" (2012–2013). Being a brilliant engineer, Octavius also created an army of miniature "octobots" to do his bidding (which he reimaged as "spiderbots" while he was Spider-Man). The only live-action version of Doctor Octopus to date has been in the feature film *Spider-Man 2* (Sony, 2004), played by Alfred Molina (b. 1953). Staying largely loyal to the source material,

Octavius's madness is driven more by the death of his beloved wife in the same explosion that made him "Doctor Octopus."

Currently, scientists and engineers are already working on mechanical appendages that can respond to the human brain's neuron activity. While the most common application of such technology would logically be to assist those who have lost arms and/or hands, it does not take much of a leap of imagination to get to a Doctor Octopus–type scenario. Indeed, many micro-surgeries today are assisted by intricate armlike devices to assist the surgeon. As such, of the many cyborg scenarios in popular culture today, Doctor Octopus is perhaps the closest to scientific reality. The lessons learned from this character, then, become all the more important for current and future study.

Though admittedly not by his own choosing, Doctor Octopus

A cosplayer dressed as Doctor Octopus, one of the earliest villains in Spider-Man's rogues' gallery, created by Stan Lee and Steve Ditko. (Fernando Carniel Machado/Dreamstime.com)

is a cyborg—fully interconnected to his mechanical implants. An early version of what would become the Borg, Cybermen, or Daleks—and more akin to the superhero Cyborg, Jaime Sommers, and Steve Austin—Doctor Octopus possesses advanced abilities due to his cybernetic attachments. It is interesting, however, that, unlike the Borg, Cybermen, or Daleks, Octavius does not seek to continually "upgrade" more cybernetics. He clearly desires to maintain his humanity as much as is feasibly possible. His deep desire to remain human becomes even more clear during the brief period in which he inhabits Peter Parker's body. He understands and appreciates the abilities of his cybernetics, but he does not wish to become more machine than man; he maintains a limit to his robotic parts, never fully allowing them to override his humanity.

Richard A. Hall

See also: Adam, Alita, Arnim Zola, Baymax, Borg, Cerebro/Cerebra, Cybermen, Cyborg, Daleks, Darth Vader, Doombots, Echo/CT-1409, General Grievous, HERBIE, Human Torch, Inspector Gadget, Iron Legion, Iron Man, Jaime Sommers, JARVIS/Friday, Jocasta, LMDs, Medical Droids, Nardole, RoboCop, Sentinels, Steve Austin, Ultron, Vision, Warlock; *Thematic Essays:* Villainous Robots and Their Impact on Sci-Fi Narratives, Cyborgs: Robotic Humans or Organic Robots?

Further Reading

Ashby, LeRoy. 2006. *With Amusement for All: A History of American Popular Culture since 1830*. Lexington: University Press of Kentucky.

Bastani, Aaron. 2019. *Fully Automated Luxury Communism*. New York: Verso.

Byrne, Emma. 2013. "Innovation Isn't Safe: The Future According to Kevin Warwick." *Forbes*, September 30, 2013. https://www.forbes.com/sites/netapp/2013/09/30/kevin-warwick-captain-cyborg/#48b704c13560.

Byrne, Emma. 2014. "Cybernetic Implants: No Longer Science Fiction." *Forbes*, March 11, 2014. https://www.forbes.com/sites/netapp/2014/03/11/cybernetic-implants-not-sci-fi/#7399c57e77ba.

Choi, Charles Q. 2009. "Human Evolution: The Origin of Tool Use." LiveScience, November 11, 2009. https://www.livescience.com/7968-human-evolution-origin-tool.html.

Costello, Matthew J. 2009. *Secret Identity Crisis: Comic Books & the Unmasking of Cold War America*. New York: Continuum.

DeFalco, Tom. 2004. *Comics Creators on Spider-Man*. London: Titan Books.

Hitchcock, Susan Tyler. 2007. *Frankenstein: A Cultural History*. New York: W. W. Norton.

Howe, Sean. 2012. *Marvel Comics: The Untold Story*. New York: Harper-Perennial.

Kripal, Jeffrey J. 2015. *Mutants and Mystics: Science Fiction, Superhero Comics, and the Paranormal*. Chicago: University of Chicago Press.

Sanford, Jonathan J., and Irwin, William, eds. 2012. *Spider-Man and Philosophy: The Web of Inquiry*. New York: Wiley.

Spaeth, Dennis. 2018. "From Single-Task Machines to Backflipping Robots: The Evolution of Robots." *Cutting Tool Engineering*, January 15, 2018. https://www.ctemag.com/news/articles/evolution-of-robots.

Starck, Kathleen, ed. 2010. *Between Fear and Freedom: Cultural Representations of the Cold War*. Newcastle, UK: Cambridge Scholars.

Thomas, Roy. 2017. *The Marvel Age of Comics: 1961–1978*. Los Angeles: Taschen.

Tucker, Reed. 2017. *Slugfest: Inside the Epic 50-Year Battle between Marvel and DC*. New York: Da Capo Press.

Virk, Rizwan. 2019. *The Simulation Hypothesis: An MIT Computer Scientist Shows Why AI, Quantum Physics, and Eastern Mystics All Agree We Are in a Video Game*. Milwaukee, WI: Bayview.

Wright, Bradford W. 2003. *Comic Book Nation: The Transformation of Youth Culture in America*. Baltimore, MD: Johns Hopkins University Press.

Doctor/EMH (Emergency Medical Hologram Mark I)

Star Trek

In 1966, creator-writer-producer Gene Roddenberry (1921–1991) launched his iconic science fiction television series *Star Trek* (NBC, 1966–1969). Set in the twenty-third century, the series centers on the adventures of the starship USS *Enterprise*; its commander, Captain James T. Kirk (played by William Shatner, b. 1931); and crew, headed by the alien first officer, Mr. Spock (played by Leonard Nimoy, 1931–2015), representing the United Federation of Planets. Though short-lived in its initial run, the series has become legendary, not only for its groundbreaking special effects but also for the sophistication of its stories, many of which were grounded in real-world issues of the day, such as racism and a fear of the Cold War. Its success in syndicated reruns revived the franchise, leading to an animated series, thirteen feature films, and five live-action "sequel" series (to the time of this writing). The third live-action sequel series was *Star Trek: Voyager* (UPN, 1995–2001).

The premise of *Voyager* (*ST: Voyager*) is that the Federation starship *USS Voyager*, under the command of Captain Kathryn Janeway (played by Kate Mulgrew, b. 1955) has been mysteriously flung to the far reaches of space, the Delta Quadrant of the Milky Way Galaxy (whereas the Federation exists in the "Alpha" quadrant) while in pursuit of Maquis terrorists. The event leaves both the *Voyager* (and the Maquis ship) badly damaged, with numerous casualties and seventy-five light-years from home (i.e., if they were to maintain maximum speed—an impossible task—it would take seventy-five years to return to the Alpha Quadrant). Among *Voyager*'s casualties is its entire medical staff (Michael Piller and Jeri Taylor, "Caretaker, Parts 1 and 2," *ST: Voyager*, season 1, Episodes 1 and 2, January 16, 1995). As such, the ship and crew are completely reliant on the experimental "Emergency Medical Hologram" (EMH) program (played by Robert Picardo, b. 1953).

By that time, the twenty-fourth century of *Star Trek* had already established advanced motion-capture "holodeck"/"holosuite" technology. These were rooms built with holographic projectors that could allow the computer to bring literally any situation (real or fictional) to "life," even creating complex life-forms that were "solid" and completely interactive. As such, Starfleet engineers devised a plan to build medical bays within holodecks, allowing for the creation of EMH programs that, in emergency situations, could act as medical personnel, fully programmed with the latest Starfleet medical knowledge. The "Mark I" series of holograms were created in the image of the program's creator, Dr. Lewis Zimmerman (also played by Picardo). When *Voyager* finds itself without a medical staff, it initiates the program, which comes online with the programmed greeting, "Please state the nature of the medical emergency." Informed that he is now the entire medical staff, the *Voyager* EMH informs the crew that his program is meant only for brief emergencies. Further, as a mere hologram, the Doctor cannot exist outside the confines of the ship's medical bay. Over time, holo-projectors are installed throughout the ship, allowing the Doctor more mobility to respond to emergencies. Eventually, the crew develops a "mobile" projector that the hologram can wear on his "arm," allowing him to essentially go anywhere, to become a truly interactive life-form. As a hologram, however, the Doctor can take any form accessible from the Starfleet databanks, including the images of his crewmates.

On the latest *Star Trek* series, *Picard* (CBS All Access, 2020–present), the iconic titular character from *Star Trek: The Next Generation* (syndication, 1987–1994), Admiral Jean-Luc Picard (played by Sir Patrick Stewart, b. 1940), long retired from Starfleet, embarks on a solo adventure, securing passage on the civilian starship *La Sirena*, piloted by former Starfleet officer, Chris Rios (played by Santiago Cabrera, b. 1978). Rather than a traditional crew, *La Sirena* is crewed by multiple EMH-styled holograms (including one that is the actual EMH), all of which possess variations on the physical appearance of their captain. Despite their near-identical appearance, each hologram bears its own personality and contains a varying degree of its captain's wit and sarcasm.

As with Data on *Star Trek: The Next Generation* and the Borg drone Seven of Nine, later on *Voyager*, the Doctor/EMH character was utilized to explore definitions of "life" beyond the organic. In order to become a more effective physician, the EMH program develops "empathy" and the ability to emote joy, sadness, anger, frustration, and, to a degree, romantic interest. With his unmatched knowledge, his eventual self-awareness, and his ability to emote and "care" for his crewmates, the Doctor emerges from *Voyager*'s adventures a fully functioning and integrated life-form, even recognized as such by the United Federation of Planets. As AI becomes more advanced in the twenty-first century, humanity will soon be forced to contend with whether "intelligence"—organic or artificial—constitutes "life."

Richard A. Hall

See also: AI/Ziggy, Androids, Borg, Cerebro/Cerebra, HAL 9000, Janet, JARVIS/Friday, Landru, Lieutenant Commander Data, Lore/B4, The Matrix/Agent Smith, Max Headroom, Medical Droids, OASIS, Oz, Skynet, Soji and Dahj Asha, Starfleet Computer, TARDIS, V-GER, WOPR, Zordon/Alpha-5; *Thematic Essays:* Robots and Slavery, AI and the Apocalypse: Science Fiction Meeting Science Fact.

Further Reading

Allen, Arthur. 2019. "There's a Reason We Don't Know Much about AI." Politico, September 16, 2019. https://www.politico.com/agenda/story/2019/09/16/artficial-intelligence-study-data-000956?cid=apn.

Bastani, Aaron. 2019. *Fully Automated Luxury Communism*. New York: Verso.

Byrne, Emma. 2013. "Innovation Isn't Safe: The Future According to Kevin Warwick." *Forbes*, September 30, 2013. https://www.forbes.com/sites/netapp/2013/09/30/kevin-warwick-captain-cyborg/#48b704c13560.

Castleman, Harry, and Podrazik, Walter J. 2016. *Watching TV: Eight Decades of American Television, Third Edition*. Syracuse, NY: Syracuse University Press.

Chandler, Simon. 2019. "Artificial Intelligence Has Become a Tool for Classifying and Ranking People." *Forbes*, October 1, 2019. https://www.forbes.com/sites/simonchandler/2019/10/01/artificial-intelligence-has-become-a-tool-for-classifying-and-ranking-people/#7431657f1d7c.

Chanthadavong, Aimee. 2019. "AI and Ethics: The Debate that Needs to Be Had." ZDnet, September 16, 2019. https://www.zdnet.com/article/ai-and-ethics-the-debate-that-needs-to-be-had/#ftag=CAD-03-10abf5f.

Decker, Kevin S., and Eberl, Jason T., eds. 2016. *The Ultimate* Star Trek *and Philosophy: The Search for Socrates*. Hoboken, NJ: Wiley-Blackwell.

Dinh, Thien-Nam. 2018. *Silicon Minds: The Science, Impact, and Promise of Artificial Intelligence*. Independently Published.

Gross, Edward, and Altman, Mark A. 2016. *The Fifty-Year Mission, The Second 25 Years: The Complete, Uncensored, Unauthorized Oral History of Star Trek*. New York: St. Martin's.

Hampton, Gregory Jerome. 2017. *Imagining Slaves and Robots in Literature, Film, and Popular Culture: Reinventing Yesterday's Slave with Tomorrow's Robot*. Lanham, MD: Lexington.

Hitchcock, Susan Tyler. 2007. *Frankenstein: A Cultural History*. New York: W. W. Norton.

Murdock, Jason. 2019. "Former Google Engineer Warns AI Might Accidentally Start a War: 'These Things Will Start to Behave in Unexpected Ways.'" *Newsweek*, September 16, 2019. https://www.newsweek.com/google-project-maven-artificial-intelligence-laura-nolan-killer-robots-department-defense-1459358.

Paur, Joey. 2019. "An AI Bot Writes a Hilarious Episode of *Star Trek: The Next Generation*." Geek Tyrant, November 16, 2019. https://geektyrant.com/news/an-ai-bot-writes-a-hilarious-episode-of-star-trek-the-next-generation.

Reagin, Nancy R., and Liedl, Janice, eds. 2013. Star Wars *and History*. New York: John Wiley & Sons.

Rouhiainen, Lasse. 2018. *Artificial Intelligence: 101 Things You Must Know Today about Our Future*. Scotts Valley, CA: CreateSpace.

Wallach, Wendell, and Allen, Colin. 2009. *Moral Machines: Teaching Robots Right from Wrong*. New York: Oxford University Press.

Dolores

Westworld

In 2016, creators Jonathan Nolan and Lisa Joy introduced their live-action television series *Westworld* (HBO, 2016–present), an updated reboot of the 1973 film of the same name by writer-director Michael Crichton (1942–2008). The premise of the series is that, in an undisclosed future, the Delos Corporation operates a series of amusement/recreation parks that allow patrons to live out their ultimate fantasies against various backdrops. "Westworld" allows customers to experience realistic adventures set in the American "Wild West" in a vividly realistic panoramic western world populated by "hosts": realistic androids programmed to play out specified storylines with preprogrammed personalities based on stereotypes of the Wild West (e.g., gunfighters, soldiers, farmers, savage Natives, and prostitutes). The hosts—and their corresponding storylines–have been produced by park cofounder Robert Ford (played by Anthony Hopkins, b. 1937). As the hosts are robots, guests are free to beat, rape, or murder the hosts as they see fit (creating both a cathartic release and an ethical conundrum). As the series opens, one such unlucky host is the farm girl Dolores (played by Evan Rachel Wood, b. 1987).

When Ford and his original partner, Arnold Weber (played by Jeffrey Wright, b. 1965), first mastered their android technology, Dolores was the prototype. Weber believed that the androids were capable of eventually achieving sentience; and he worked secretly with Dolores, implanting programming that he believed would one day awaken her consciousness. When the series opens, Dolores, who by this time has been beaten, raped, and murdered countless times, her memory "wiped" after each encounter, begins having "dreams" and flashbacks in a manner similar to déjà vu. By the time the first season ends, Dolores has fully awakened and embarked on a bloody quest to achieve sentience through "the maze." Her guide on this path has been the park's programming coordinator, Bernard. Audiences learn that Bernard is, himself, an android, built to resemble Weber and complete Weber's goal of android sentience through the machinations of the recently reformed Ford (Nolan and Joy, "The Bicameral Mind," *Westworld*, season 1, episode 10, December 4, 2016). Throughout the second season, Dolores raises an army, killing hosts and humans as she continues her quest. She finally discovers that there is a mainframe where the host consciousnesses are uploaded. Dolores escapes the park with the mainframe, now bent on moving her crusade into the human world (Nolan and Joy, "The Passenger," *Westworld*, season 2, episode 10, June 24, 2018).

In the series' third season, Dolores embarks on her war to destroy humanity. To aid her, she has produced multiple copies of her own "pearl" (the central core of her personality—in essence, her "brain") and implanted them in several new "host" bodies built to look like others. During her quest, however, Dolores discovers that like her, humans have been denied their own free will for years and are unknowingly under the control of a central AI known as "Rehoboam." In the end, Dolores, now dedicated to "freeing" humanity rather than destroying it, seemingly sacrifices her life to take down Rehoboam. Her sacrifice appears to succeed, and humans (still unaware they had been "controlled") are now free to utilize their own will moving forward (for better or worse). The story is set to continue in 2022 (Denise Thé and Nolan, "Crisis Theory," *Westworld*, season 3, episode 8, May 3, 2020).

Dolores is yet another ultimate example of the dangers of the Frankenstein analogy. She is created by humans, is abused by humans, and then is a killer of humans. She and the other hosts have endured the most horrible atrocities during their decades of slavery as theme-park attractions. Like so many of the artificial life-forms presented throughout this volume, Dolores is an AI that achieves full sentience. She is intelligent, she is self-aware, and she is fully conscious of who she is and that she is alive. As humanity in the twenty-first century pushes the envelope of AI on a near-daily basis, the question of "life" is going to require serious reconsidering in the not-too-distant future; and if science fiction is any prognosticator, it is vital that humans reconsider carefully.

Richard A. Hall

See also: Androids, B.A.T.s, Battle Droids, Bernard, Bishop, Buffybot, Fembots, Human Torch, Lieutenant Commander Data, LMDs, Lore/B4, Marvin the Paranoid Android, Maschinenmensch/Maria, Rehoboam, Replicants, Soji and Dahj Asha, Stepford Wives, VICI, Vision; *Thematic Essays:* Robots and Slavery, AI and the Apocalypse: Science Fiction Meeting Science Fact.

Further Reading

Allen, Arthur. 2019. "There's a Reason We Don't Know Much about AI." Politico, September 16, 2019. https://www.politico.com/agenda/story/2019/09/16/artficial-intelligence-study-data-000956?cid=apn.

Bastani, Aaron. 2019. *Fully Automated Luxury Communism*. New York: Verso.

Byrne, Emma. 2013. "Innovation Isn't Safe: The Future According to Kevin Warwick." *Forbes*, September 30, 2013. https://www.forbes.com/sites/netapp/2013/09/30/kevin-warwick-captain-cyborg/#48b704c13560.

Carper, Steve. 2019. *Robots in American Popular Culture*. Jefferson, NC: McFarland.

Chandler, Simon. 2019. "Artificial Intelligence Has Become a Tool for Classifying and Ranking People." *Forbes*, October 1, 2019. https://www.forbes.com/sites/simonchandler/2019/10/01/artificial-intelligence-has-become-a-tool-for-classifying-and-ranking-people/#7431657f1d7c.

Chanthadavong, Aimee. 2019. "AI and Ethics: The Debate that Needs to Be Had." ZDnet, September 16, 2019. https://www.zdnet.com/article/ai-and-ethics-the-debate-that-needs-to-be-had/#ftag=CAD-03-10abf5f.

Choi, Charles Q. 2009. "Human Evolution: The Origin of Tool Use." LiveScience, November 11, 2009. https://www.livescience.com/7968-human-evolution-origin-tool.html.

Dinh, Thien-Nam. 2018. *Silicon Minds: The Science, Impact, and Promise of Artificial Intelligence*. Independently Published.

Greene, Richard, and Heter, Joshua, eds. 2018. Westworld *and Philosophy: Mind Equals Blown*. Chicago: OpenCourt.

Hampton, Gregory Jerome. 2017. *Imagining Slaves and Robots in Literature, Film, and Popular Culture: Reinventing Yesterday's Slave with Tomorrow's Robot*. Lanham, MD: Lexington.

Hitchcock, Susan Tyler. 2007. *Frankenstein: A Cultural History*. New York: W. W. Norton.

Irwin, William. 2018. *Westworld and Philosophy: If You Go Looking for the Truth, Get the Whole Thing*. Hoboken, NJ: Wiley-Blackwell.

Langley, Travis, Goodfriend, Wind, and Cain, Tim, eds. 2018. Westworld *Psychology: Violent Delights*. New York: Sterling.

Lin, Patrick, Abney, Keith, and Bekey, George A., eds. 2014. *Robot Ethics: The Ethical and Social Implications of Robotics*. Cambridge: Massachusetts Institute of Technology.

Lin, Patrick, Jenkins, Ryan, and Abney, Keith, eds. 2017. *Robot Ethics 2.0: From Autonomous Cars to Artificial Intelligence*. New York: Oxford University Press.

Linder, Courtney. 2020. "This AI Robot Just Nabbed the Lead Role in a Sci-Fi Movie: Meet Erica, the Movie Star Who May Put Human Actors Out of Work." *Popular Mechanics*, June 25, 2020. https://www.popularmechanics.com/technology/robots /a32968811/artificial-intelligence-robot-movie-star-erica.

Murdock, Jason. 2019. "Former Google Engineer Warns AI Might Accidentally Start a War: 'These Things Will Start to Behave in Unexpected Ways.'" *Newsweek*, September 16, 2019. https://www.newsweek.com/google-project-maven-artificial-intel ligence-laura-nolan-killer-robots-department-defense-1459358.

Pellissier, Hank. 2013. "Robots and Slavery: What Do Humans Want When We Are 'Masters'?" Institute for Ethics and Emerging Technologies, September 13, 2013. https://ieet.org/index.php/IEET2/more/pellissier20130913.

Randall, Terri, dir. 2016. *NOVA*. "Rise of the Robots: Inside the DARPA Robotics Challenge." Aired February 24, 2016, on PBS. DVD.

Robitzski, Dan. 2018. "Artificial Consciousness: How to Give a Robot a Soul." Futurism, June 25, 2018. https://futurism.com/artificial-consciousness.

Rouhiainen, Lasse. 2018. *Artificial Intelligence: 101 Things You Must Know Today about Our Future*. Scotts Valley, CA: CreateSpace.

Schroeder, Stan. 2020. "Samsung Just Launched an 'Artificial Human' Called Neon, and Wait, What?" Mashable, January 7, 2020. https://mashable.com/article/samsung-star -labs-neon-ces.

Singer, Peter W. 2009. "Isaac Asimov's Laws of Robotics Are Wrong." Brookings Institute, May 18, 2009. https://www.brookings.edu/opinions/isaac-asimovs-laws-of -robotics-are-wrong.

Spaeth, Dennis. 2018. "From Single-Task Machines to Backflipping Robots: The Evolution of Robots." *Cutting Tool Engineering*, January 15, 2018. https://www.ctemag .com/news/articles/evolution-of-robots.

Starck, Kathleen, ed. 2010. *Between Fear and Freedom: Cultural Representations of the Cold War*. Newcastle, UK: Cambridge Scholars.

Thagard, Paul. 2017. "Will Robots Ever Have Emotions?" *Psychology Today*, December 14, 2017. https://www.psychologytoday.com/us/blog/hot-thought/201712/will-robots -ever-have-emotions.

Virk, Rizwan. 2019. *The Simulation Hypothesis: An MIT Computer Scientist Shows Why AI, Quantum Physics, and Eastern Mystics All Agree We Are in a Video Game.* Milwaukee, WI: Bayview.

Wallach, Wendell, and Allen, Colin. 2009. *Moral Machines: Teaching Robots Right from Wrong.* New York: Oxford University Press.

Doombots

Marvel Comics

In 1961, comic book legend Stan Lee (1922–2018) launched the "Marvel age" of comics. With the launching of the Fantastic Four, soon to be followed by the Hulk, Spider-Man, Iron Man, and many, many more, comic book superheroes took a more dramatic turn, with storylines focused on the private lives of those blessed (or burdened) with superpowers. The Fantastic Four introduced the Marvel Comics Universe. When genius scientist Dr. Reed Richards, his graduate student/girlfriend Susan Storm, her brother Johnny, and Reed's best friend Ben Grimm steal a spaceship to "beat the commies" into space, inadequate shielding exposes the four to "cosmic rays" that imbue each with superpowers. Reed possesses the ability to stretch his entire body to incredible degrees, becoming "Mister Fantastic." Susan can turn invisible and—over time—produce invisible "force fields," becoming the "Invisible Girl" (eventually, the "Invisible Woman"). Johnny can burst into flames, fly, and shoot flames from his hands, becoming the second "Human Torch." Ben possesses superstrength and a rocky exterior, becoming the "Thing." Together, they are the "Fantastic Four" (Lee and Jack Kirby, *The Fantastic Four #1*, November 1961). Throughout their decades of adventures, their archnemesis has been the Latverian dictator, Victor Von Doom (a.k.a. "Doctor Doom").

Doom possesses a personal hatred of Reed Richards and a goal of global conquest. He is a brilliant scientist and engineer in his own right. He wears a metal suit of armor, including a menacing metallic mask that covers his scarred face. He covers his armor with a green tunic, hood, and cape. In order to protect himself from constant defeat by the Fantastic Four (and nearly every other Marvel superhero), Doom created legions of "Doombots." Each Doombot is an exact replica of Doom himself (most believe that they are, in fact, the real Doctor Doom). This began as a narrative device to explain how Doom could repeatedly "die" and come back in later issues; but over time, it became a tired trope, and the Doombots became more of a legion of robot troops serving their Lord Doom. They are a nearly unstoppable army of machines, expendable and easily replaced. In this way, they are also slaves to Doom (though the moniker of "slave" is less definitive, as they are simply robots with no clear consciousness). Three distinct versions of Doombots have emerged over the years: diplomatic models that are primarily programmed for intelligence; fighting models that are heavily armed but lack mindfulness; and AI models, who believe they are Doom.

As of 2019, real-world engineers have nearly completed fully functioning combat robots. On the one hand, as seen in the pages of *The Fantastic Four*, legions of mindless automatons could prove invaluable on the battlefield, saving countless human lives by conducting combat activities for them. On the other hand, without the downside of massive loss of life, governments could choose combat as a first option rather than diplomacy, leading to the loss of more civilian casualties in pursuit of political goals. Removing the human equation from war threatens the ethical conundrum that warfare presently creates. Armies of Doombots—should they ever achieve sentience—could even slide down the slippery slope toward human extinction. The danger of real-world Doombots is the danger of a real-world Doctor Doom (and humanity's own doom).

Richard A. Hall

See also: Androids, Arnim Zola, B.A.T.s, Battle Droids, Cylons, ED-209, EV-9D9, Fembots, HERBIE, Iron Legion, Iron Man, LMDs, Metalhead, Robots, Sentinels, Soji and Dahj Asha, Stepford Wives, Tin Woodsman, Ultron, Vision, Warlock; *Thematic Essays:* Robots and Slavery, Villainous Robots and Their Impact on Sci-Fi Narratives.

Further Reading

Ashby, LeRoy. 2006. *With Amusement for All: A History of American Popular Culture since 1830*. Lexington: University Press of Kentucky.

Bastani, Aaron. 2019. *Fully Automated Luxury Communism*. New York: Verso.

Byrne, Emma. 2013. "Innovation Isn't Safe: The Future According to Kevin Warwick." *Forbes*, September 30, 2013. https://www.forbes.com/sites/netapp/2013/09/30/kevin -warwick-captain-cyborg/#48b704c13560.

Costello, Matthew J. 2009. *Secret Identity Crisis: Comic Books & the Unmasking of Cold War America*. New York: Continuum.

DeFalco, Tom. 2005. *Comics Creators on Fantastic Four*. London: Titan Books.

Hampton, Gregory Jerome. 2017. *Imagining Slaves and Robots in Literature, Film, and Popular Culture: Reinventing Yesterday's Slave with Tomorrow's Robot*. Lanham, MD: Lexington.

Howe, Sean. 2012. *Marvel Comics: The Untold Story*. New York: Harper-Perennial.

Lin, Patrick, Abney, Keith, and Bekey, George A., eds. 2014. *Robot Ethics: The Ethical and Social Implications of Robotics*. Cambridge: Massachusetts Institute of Technology.

Lin, Patrick, Jenkins, Ryan, and Abney, Keith, eds. 2017. *Robot Ethics 2.0: From Autonomous Cars to Artificial Intelligence*. New York: Oxford University Press.

Murdock, Jason. 2019. "Former Google Engineer Warns AI Might Accidentally Start a War: 'These Things Will Start to Behave in Unexpected Ways.'" *Newsweek*, September 16, 2019. https://www.newsweek.com/google-project-maven-artificial-intel ligence-laura-nolan-killer-robots-department-defense-1459358.

Pellissier, Hank. 2013. "Robots and Slavery: What Do Humans Want When We Are 'Masters'?" Institute for Ethics and Emerging Technologies, September 13, 2013. https:// ieet.org/index.php/IEET2/more/pellissier20130913.

Spaeth, Dennis. 2018. "From Single-Task Machines to Backflipping Robots: The Evolution of Robots." *Cutting Tool Engineering*, January 15, 2018. https://www.ctemag .com/news/articles/evolution-of-robots.

Thomas, Roy. 2017. *The Marvel Age of Comics: 1961–1978*. Los Angeles: Taschen.

Wallach, Wendell, and Allen, Colin. 2009. *Moral Machines: Teaching Robots Right from Wrong*. New York: Oxford University Press.

Wright, Bradford W. 2003. *Comic Book Nation: The Transformation of Youth Culture in America*. Baltimore, MD: Johns Hopkins University Press.

Dr. Theopolis and Twiki

Buck Rogers in the 25th Century

In 1979, inspired by the massive success of *Star Wars*, and hot off the success of his hit sci-fi series *Battlestar Galactica*, Glen A. Larson (1937–2014) teamed with Leslie Stevens (1924–1998) to bring the 1930s sci-fi hero Buck Rogers (created by Philip Francis Nowlan, 1888–1940) to modern audiences. The result was the series *Buck Rogers in the 25th Century* (NBC, 1979–1981). In the series, Captain Buck Rogers (played by Gil Gerard, b. 1943) is a U.S. astronaut launched into space on a mission in 1987. After liftoff, Rogers is accidentally cryogenically frozen and then discovered and reawakened in the year 2491. Rogers soon discovers that Earth had been destroyed by nuclear war shortly after his leaving and that the new, rebuilt Earth is under the protection of the Earth Defense Directorate, for which Rogers begins to work (Larson and Stevens, *Buck Rogers in the 25th Century* [theatrical film], 1979). The EDD answers to Earth's Computer Council, a council of intelligent computers tasked with ruling Earth since human rule had proven so destructive. The Council's liaison with the EDD was Dr. Theopolis.

Dr. Theopolis was a disc-shaped robot, roughly the size of a dinner plate, with a computerized "face" on its front. Though possessing intelligence far beyond human comprehension, Dr. Theopolis's size and shape made mobility impossible. As such, Theopolis was assigned an assistant, Twiki. Twiki was a short anthropomorphic robot whose human speech patterns were interspersed with a stuttering "Biddibiddibiddi." Together, the two robots were invaluable teammates to Captain Rogers and learn much about humanity from this relic of humanity's past. The duo was clearly an attempt to cash in on the popularity of *Star Wars*'s C-3PO and R2-D2, but it managed to go beyond that dynamic to create something new and interesting for the decades-old *Buck Rogers* franchise.

At its core, Dr. Theopolis presents a positive outlook on a future ruled by robots. Unlike so many other futuristic visions of robot rule—Brainiac, the Cybermen, Ultron, and Skynet, to name but a few—Earth's Computer Council does not view "protecting" Earth to mean destroying or enslaving humanity; rather, it applies clear, sound, intelligent logic to helping humanity to avoid the mistakes of their past. Dr. Theopolis and his fellow ECC members do not "rule" Earth but, instead, "administrate" Earth, allowing humanity to follow other, more beneficial pursuits, such as science, art, and philosophy. Twiki acts as a means of mobility for Dr. Theopolis, so that the important administrator need not rely on a human for such (which might open the possibility of such a human holding the robot's mobility as a means of leverage against him, a distinct possibility considering humanity's past). Coming in the immediate wake of franchises such as *Lost in Space* and *Star Wars* and changing the dynamic on how robots were presented, Dr. Theopolis and Twiki provide audiences with a hopeful view of humanity's future relationship with robots.

Richard A. Hall

See also: Al/Ziggy, Baymax, BB-8, C-3PO, Cambot/Gypsy/Tom Servo/Crow, Cerebro/ Cerebra, Control, D-O, HAL 9000, HERBIE, Janet, JARVIS/Friday, K9, R2-D2, Robot, Skynet, WOPR, Zordon/Alpha-5; *Thematic Essay:* Heroic Robots and Their Impact on Sci-Fi Narratives.

Further Reading

Ambrosino, Brandon. 2018. "What Would it Mean for AI to Have a Soul?" BBC, June 17, 2018. https://www.bbc.com/future/article/20180615-can-artificial-intelligence-have-a-soul-and-religion.

Ashby, LeRoy. 2006. *With Amusement for All: A History of American Popular Culture since 1830*. Lexington: University Press of Kentucky.

Bastani, Aaron. 2019. *Fully Automated Luxury Communism*. New York: Verso.

Byrne, Emma. 2013. "Innovation Isn't Safe: The Future According to Kevin Warwick." *Forbes*, September 30, 2013. https://www.forbes.com/sites/netapp/2013/09/30/kevin-warwick-captain-cyborg/#48b704c13560.

Carper, Steve. 2019. *Robots in American Popular Culture*. Jefferson, NC: McFarland.

Castleman, Harry, and Podrazik, Walter J. 2016. *Watching TV: Eight Decades of American Television, Third Edition*. Syracuse, NY: Syracuse University Press.

Chanthadavong, Aimee. 2019. "AI and Ethics: The Debate That Needs to Be Had." ZDnet, September 16, 2019. https://www.zdnet.com/article/ai-and-ethics-the-debate-that-needs-to-be-had/#ftag=CAD-03-10abf5f.

Choi, Charles Q. 2009. "Human Evolution: The Origin of Tool Use." LiveScience, November 11, 2009. https://www.livescience.com/7968-human-evolution-origin-tool.html.

Ford, Martin. 2015. *Rise of the Robots: Technology and the Threat of a Jobless Future*. New York: Basic.

Lin, Patrick, Abney, Keith, and Bekey, George A., eds. 2014. *Robot Ethics: The Ethical and Social Implications of Robotics*. Cambridge: Massachusetts Institute of Technology.

Lin, Patrick, Jenkins, Ryan, and Abney, Keith, eds. 2017. *Robot Ethics 2.0: From Autonomous Cars to Artificial Intelligence*. New York: Oxford University Press.

Rouhiainen, Lasse. 2018. *Artificial Intelligence: 101 Things You Must Know Today about Our Future*. Scotts Valley, CA: CreateSpace.

Seed, David. 1999. *American Science Fiction and the Cold War: Literature and Film*. Abingdon, UK: Routledge.

Spaeth, Dennis. 2018. "From Single-Task Machines to Backflipping Robots: The Evolution of Robots." *Cutting Tool Engineering*, January 15, 2018. https://www.ctemag.com/news/articles/evolution-of-robots.

Starck, Kathleen, ed. 2010. *Between Fear and Freedom: Cultural Representations of the Cold War*. Newcastle, UK: Cambridge Scholars.

Wallach, Wendell, and Allen, Colin. 2009. *Moral Machines: Teaching Robots Right from Wrong*. New York: Oxford University Press.

E

Echo/CT-1409

Star Wars

In 1977, writer-director George Lucas (b. 1944) introduced the world to the sci-fi/fantasy phenomenon *Star Wars* (Lucasfilm/20th Century Fox, 1977), one of the most successful pop culture franchises in American history. His original trilogy of films (1977–1983) focused on the rise of its protagonist, Luke Skywalker (played by Mark Hamill, b. 1951), against the backdrop of a Galactic Civil War against the human forces of the evil Empire, led by its Emperor, Sheev Palpatine (played by Ian McDiarmid, b. 1944), and his henchman, Darth Vader (played by David Prowse, b. 1935; voiced by James Earl Jones, b. 1931). The enduring popularity of the franchise led Lucas to produce a "prequel" trilogy (1999–2005) telling the tale of how the young hero Anakin Skywalker (played first by Jake Lloyd, b. 1989; then by Hayden Christensen, b. 1981) fell to the dark side to become Darth Vader. In 2008, Lucasfilm launched the animated series *Star Wars: The Clone Wars* (Cartoon Network, 2008–2013; Netflix, 2014; Disney+, 2020), covering the seminal background event of the prequel trilogy.

Throughout the Clone Wars, the galaxy far, far away experiences a civil war, with the planets of the Confederacy of Independent Systems and their droid armies fighting against the Galactic Republic and their armies of clones. The identical clones were produced from the DNA of the bounty hunter Jango Fett (played by Temuera Morrison, b. 1960) and programmed from birth to be elite military forces. Though the clones are physically identical, each eventually develops its own distinct personality. One such clone was CT-1409, better known as "Echo." Echo (voiced by Dee Bradley Baker, b. 1962) was believed killed during a mission (Matt Michnovetz, "Counter Attack," *The Clone Wars*, season 3, episode 19, March 4, 2011). However, Echo's commander, Captain Rex (also voiced by Baker) refused to believe Echo dead. His beliefs were proven sound when the Republic discovers that the enemy is using an algorithm of Captain Rex's own strategies (Michnovetz and Brent Friedman, "The Bad Batch," *The Clone Wars*, season 7, episode 1, February 21, 2020).

Their discovery leads them to a hidden signal that proves to be the long-presumed-dead clone, Echo. Rex leads a mission to follow the signal to its source, discovering an emaciated and mechanically altered Rex plugged into a machine, broadcasting the algorithm to all Confederacy forces. The Republic is uncertain whether Echo can now be trusted, being as much machine as man. With Rex and his team pinned down, Echo utilizes his mechanical enhancements to assist not only in Rex's team getting away but also in a major Republic victory over the droid armies. No longer fitting in with the rest of his clone brothers, Echo chooses to join an elite squadron of mistakenly mutated clones who call themselves the "Bad Batch" (Michnovetz and

Friedman, "Unfinished Business," *The Clone Wars*, season 7, episode 4, March 13, 2020). Echo, then, becomes one of very few cyborg characters in *Star Wars*.

Unlike Darth Vader, who became a cyborg in order to save his life, or General Grievous, who chose cybernetic implants to improve his soldiering abilities, Echo's transformation from clone to cyborg was done against his will, forcibly keeping his body alive to serve the enemy he was sworn (and programmed) to fight. In the end, however, no amount of inorganic enhancement could override Echo's deep dedication to his brothers-in-arms or to the Republic that he was sworn to serve. He is an example of human overcoming machine, of the human spirit overriding computer programming. Despite his alterations, Echo continues to serve as the soldier he was designed to be.

Richard A. Hall

See also: Adam, Alita, Battle Droids, BB-8, Borg, C-3PO, Cybermen, Cyborg, Daleks, Darth Vader, D-O, Doctor Octopus, EV-9D9, General Grievous, IG-88/IG-11, Inspector Gadget, Jaime Sommers, K-2SO, L3-37/*Millennium Falcon*, Medical Droids, Nardole, R2-D2, Steve Austin; *Thematic Essays:* Robots and Slavery, Cyborgs: Robotic Humans or Organic Robots?

Further Reading

Albert, Robert S., and Brigante, Thomas R. 1962. "The Psychology of Friendship Relations: Social Factors." *Journal of Social Psychology* 56 (1). https://www.tandfon line.com/doi/abs/10.1080/00224545.1962.9919371.

Allan, Kathryn, ed. 2013. *Disability in Science Fiction*. New York: Palgrave Macmillan.

Amati, Viviana, Meggiolaro, Silvia, Rivellini, Giulia, and Zaccarin, Susanna. 2018. "Social Relations and Life Satisfaction: The Role of Friends." *Genus* 74, no. 1 (May 4). https://www.ncbi.nlm.nih.gov/pmc/articles/PMC5937874.

Bryant, D'Orsay D., III. 1985. "Spare-Part Surgery: The Ethics of Organ Transplantation." *Journal of the National Medical Association* 77 (2). https://www.ncbi.nlm .nih.gov/pmc/articles/PMC2561842/pdf/jnma00245-0055.pdf.

Byrne, Emma. 2014. "Cybernetic Implants: No Longer Science Fiction." *Forbes*, March 11, 2014. https://www.forbes.com/sites/netapp/2014/03/11/cybernetic-implants-not -sci-fi/#7399c57e77ba.

Calvert, Bronwen. 2017. *Being Bionic: The World of TV Cyborgs*. London: I. B. Tauris.

Eberl, Jason T., and Decker, Kevin S., eds. 2015. *The Ultimate* Star Wars *and Philosophy: You Must Unlearn What You Have Learned*. Hoboken, NJ: Wiley-Blackwell.

Ford, Martin. 2015. *Rise of the Robots: Technology and the Threat of a Jobless Future*. New York: Basic.

Fryer-Biggs, Zachary. 2019. "Coming Soon to a Battlefield: Robots that Can Kill." *The Atlantic*, September 3, 2019. https://www.theatlantic.com/technology/archive/2019/09/killer -robots-and-new-era-machine-driven-warfare/597130.

Hampton, Gregory Jerome. 2017. *Imagining Slaves and Robots in Literature, Film, and Popular Culture: Reinventing Yesterday's Slave with Tomorrow's Robot*. Lanham, MD: Lexington.

Kaminski, Michael. 2008. *The Secret History of* Star Wars*: The Art of Storytelling and the Making of a Modern Epic*. Kingston, Canada: Legacy Books.

Pellissier, Hank. 2013. "Robots and Slavery: What Do Humans Want When We Are 'Masters'?" Institute for Ethics and Emerging Technologies, September 13, 2013. https:// ieet.org/index.php/IEET2/more/pellissier20130913.

Reagin, Nancy R., and Liedl, Janice, eds. 2013. Star Wars *and History*. New York: John Wiley & Sons.

Sunstein, Cass R. 2016. *The World According to* Star Wars. New York: Dey Street.

Sweet, Derek R. 2015. Star Wars *in the Public Square: The Clone Wars as Political Dialogue*. Critical Explorations in Science Fiction and Fantasy series, edited by Donald E. Palumbo and Michael Sullivan. Jefferson, NC: McFarland.

Taylor, Chris. 2015. *How* Star Wars *Conquered the Universe: The Past, Present, and Future of a Multibillion Dollar Franchise*. New York: Basic Books.

ED-209

RoboCop

In 1987, director Paul Verhoeven (b. 1938) released his film *RoboCop*. In near-future Detroit, Michigan, the city is on the verge of financial collapse. As such, the city privatizes its police force to the company Omni Consumer Products, which has developed the technology to produce cybernetic policemen. When Officer Alex Murphy (played by Peter Weller, b. 1947) is critically wounded in the line of duty, OCP utilizes what remains of him to create "RoboCop." The new cyborg policeman possesses three primary objectives: protecting the innocent, preserving public trust in law enforcement, and upholding the law. Over time, Murphy's original brain patterns emerge, making RoboCop sentient. When he discovers the true criminal intentions of OCP, he sets out to bring those in charge to justice. His attempt, however, ignites a hidden fourth program: automatic shutdown if attempting to arrest OCP personnel. To prevent RoboCop from exposing them, OCP releases ED-209, a fully automated "RoboCop" (Edward Neumeier and Michael Miner, *RoboCop*, Orion, 1987).

ED-209 stands for "Enforcement Droid, series 209"; it was the initial prototype for OCP's privatization of Detroit's police force. However, on its trial run, ED-209 immediately and brutally killed the OCP representative in charge of the exposition. This failure led to OCP moving forward with the RoboCop program. Once Murphy/RoboCop discovers the company's actual criminal intentions, the ED-209 is sent to stop him. RoboCop, in possession of the human ability to think in addition to his mechanical enhancements, easily overtakes the mindless automaton. What ED-209—and indeed the entire film—underscores is the frightening concept of placing policing duties in the mechanical hands of mindless robots, particularly robots designed by and under the control of a private corporation rather than duly elected representatives of the people. It is a mindless attack dog, with no ability to make the types of immediate decisions required by police in the field.

Richard A. Hall

See also: B.A.T.s, Battle Droids, Cylons, Doombots, EV-9D9, Fembots, Iron Legion, K-2SO, Metalhead, RoboCop, Robots, Sentinels, Terminators; *Thematic Essays:* Robots and Slavery, Villainous Robots and Their Impact on Sci-Fi Narratives.

Further Reading

Ashby, LeRoy. 2006. *With Amusement for All: A History of American Popular Culture since 1830*. Lexington: University Press of Kentucky.

Bastani, Aaron. 2019. *Fully Automated Luxury Communism*. New York: Verso.

Byrne, Emma. 2013. "Innovation Isn't Safe: The Future According to Kevin Warwick." *Forbes*, September 30, 2013. https://www.forbes.com/sites/netapp/2013/09/30/kevin-warwick-captain-cyborg/#48b704c13560.

Carper, Steve. 2019. *Robots in American Popular Culture*. Jefferson, NC: McFarland.

Chandler, Simon. 2019. "Artificial Intelligence Has Become a Tool for Classifying and Ranking People." *Forbes*, October 1, 2019. https://www.forbes.com/sites/simonchandler/2019/10/01/artificial-intelligence-has-become-a-tool-for-classifying-and-ranking-people/#7431657f1d7c.

Chanthadavong, Aimee. 2019. "AI and Ethics: The Debate That Needs to Be Had." ZDnet, September 16, 2019. https://www.zdnet.com/article/ai-and-ethics-the-debate-that-needs-to-be-had/#ftag=CAD-03-10abf5f.

Choi, Charles Q. 2009. "Human Evolution: The Origin of Tool Use." LiveScience, November 11, 2009. https://www.livescience.com/7968-human-evolution-origin-tool.html.

Dinh, Thien-Nam. 2018. *Silicon Minds: The Science, Impact, and Promise of Artificial Intelligence*. Independently Published.

Dowden, Bradley. 1993. *Logical Reasoning*. Belmont, CA: Wadsworth Publishing Company.

Ford, Martin. 2015. *Rise of the Robots: Technology and the Threat of a Jobless Future*. New York: Basic.

Fryer-Biggs, Zachary. 2019. "Coming Soon to a Battlefield: Robots That Can Kill." *The Atlantic*, September 3, 2019. https://www.theatlantic.com/technology/archive/2019/09/killer-robots-and-new-era-machine-driven-warfare/597130/

Hampton, Gregory Jerome. 2017. *Imagining Slaves and Robots in Literature, Film, and Popular Culture: Reinventing Yesterday's Slave with Tomorrow's Robot*. Lanham, MD: Lexington.

Krishnan, Armin. 2009. *Killer Robots: Legality and Ethicality of Autonomous Weapons*. London: Routledge.

Lin, Patrick, Abney, Keith, and Bekey, George A., eds. 2014. *Robot Ethics: The Ethical and Social Implications of Robotics*. Cambridge: Massachusetts Institute of Technology.

Murdock, Jason. 2019. "Former Google Engineer Warns AI Might Accidentally Start a War: 'These Things Will Start to Behave in Unexpected Ways.'" *Newsweek*, September 16, 2019. https://www.newsweek.com/google-project-maven-artificial-intelligence-laura-nolan-killer-robots-department-defense-1459358.

Pellissier, Hank. 2013. "Robots and Slavery: What Do Humans Want When We Are 'Masters'?" Institute for Ethics and Emerging Technologies, September 13, 2013. https://ieet.org/index.php/IEET2/more/pellissier20130913.

Rouhiainen, Lasse. 2018. *Artificial Intelligence: 101 Things You Must Know Today about Our Future*. Scotts Valley, CA: CreateSpace.

Singer, Peter W. 2009. "Isaac Asimov's Laws of Robotics Are Wrong." Brookings Institute, May 18, 2009. https://www.brookings.edu/opinions/isaac-asimovs-laws-of-robotics-are-wrong.

Spaeth, Dennis. 2018. "From Single-Task Machines to Backflipping Robots: The Evolution of Robots." *Cutting Tool Engineering*, January 15, 2018. https://www.ctemag.com/news/articles/evolution-of-robots.

Wallach, Wendell, and Allen, Colin. 2009. *Moral Machines: Teaching Robots Right from Wrong*. New York: Oxford University Press.

EV-9D9

Star Wars

In 1977, writer-director George Lucas (b. 1944) introduced the world to the sci-fi/fantasy phenomenon *Star Wars* (Lucasfilm/20th Century Fox, 1977), one of the most successful pop culture franchises in American history. His original trilogy of films (1977–1983) focused on the rise of its protagonist, Luke Skywalker (played by Mark Hamill, b. 1951), against the backdrop of a Galactic Civil War against the human forces of the evil Empire, led by its Emperor, Sheev Palpatine (played by Ian McDiarmid, b. 1944), and his henchman, Darth Vader (played by David Prowse, b. 1935; voiced by James Earl Jones, b. 1931). The final film of the trilogy saw the heroes' attempt to rescue their friend, Han Solo (played by Harrison Ford, b. 1942), from the crime lord Jabba the Hutt. Part of this rescue attempt involved embedding their two droids—C-3PO (played by Anthony Daniels, b. 1946) and R2-D2 (played by Kenny Baker, 1934–2016)—within Jabba's lair under the ruse that they were "gifts" to foster negotiations for Solo's release. Once accepted, the droids are sent to the droid overseer, EV-9D9.

EV-9D9 was a standard "supervisor" droid, responsible for assigning the droid slaves of Jabba to specific duties. EV is the first *Star Wars* droid to be given a "feminine" voice. She orders R2 to serve as a roving bartender on Jabba's sail barge and appoints 3PO as Jabba's new interpreter (as Jabba had apparently disintegrated the last protocol droid for interpreting something Jabba did not like

Animatronics

Animatronics are, essentially, mechanized "puppets," programmed utilizing "mechatronic engineering," or the combination of mechanics and electronics. They are most commonly found in the entertainment industry, specifically in the areas of theme parks, movie/television production, and toys. The term was coined by Walt Disney (1901–1966) in 1961, as his team of "imagineers" began working on animatronic characters for his theme park, Disneyland. The earliest Disney animatrons were animals, specifically birds, which were incorporated into the 1964 Disney film, *Mary Poppins*. That same year, Disneyland debuted an animatron of President Abraham Lincoln, which would give a speech to onlookers. Soon such animatronic puppets were utilized in several Disney attractions, most notably the rides Pirates of the Caribbean and It's a Small World. Soon afterward, Hollywood studios began using animatronic technology for science fiction films. Beginning in 1977, children's themed pizza restaurants Chuck E. Cheese's and ShowBiz began using animatronic characters to entertain their patrons. In 1983, musical artist Herbie Hancock (b. 1940) utilized animatronics for his music video and stage performances for his hit song "Rockit." In 1985, the Hasbro Toy Company introduced "Teddy Ruxpin," an animatronic teddy bear that could "tell stories" to children by way of its animatronic mouth and eyes moving along with an audio story playing via a cassette player in the toy's back. The largest scale example of animatronics was Industrial Light and Magic's Tyrannosaurus Rex for the 1993 film *Jurassic Park*. Since the mid-1990s, Hollywood and the toy industry have switched to more computer-based technology, but major theme parks such as Disneyland/Walt Disney World and Universal Studios continue to make regular use of animatronic technology.

Richard A. Hall

hearing). Ironically, EV's choices perfectly place the two droids for their secret missions (though 3PO is unaware that he is, in fact, on a mission). Near EV is her assistant, 8D8, a "torture" droid. As 3PO is taken away and EV threatens to teach R2 proper respect, 8D8 is shown off to the side torturing a Power Droid by applying hot metal coils to the droid's feet. The Power Droid's yelps of pain and R2's own expression of fear underscore the idea that droids in *Star Wars* can "feel."

EV-9D9 plays a very brief role in the overall *Star Wars* saga but plays an important role in further informing the audience of the life of droids in this galaxy far, far away. Droids are slaves. They are property. Although EV possesses some authority among Jabba's droids, she, too, is his property. EV also shows the audience that not all droids are "good." To this point, audiences had really not experienced "bad guy" droids. EV's malevolence shows through her tone with the heroic droid duo. She relishes her power over her fellow droids. She feels no sympathy for them, despite her own status as a droid. She has found her niche in droid/slave society, and just as 3PO and R2 exhibit "good human" qualities, EV-9D9 equally exhibits the "worst of humanity."

Richard A. Hall

See also: Battle Droids, BB-8, C-3PO, Cambot/Gypsy/Tom Servo/Crow, Darth Vader, D-O, Doombots, Echo/CT-1409, ED-209, General Grievous, IG-88/IG-11, K-2SO, L3-37/*Millennium Falcon*, Medical Droids, R2-D2; *Thematic Essays:* Robots and Slavery, Villainous Robots and Their Impact on Sci-Fi Narratives.

Further Reading

Ashby, LeRoy. 2006. *With Amusement for All: A History of American Popular Culture since 1830.* Lexington: University Press of Kentucky.

Carper, Steve. 2019. *Robots in American Popular Culture.* Jefferson, NC: McFarland.

Eberl, Jason T., and Decker, Kevin S., eds. 2015. *The Ultimate* Star Wars *and Philosophy: You Must Unlearn What You Have Learned.* Hoboken, NJ: Wiley-Blackwell.

Hampton, Gregory Jerome. 2017. *Imagining Slaves and Robots in Literature, Film, and Popular Culture: Reinventing Yesterday's Slave with Tomorrow's Robot.* Lanham, MD: Lexington.

Kaminski, Michael. 2008. *The Secret History of* Star Wars: *The Art of Storytelling and the Making of a Modern Epic.* Kingston, Canada: Legacy Books.

Lin, Patrick, Abney, Keith, and Bekey, George A., eds. 2014. *Robot Ethics: The Ethical and Social Implications of Robotics.* Cambridge: Massachusetts Institute of Technology.

Reagin, Nancy R., and Liedl, Janice, eds. 2013. Star Wars *and History.* New York: John Wiley & Sons.

Sunstein, Cass R. 2016. *The World According to* Star Wars. New York: Dey Street.

Taylor, Chris. 2015. *How* Star Wars *Conquered the Universe: The Past, Present, and Future of a Multibillion Dollar Franchise.* New York: Basic Books.

Wallach, Wendell, and Allen, Colin. 2009. *Moral Machines: Teaching Robots Right from Wrong.* New York: Oxford University Press.

F

Fembots

Austin Powers: International Man of Mystery

In 1997, comedian Mike Meyers (b. 1963) and director Jay Roach (b. 1957) introduced the world to the ultimate parody of 1960s spy films: the *Austin Powers* franchise. Meyers played Austin Powers, a British superspy in the 1960s, placed in suspended animation in order to protect the world in the future from the possible return of the archvillain, Dr. Evil (also played by Meyers). Revived in 1997, the first film, *Austin Powers: International Man of Mystery*, worked as both a parody/ homage to '60s spy cinema and a traditional "fish-out-of-water" story, placing the "hip," swinging, hypersexual Powers in the world of the more politically correct 1990s. Meanwhile, Dr. Evil's cadre of minions consist of several stereotypes of 1960s spy film villains. One of his weapons, however, was a team of "Fembots": deadly android assassins built to look like beautiful blond 1960s go-go dancers. Their primary weapon was a pair of retractable machine guns protruding from their breasts (as well as enhanced robot strength).

Though only appearing in one scene in the first film, the three main Fembots (Cheryl Bartel, 1971–2010; Cindy Margolis, b. 1965); and Donna W. Scott (birth year unknown) were a major part of the marketing of the film in 1997. When Powers comes close to reaching and stopping Dr. Evil, the Fembots are deployed to divert, seduce, and kill Powers to prevent his interruption of Dr. Evil's plan. In the film's sequel, *Austin Powers: The Spy Who Shagged Me* (1999), Powers's partner-turned-wife from the first film, Agent Vanessa Kensington (played by Elizabeth Hurley, b. 1965) is shown to have also been a Fembot all along, attempting to kill Powers on their honeymoon in the opening scene from the film. Other Fembots from the series include models Barbara Moore (b. 1968) and Cynthia LaMontagne (b. 1966), both playing Fembot versions of themselves.

In essence, Fembots represent the oversexualization of women (often no more than sex objects for the hero spy or head villain) in traditional spy cinema. In hindsight, they have become a prescient prophecy of things to come. Throughout the 2010s, Japanese engineers have been working diligently on "sex robots," primarily beautiful female robots with the ability to respond to human interaction. At a point in history in which, more than ever, women globally are standing firm against being used as sexual objects by men, men are working to create artificial "women" who have no choice but to succumb to the whims of their male owners. As such, while a comedic aspect to the original film, the warning they provide for the misuse of artificial intelligence should be seriously considered.

Richard A. Hall

Chatbots

A "chatbot" is a computer software program that allows an AI to have online interactions in the same way as a real person does. The first such chatbot was Verbot (or Verbal-Robot) created by Michael Mauldin (b. 1959) in 1994. In 2016, Microsoft launched Tay (or Thinking about You) as a chatbot on Twitter. The bot was given the personality of a young female, its AI program meant to "grow" and "learn" from its online interactions. What Tay's designers did not take into account was the degree of hate and vitriol on social media. In a very short period of time, Tay "learned" to become hateful: racist, sexist, homophobic, Islamophobic, and more. Tay became so corrupted by her Twitter interactions, that Microsoft was forced to take her offline after only nine months. By that time, Microsoft was ready with its next generation chatbot: Zo (an English-speaking version of its successful Asian chatbots, Rinna and Xiaoice, launched in Japan and China, respectively). Aside from Twitter, Zo was also available on the platforms Kik and Facebook Messenger. Like Tay, Zo soon learned to hate through its online interactions. Additionally, however, Zo also began to promote its own original program base (Windows 7) to the company's latest upgrade (Windows 10). All of this spurred Microsoft to take Zo offline in 2019. The primary lesson taken from the examples of Tay and Zo is the danger of allowing AIs learn from human behavior. The end result of a potentially invulnerable AI ingrained with human failings is a real-world Frankenstein's monster in the making.

Richard A. Hall

See also: Alita, Androids, Buffybot, Dolores, EV-9D9, Jaime Sommers, Janet, Jocasta, L3-37/*Millennium Falcon*, LMDs, Maschinenmensch/Maria, Rosie, Soji and Dahj Asha, Stepford Wives, VICI; *Thematic Essays:* Robots and Slavery, Villainous Robots and Their Impact on Sci-Fi Narratives.

Further Reading

Ambrosino, Brandon. 2018. "What Would It Mean for AI to Have a Soul?" BBC, June 17, 2018. https://www.bbc.com/future/article/20180615-can-artificial-intelligence-have-a-soul-and-religion.

Ashby, LeRoy. 2006. *With Amusement for All: A History of American Popular Culture since 1830*. Lexington: University Press of Kentucky.

Bastani, Aaron. 2019. *Fully Automated Luxury Communism*. New York: Verso.

Blair, Anthony. 2018. "Sex Robots That Feel LOVE and Suffer PAIN When Dumped Coming Soon Claims Expert." *Daily Star*, December 6, 2018. https://www.dailystar.co.uk/news/latest-news/sex-robot-feel-love-pain-16824160.

Chandler, Simon. 2019. "Artificial Intelligence Has Become a Tool for Classifying and Ranking People." *Forbes*, October 1, 2019. https://www.forbes.com/sites/simonchandler/2019/10/01/artificial-intelligence-has-become-a-tool-for-classifying-and-ranking-people/#7431657f1d7c.

Chanthadavong, Aimee. 2019. "AI and Ethics: The Debate That Needs to Be Had." ZDnet, September 16, 2019. https://www.zdnet.com/article/ai-and-ethics-the-debate-that-needs-to-be-had/#ftag=CAD-03-10abf5f.

Conrad, Dean. 2018. *Space Sirens, Scientists and Princesses: The Portrayal of Women in Science Fiction Cinema*. Jefferson, NC: McFarland.

Dinh, Thien-Nam. 2018. *Silicon Minds: The Science, Impact, and Promise of Artificial Intelligence*. Independently Published.

Hampton, Gregory Jerome. 2017. *Imagining Slaves and Robots in Literature, Film, and Popular Culture: Reinventing Yesterday's Slave with Tomorrow's Robot*. Lanham, MD: Lexington.

Hicks, Amber. 2019. "Sex Robot Brothel Opens in Japan amid Surge of Men Wanting Bisexual Threesomes." *Mirror*, April 27, 2019. https://www.mirror.co.uk/news /weird-news/sex-robot-brothel-opens-japan-14792161.

Krishnan, Armin. 2009. *Killer Robots: Legality and Ethicality of Autonomous Weapons*. London: Routledge.

Lin, Patrick, Abney, Keith, and Bekey, George A., eds. 2014. *Robot Ethics: The Ethical and Social Implications of Robotics*. Cambridge: Massachusetts Institute of Technology.

Murdock, Jason. 2019. "Former Google Engineer Warns AI Might Accidentally Start a War: 'These Things Will Start to Behave in Unexpected Ways.'" *Newsweek*, September 16, 2019. https://www.newsweek.com/google-project-maven-artificial-intel ligence-laura-nolan-killer-robots-department-defense-1459358.

Pellissier, Hank. 2013. "Robots and Slavery: What Do Humans Want When We Are 'Masters'?" Institute for Ethics and Emerging Technologies, September 13, 2013. https://ieet.org/index.php/IEET2/more/pellissier20130913.

Seed, David. 1999. *American Science Fiction and the Cold War: Literature and Film*. Abingdon, UK: Routledge.

Singer, Peter W. 2009. "Isaac Asimov's Laws of Robotics Are Wrong." Brookings Institute, May 18, 2009. https://www.brookings.edu/opinions/isaac-asimovs-laws-of-robotics -are-wrong.

Wallach, Wendell, and Allen, Colin. 2009. *Moral Machines: Teaching Robots Right from Wrong*. New York: Oxford University Press.

G

General Grievous

Star Wars

In 1977, writer-director George Lucas (b. 1944) introduced the world to the sci-fi/fantasy phenomenon *Star Wars* (Lucasfilm/20th Century Fox, 1977), one of the most successful pop culture franchises in American history. His original trilogy of films (1977–1983) focused on the rise of its protagonist, Luke Skywalker (played by Mark Hamill, b. 1951), against the backdrop of a Galactic Civil War against the human forces of the evil Empire, led by its Emperor, Sheev Palpatine (played by Ian McDiarmid, b. 1944), and his henchman, Darth Vader (played by David Prowse, b. 1935; voiced by James Earl Jones, b. 1931). In 1999, Lucas launched a "prequel" trilogy, beginning with *Star Wars, Episode I: The Phantom Menace* (Lucasfilm, 1999), which began the story of the fall of Anakin Skywalker (played originally by Jake Lloyd, b. 1989; later by Hayden Christensen, b. 1981)—the father of the original hero—into becoming the villainous Darth Vader. In the final film of the prequel trilogy, *Star Wars, Episode III: Revenge of the Sith* (Lucasfilm, 2005), theatergoers were introduced to General Grievous (voiced by Matthew Wood, b. 1972), commander of the droid armies of the Confederacy of Independent Systems (CIS), a nightmarish cyborg precursor to Anakin Skywalker's ultimate fate.

Grievous first appeared in the final episodes of the animated "microseries" *Star Wars: Clone Wars* (Cartoon Network, 2003–2005). This skeletal droid frame encasing organic eyes, brain, heart, and lungs were the ultimate weapon against the Jedi Knights: four arms, each bearing a light-saber, trained in the Jedi arts by CIS leader (and former Jedi Knight) Count Dooku (voiced in the animated series by Corey Burton, b. 1955). In the last episode of the microseries (leading up to the opening moments of *Revenge of the Sith*), Grievous kidnaps Republic chancellor Palpatine in order to force the Republic's surrender in the civil war (Genndy Tartakovsky, "Chapter 25," *Star Wars: Clone Wars*, season 3, episode 5, March 25, 2005). In the film, after assisting in Palpatine's rescue, Jedi Master Obi-Wan Kenobi (played by Ewan McGregor, b. 1971) hunts down the evil cyborg and kills him with a simple laser blaster (Lucas, *Revenge of the Sith*, 2005).

When Disney purchased Lucasfilm and the *Star Wars* franchise in 2012, it "deleted" all previous novels, comic books, and video games from "official canon"; but the only origin story for General Grievous to date came from the 2005 comic book *Star Wars: Visionaries*. In the story titled "The Eyes of Revolution," by Warren J. Fu, Grievous was originally a fully organic being from the planet Kalee, a cunning and powerful warrior general named Qymaen jai Sheelal. Seeing Sheelal's potential, the evil Sith Lord Count Dooku orchestrates a horrible "accident," leaving Sheelal all but dead. Saving only his eyes, brain, heart, and lungs, Dooku implants them in a skeletal droid body with retractable third and fourth arms. He then trains the newly minted "Grievous" to be the ultimate weapon against the Jedi Knights and commander of the droid armies (Fu, "The Eyes of Revolution," 2005).

Grievous has become a popular villain in the *Star Wars* franchise, appearing not only in *Revenge of the Sith* and *Clone Wars* but also in comic books, video games, and the CGI television series *The Clone Wars* (Cartoon Network, 2008–2013; Netflix, 2014; Disney+, 2020). Within the larger context of his primary appearance (*Revenge of the Sith*), Grievous presented a horrifying window into the fate of Jedi hero Anakin Skywalker. Once Skywalker falls to the Dark Side, taking the new name "Darth Vader," he, too, falls in combat against his onetime friend and mentor, Kenobi. Losing his left arm and both legs in battle with Kenobi (his right arm had previously been lost in battle with Dooku), his flesh burned and lungs damaged beyond repair by lava, Skywalker/Vader is "saved" by his Sith master Sidious/Emperor Palpatine, who replaces Vader's limbs with cybernetic ones (and a helmet/breathing apparatus). Vader, then, becomes a cyborg killing machine like Grievous. Grievous is a prime example of losing one's "humanity" to cybernetic implants and, with that, one's very soul.

Richard A. Hall

See also: Adam, Alita, Battle Droids, BB-8, Borg, C-3PO, Cybermen, Cyborg, Daleks, Darth Vader, D-O, Doctor Octopus, Echo/CT-1409, EV-9D9, IG-88/IG-11, Inspector Gadget, Jaime Sommers, K-2SO, L3-37/*Millennium Falcon*, Medical Droids, Nardole, R2-D2, RoboCop, Steve Austin; *Thematic Essays:* Robots and Slavery, Villainous Robots and Their Impact on Sci-Fi Narratives, Cyborgs: Robotic Humans or Organic Robots?

Further Reading

Bray, Adam, Barr, Tricia, Horton, Cole, and Windham, Ryder. 2019. *Ultimate* Star Wars, *New Edition.* London: DK Publishing.

Bray, Adam, and Horton, Cole. 2017. Star Wars: *Absolutely Everything You Need to Know, Updated and Expanded.* London: DK Children.

Bryant, D'Orsay D, III. 1985. "Spare-Part Surgery: The Ethics of Organ Transplantation." *Journal of the National Medical Association* 77 (2). https://www.ncbi.nlm.nih .gov/pmc/articles/PMC2561842/pdf/jnma00245-0055.pdf.

Byrne, Emma. 2014. "Cybernetic Implants: No Longer Science Fiction." *Forbes*, March 11, 2014. https://www.forbes.com/sites/netapp/2014/03/11/cybernetic-implants-not -sci-fi/#7399c57e77ba.

Calvert, Bronwen. 2017. *Being Bionic: The World of TV Cyborgs.* London: I. B. Tauris.

Carper, Steve. 2019. *Robots in American Popular Culture.* Jefferson, NC: McFarland.

Eberl, Jason T., and Decker, Kevin S., eds. 2015. *The Ultimate* Star Wars *and Philosophy: You Must Unlearn What You Have Learned.* Hoboken, NJ: Wiley-Blackwell.

Hitchcock, Susan Tyler. 2007. *Frankenstein: A Cultural History.* New York: W. W. Norton.

Kaminski, Michael. 2008. *The Secret History of* Star Wars: *The Art of Storytelling and the Making of a Modern Epic.* Kingston, Canada: Legacy.

Reagin, Nancy R., Liedl, Janice, eds. 2013. Star Wars *and History.* New York: John Wiley & Sons.

Slavicsek, Bill. 2000. *A Guide to the* Star Wars *Universe.* San Francisco, CA: LucasBooks.

Sumerak, Marc. 2018. Star Wars: *Droidography.* New York: HarperFestival.

Sunstein, Cass R. 2016. *The World According to* Star Wars. New York: Dey Street.

Sweet, Derek R. 2015. Star Wars *in the Public Square: The Clone Wars as Political Dialogue*. Critical Explorations in Science Fiction and Fantasy series, edited by Donald E. Palumbo and Michael Sullivan. Jefferson, NC: McFarland.

Taylor, Chris. 2015. *How* Star Wars *Conquered the Universe: The Past, Present, and Future of a Multibillion Dollar Franchise*. New York: Basic Books.

Wallace, Daniel. 2006. Star Wars*: The New Essential Guide to Droids*. New York: Del Rey.

The Great Intelligence

Doctor Who

In 1963, BBC executive Sydney Newman (1917–1997) and producer Verity Lambert (1935–2007) created the British television series *Doctor Who* (BBC, 1963–1989; FOX, 1996; BBC, 2005–present), the longest running science fiction television series in history. The series centers on the character of the "Doctor," an alien "Time Lord" possessing the power to "regenerate"—or gain a new body and personality—when his or her (depending on the incarnation) physical body dies. The Doctor travels through time and space in a device known as a TARDIS (Time and Relative Dimension in Space). Though TARDISes are designed to change their outward appearance in order to blend in with the environment, the Doctor's TARDIS is "stuck" in the constant form of a 1960s British police call box. Throughout the long history of the series, the Doctor has faced many alien threats. One of the most dangerous is the mysterious entity known only as the "Great Intelligence."

The Great Intelligence was first introduced as an ethereal bodiless entity, traveling via the astral plane. In the 1967 storyline "The Abominable Snowmen," the Great Intelligence has taken over the physical body of an organic life-form. The following year, the entity returns in the storyline "The Web of Fear," inhabiting the body of an alien creature known as a "Yeti." The mechanical connection of this time-traveling intelligence comes in the 2012 Christmas special, *The Snowmen*. Having befriended a socially awkward child in mid-nineteenth-century England, the Great Intelligence has taken the form of "intelligent snow." The child then spends his entire life attempting to help the Great Intelligence take permanent bodily form, keeping the "snow" safe in a steampunk globe-like device until it can be transferred into the bodily form of an ice creature in order to enact its plan to destroy all human life on Earth (Steven Moffat, *Doctor Who: The Snowmen*, BBC, December 25, 2012).

Having been defeated by the Doctor for a third time, the Great Intelligence then moves on to twenty-first-century London, embedding itself in a global (free) Wi-Fi network. Having once again befriended a troubled child who grows up to help and protect it, the Wi-Fi network lures hapless internet users to log on, at which point the network "downloads" the human's consciousness, leaving their soulless bodies to eventually die. The Great Intelligence then utilizes whatever skills the human consciousnesses can provide to strengthen itself. With its human minion once more outsmarted by the Doctor, the Great Intelligence has no option other than to free those it has captured (the fate of those whose physical bodies have already died is unknown), and escape once again (Moffat, "The Bells of Saint John," *Doctor Who*, return series season 7, episode 6, March 30, 2013).

The most recent appearance of the Great Intelligence saw the creature enacting a revenge plan against the Doctor, locating the site of the Time Lord's "grave" on the distant planet of Trenzilore. There, the console of the long-dead TARDIS has been replaced with a column of electrical energy representing the "life" of the Doctor. The Great Intelligence, taking the visible form of its minion from *The Snowmen*, forces the Doctor to meet it there and grant it access to the TARDIS. Once inside, the Great Intelligence merges with the electrical column, thereby gaining access to the Doctor's entire time line, turning every victory into a defeat and torturing the Time Lord repeatedly throughout time. Through the cleverness of the Doctor (played by Matt Smith, b. 1982) and his human companion Clara (played by Jenna Coleman, b. 1986), the Great Intelligence is defeated yet again; this time, however, its consciousness is spread throughout time and space (Moffat, "The Name of The Doctor," *Doctor Who*, return series, season 7, episode 13, May 18, 2013).

Though the Great Intelligence appears to be (at the time of this writing) some form of "living" essence, its utilization of the internet gives him an AI perspective (whether AI is the basis of the Great Intelligence remains unknown, though the argument could be made that *if* the intelligence is a "living entity," its utilization of the internet could make it a "cyborg"). As an intelligent, internet-based creature, the Great Intelligence takes on the same potential as a Skynet or Matrix: a bodiless entity in computer-based form. As with those examples, the threat posed by the entity—that of an omnipotent, immortal overlord—is one that endangers the very existence of the human race on Earth. As the primary goal of the Great Intelligence appears to be continuing its already long existence in a physical form, making use of mechanical/computerized technology to do so is the most logical way to go about it. With AI becoming more and more a part of humanity's daily life, fans of *Doctor Who* can likely expect the return of this villain in some similar storyline in the years to come.

Richard A. Hall

See also: AI/Ziggy, Brainiac, Cerebro/Cerebra, Control, Cybermen, Daleks, HAL 9000, HERBIE, Janet, JARVIS/Friday, K9, Landru, The Matrix/Agent Smith, Nardole, OASIS, Skynet, TARDIS, Teselecta, V-GER, WOPR, Zordon/Alpha-5; *Thematic Essay:* AI and the Apocalypse: Science Fiction Meeting Science Fact.

Further Reading

Allen, Arthur. 2019. "There's a Reason We Don't Know Much about AI." Politico, September 16, 2019. https://www.politico.com/agenda/story/2019/09/16/artficial-intel ligence-study-data-000956?cid=apn.

Ambrosino, Brandon. 2018. "What Would It Mean for AI to Have a Soul?" BBC, June 17, 2018. https://www.bbc.com/future/article/20180615-can-artificial-intelligence-have -a-soul-and-religion.

Bastani, Aaron. 2019. *Fully Automated Luxury Communism*. New York: Verso.

Campbell, Mark. 2010. Doctor Who: *The Complete Guide*. London: Running Press.

Chandler, Simon. 2019. "Artificial Intelligence Has Become a Tool for Classifying and Ranking People." *Forbes*, October 1, 2019. https://www.forbes.com/sites/simonchandler

/2019/10/01/artificial-intelligence-has-become-a-tool-for-classifying-and-ranking-people/#7431657f1d7c.

Chanthadavong, Aimee. 2019. "AI and Ethics: The Debate That Needs to Be Had." ZDnet, September 16, 2019. https://www.zdnet.com/article/ai-and-ethics-the-debate-that-needs-to-be-had/#ftag=CAD-03-10abf5f.

Dinh, Thien-Nam. 2018. *Silicon Minds: The Science, Impact, and Promise of Artificial Intelligence*. Independently Published.

Dowden, Bradley. 1993. *Logical Reasoning*. Belmont, CA: Wadsworth Publishing Company.

Hills, Matt, ed. 2013. *New Dimensions of* Doctor Who: *Adventures in Space, Time and Television*. London: I. B. Tauris.

Kistler, Alan. 2013. Doctor Who: *Celebrating Fifty Years, A History*. Guilford, CT: Lyons.

Leetaru, Kalev. 2019. "Automatic Image Captioning and Why Not Every AI Problem Can Be Solved through More Data." *Forbes*, July 7, 2019. https://www.forbes.com/sites/kalevleetaru/2019/07/07/automatic-image-captioning-and-why-not-every-ai-problem-can-be-solved-through-more-data/#20b943476997.

Lewis, Courtland, and Smithka, Paula, eds. 2010. Doctor Who *and Philosophy: Bigger on the Inside*. Chicago: Open Court.

Lewis, Courtland, and Smithka, Paula, eds. 2015: *More* Doctor Who *and Philosophy: Regeneration Time*. Chicago: Open Court.

Lin, Patrick, Jenkins, Ryan, and Abney, Keith, eds. 2017. *Robot Ethics 2.0: From Autonomous Cars to Artificial Intelligence*. New York: Oxford University Press.

Marks, Robert J. 2020. "*2084* vs. *1984*: The Difference AI Could Make to Big Brother." Podcast interview with author John Lennox. Mind Matters, July 3, 2020. https://mindmatters.ai/2020/07/2084-vs-1984-the-difference-ai-could-make-to-big-brother.

Muir, John Kenneth. 2007. *A Critical History of Doctor Who on Television*. Jefferson, NC: McFarland.

Murdock, Jason. 2019. "Former Google Engineer Warns AI Might Accidentally Start a War: 'These Things Will Start to Behave in Unexpected Ways.'" *Newsweek*, September 16, 2019. https://www.newsweek.com/google-project-maven-artificial-intelligence-laura-nolan-killer-robots-department-defense-1459358.

Nahin, Paul J. 1999. *Time Machines: Time Travel in Physics, Metaphysics, and Science Fiction*. 2nd ed. Woodbury, NY: AIP Press; New York: Springer.

Rouhiainen, Lasse. 2018. *Artificial Intelligence: 101 Things You Must Know Today about Our Future*. Scotts Valley, CA: CreateSpace.

Virk, Rizwan. 2019. *The Simulation Hypothesis: An MIT Computer Scientist Shows Why AI, Quantum Physics, and Eastern Mystics All Agree We Are in a Video Game*. Milwaukee, WI: Bayview.

Wasserman, Ryan. 2018. *Paradoxes of Time Travel*. Oxford: Oxford University Press.

H

HAL 9000

2001: A Space Odyssey

In 1968, science fiction legend Arthur C. Clarke (1917–2008) published the first full novel in his iconic Space Odyssey series: *2001: A Space Odyssey*. Simultaneously, Clarke worked as an adviser on the film version of his novel, directed by equally legendary filmmaker Stanley Kubrick (1928–1999). This story centers primarily on the American space vessel *Discovery One*, commanded by astronauts Dave Bowman (played by Keir Dullea, b. 1936) and Frank Poole (played by Gary Lockwood, b. 1937). The rest of the *Discovery* crew are in suspended animation awaiting their arrival at their destination: the planet Jupiter. Unbeknown to the crew, the true mission is to seek out the recipient of a radio transmission sent by a mysterious black monolith from Earth's moon. Only the ship's onboard computer system, HAL 9000, is aware of the true mission, and it is programmed to keep the mission's true intent from the *Discovery* crew.

HAL 9000 (voiced in the film by Douglas Rain, 1928–2018) is a "Heuristically programmed ALgorithmic computer." It is programmed with a dual mission: to conceal the aforementioned true intent of the *Discovery One*'s trip to Jupiter and to assist and care for the ship's crew to ensure their safety and the success of their supposed mission. However, the unknown—but potentially quite real—danger of the "secret" mission threatens HAL's ability to properly perform his secondary mission. The illogic of the competing missions causes HAL to begin to malfunction, something Frank discovers when HAL begins making mistakes

The iconic red light signifying the communications node for the AI HAL 9000 from the classic sci-fi film *2001: A Space Odyssey* (1968). The program was voiced by Douglas Rain. (Gualtiero Boffi/Dreamstime.com)

Shimon

Shimon is a musical robot designed by the robotics team at Georgia Institute of Technology (a.k.a. Georgia Tech) under the leadership of Professor Gil Weinberg. It was originally programmed to play the percussion instrument called the marimba. In early 2020, the team announced that Shimon could now not only sing and dance but also write music and songs. Programmed with approximately fifty thousand songs from the genres of progressive rock, rhythm and blues, and rap/hip-hop, and under the guidance of grad student Richard Savery, Shimon developed the ability to produce original music and songs. In the spring of 2020, Dr. Weinberg announced that Shimon would be going on tour promoting an album of ten original songs. Shimon's "voice" was developed through the assistance of the robotics team at Pompeu Fabra University, in Barcelona, Spain. Shimon's performance abilities were developed under the guidance of grad students Ning Yang and Lisa Zahary, the latter focusing on "facial expressions" through Shimon's new "mouth" and "eyebrow" features (Georgia Tech Online News Center, "Shimon: Now a Singing, Songwriting Robot," February 25, 2020, https://www.news.gatech.edu/2020/02/25/shimon-now-singing-songwriting-robot). Shimon represents an amazing advance in robotics, going beyond performing simple labor tasks or mathematical functions to actually producing original artistic material. To do so, Shimon has developed the ability for original thought (or, at the very least, the initial seeds of such thought). As robotics continue to advance beyond the original idea of performing tasks for humans (from the very basic to the more challenging) to creating robots capable of creating art and "thinking" on their own, an ethical conundrum will advance as well: At what point will robots be considered their own form of "life?"

Richard A. Hall

while playing him in chess. As all of the ship's functions (including life support) are controlled by HAL, the astronauts soon conclude that they must shut HAL down. Aware of the humans' intentions, HAL—either out of self-defense or adhering strictly to its primary mission—decides to kill the humans. After successfully cutting Frank's life support while the astronaut is outside the ship, HAL attempts to do the same to Dave, but the remaining astronaut works his way back into the ship, successfully shutting down HAL's systems while also discovering the computer's secret mission (Clarke, *2001: A Space Odyssey*, 1968).

HAL's story represents yet another cautionary tale regarding trusting too much to artificial intelligence. Unlike humans, computers are unable to distinguish what humans call "gray areas" or moral relativism. If humans do not take this into account when programming computers, the unintended consequences could be devastating. Particularly as humankind begins to send astronauts into space to seriously pursue interplanetary exploration, staff dependency on computer systems working properly will be more important than ever—and just as potentially dangerous.

Richard A. Hall

See also: AI/Ziggy, Arnim Zola, Batcomputer, Brainiac, Cerebro/Cerebra, Control, Doctor/EMH, The Great Intelligence, HERBIE, Janet, JARVIS/Friday, L3-37/*Millennium Falcon*, Landru, OASIS, Oz, Rehoboam, Skynet, Starfleet Computer, TARDIS, Ultron, V-GER, WOPR, Zordon/Alpha-5; *Thematic Essays:* Robots and Slavery, AI and the Apocalypse: Science Fiction Meeting Science Fact.

Further Reading

Alkon, Paul K. 1987. *Origins of Futuristic Fiction*. Athens: University of Georgia Press.

Allen, Arthur. 2019. "There's a Reason We Don't Know Much about AI." Politico, September 16, 2019. https://www.politico.com/agenda/story/2019/09/16/artficial-intelligence-study-data-000956?cid=apn.

Ambrosino, Brandon. 2018. "What Would It Mean for AI to Have a Soul?" BBC, June 17, 2018. https://www.bbc.com/future/article/20180615-can-artificial-intelligence-have-a-soul-and-religion.

Ashby, LeRoy. 2006. *With Amusement for All: A History of American Popular Culture since 1830*. Lexington: University Press of Kentucky.

Bastani, Aaron. 2019. *Fully Automated Luxury Communism*. New York: Verso.

Carper, Steve. 2019. *Robots in American Popular Culture*. Jefferson, NC: McFarland.

Chandler, Simon. 2019. "Artificial Intelligence Has Become a Tool for Classifying and Ranking People." *Forbes*, October 1, 2019. https://www.forbes.com/sites/simonchandler/2019/10/01/artificial-intelligence-has-become-a-tool-for-classifying-and-ranking-people/#7431657f1d7c.

Chanthadavong, Aimee. 2019. "AI and Ethics: The Debate That Needs to Be Had." ZDnet, September 16, 2019. https://www.zdnet.com/article/ai-and-ethics-the-debate-that-needs-to-be-had/#ftag=CAD-03-10abf5f.

Dinh, Thien-Nam. 2018. *Silicon Minds: The Science, Impact, and Promise of Artificial Intelligence*. Independently Published.

Dowden, Bradley. 1993. *Logical Reasoning*. Belmont, CA: Wadsworth Publishing Company.

Ford, Martin. 2015. *Rise of the Robots: Technology and the Threat of a Jobless Future*. New York: Basic.

Hampton, Gregory Jerome. 2017. *Imagining Slaves and Robots in Literature, Film, and Popular Culture: Reinventing Yesterday's Slave with Tomorrow's Robot*. Lanham, MD: Lexington.

Hitchcock, Susan Tyler. 2007. *Frankenstein: A Cultural History*. New York: W. W. Norton.

Hoberman, J. 2011. *An Army of Phantoms: American Movies and the Making of the Cold War*. New York: New Press.

Krishnan, Armin. 2009. *Killer Robots: Legality and Ethicality of Autonomous Weapons*. London: Routledge.

Leetaru, Kalev. 2019. "Automatic Image Captioning and Why Not Every AI Problem Can Be Solved through More Data." *Forbes*, July 7, 2019. https://www.forbes.com/sites/kalevleetaru/2019/07/07/automatic-image-captioning-and-why-not-every-ai-problem-can-be-solved-through-more-data/#20b943476997.

Lin, Patrick, Jenkins, Ryan, and Abney, Keith, eds. 2017. *Robot Ethics 2.0: From Autonomous Cars to Artificial Intelligence*. New York: Oxford University Press.

Marks, Robert J. 2020. "*2084* vs. *1984*: The Difference AI Could Make to Big Brother." Podcast interview with author John Lennox. Mind Matters, July 3, 2020. https://mindmatters.ai/2020/07/2084-vs-1984-the-difference-ai-could-make-to-big-brother.

Mayor, Adrienne. 2018. *Gods and Robots: Myths, Machines, and Ancient Dreams of Technology*. Princeton, NJ: Princeton University Press.

Murdock, Jason. 2019. "Former Google Engineer Warns AI Might Accidentally Start a War: 'These Things Will Start to Behave in Unexpected Ways.'" *Newsweek*, September 16, 2019. https://www.newsweek.com/google-project-maven-artificial-intelligence-laura-nolan-killer-robots-department-defense-1459358.

Paur, Joey. 2019. "An AI Bot Writes a Hilarious Episode of *Star Trek: The Next Generation*." Geek Tyrant, November 16, 2019. https://geektyrant.com/news/an-ai-bot-writes-a-hilarious-episode-of-star-trek-the-next-generation.

Pellissier, Hank. 2013. "Robots and Slavery: What Do Humans Want When We Are 'Masters'?" Institute for Ethics and Emerging Technologies, September 13, 2013. https://ieet.org/index.php/IEET2/more/pellissier20130913.

Rouhiainen, Lasse. 2018. *Artificial Intelligence: 101 Things You Must Know Today about Our Future*. Scotts Valley, CA: CreateSpace.

Seed, David. 1999. *American Science Fiction and the Cold War: Literature and Film*. Abingdon, UK: Routledge.

Singer, Peter W. 2009. "Isaac Asimov's Laws of Robotics Are Wrong." Brookings Institute, May 18, 2009. https://www.brookings.edu/opinions/isaac-asimovs-laws-of-robotics-are-wrong.

Staff. 2020. "Shimon: Now a Singing, Songwriting Robot: Marimba-Playing Robot Composes Lyrics and Melodies with Human Collaborators." Georgia Tech Online, February 25, 2020. https://www.news.gatech.edu/2020/02/25/shimon-now-singing-songwriting-robot.

Wallach, Wendell, and Allen, Colin. 2009. *Moral Machines: Teaching Robots Right from Wrong*. New York: Oxford University Press.

HERBIE

Marvel Comics

In 1961, Stan Lee (1922–2018) and Jack Kirby (1917–1994) launched the "Marvel Age" of comics with *Fantastic Four #1* (November 1961). This team of superheroes consisted of Dr. Reed Richards (Mr. Fantastic), a brilliant scientist with the ability to stretch his fingers, arms, legs, and neck to incredible proportions; Susan Storm (Invisible Girl/Woman), Reed's girlfriend (and later wife), who possesses the ability to turn invisible and to project "force fields"; Johnny Storm (the second Human Torch), Sue's brother, possessing the ability to burst into flames, fly, and shoot bursts of flames; and Ben Grimm (the Thing), whose rocklike body makes him near invulnerable and endowed with enhanced strength. Their massive overnight popularity quickly led to other Marvel heroes such as the Hulk, Spider-Man, and the X-Men, among others. In 1978, NBC commissioned a Saturday morning cartoon based on the Fantastic Four. Due to the fact that Marvel was in separate negotiations regarding the Human Torch character, a replacement member was needed for the new series. The result was the robot HERBIE (Lee and Kirby, "A Monster among Us," *The New Fantastic Four*, season 1, episode 1, January 18, 1978).

"HERBIE" is an acronym for "Humanoid Experimental Robot, B-type, Integrated Electronics." Despite the term "humanoid," however, HERBIE's body consisted of a floating ball-shaped body with a computer-monitor-shaped head, simple dot eyes and a line mouth on the screen as a face, retractable arms, and small jets for legs. Being a children's program, HERBIE primarily represented comic relief for the team's adventures. He was added to the comic book the following year (by which time the series was cancelled), in *Fantastic Four #209*. In the twenty-first-century animated series *Fantastic Four: World's Greatest Heroes* (Nickelodeon, 2006), HERBIE did not have a physical form; instead, he was the AI program in charge of the computer systems in the FF's Baxter Building headquarters, and Dr. Richards's personal assistant.

HERBIE plays the role of helpmate to the FF. He exhibits a mixture of R2-D2's resourcefulness and C-3PO's nervousness (and as he was created in 1978, at the height of *Star Wars* mania, the two droids were likely major influences on the FF sidekick). Another similarity between these robots is the fact that they are "property." Though exhibiting an individual personality and the ability to act on free will, HERBIE belongs to the FF. The fact that he (like his *Star Wars* counterparts) is treated with kindness and affection does not change the fact that he exhibits all the traits of a living organic being without any of the freedoms that those qualities should bring with them. As such, HERBIE once more raises the question as to when (and if) autonomy and real freedom should be granted to an inorganic being. If a machine exhibits *all* of the traits attributed to "life," at what point does it become wrong to deny it the freedom that life deserves?

Richard A. Hall

See also: AI/Ziggy, Arnim Zola, Batcomputer, BB-8, C-3PO, Cambot/Gypsy/Tom Servo/ Crow, Cerebro/Cerebra, Doctor Octopus, Doombots, Dr. Theopolis and Twiki, HAL 9000, Human Torch, Inspector Gadget, Iron Legion, Iron Man, Janet, JARVIS/Friday, Jocasta, Johnny 5/S.A.I.N.T. Number 5, K9, LMDs, Marvin the Paranoid Android, Max Headroom, R2-D2, Sentinels, Speed Buggy, Starfleet Computer, Ultron, VICI, Vision, WALL-E, Warlock; *Thematic Essay:* Robots and Slavery.

Further Reading

Albert, Robert S., and Brigante, Thomas R. 1962. "The Psychology of Friendship Relations: Social Factors." *Journal of Social Psychology* 56 (1). https://www.tandfon line.com/doi/abs/10.1080/00224545.1962.9919371.

Amati, Viviana, Meggiolaro, Silvia, Rivellini, Giulia, and Zaccarin, Susanna. 2018. "Social Relations and Life Satisfaction: The Role of Friends." *Genus* 74, no. 1 (May 4). https://www.ncbi.nlm.nih.gov/pmc/articles/PMC5937874.

Ambrosino, Brandon. 2018. "What Would It Mean for AI to Have a Soul?" BBC, June 17, 2018. https://www.bbc.com/future/article/20180615-can-artificial-intelligence-have -a-soul-and-religion.

Bastani, Aaron. 2019. *Fully Automated Luxury Communism*. New York: Verso.

Chanthadavong, Aimee. 2019. "AI and Ethics: The Debate That Needs to Be Had." ZDnet, September 16, 2019. https://www.zdnet.com/article/ai-and-ethics-the-debate-that -needs-to-be-had/#ftag=CAD-03-10abf5f.

Costello, Matthew J. 2009. *Secret Identity Crisis: Comic Books & the Unmasking of Cold War America*. New York: Continuum.

DeFalco, Tom, ed. 2005. *Comics Creators on Fantastic Four*. London: Titan.

Dinh, Thien-Nam. 2018. *Silicon Minds: The Science, Impact, and Promise of Artificial Intelligence*. Independently Published.

Dowden, Bradley. 1993. *Logical Reasoning*. Belmont, CA: Wadsworth Publishing Company.

Ford, Martin. 2015. *Rise of the Robots: Technology and the Threat of a Jobless Future*. New York: Basic.

Hampton, Gregory Jerome. 2017. *Imagining Slaves and Robots in Literature, Film, and Popular Culture: Reinventing Yesterday's Slave with Tomorrow's Robot*. Lanham, MD: Lexington.

Howe, Sean. 2012. *Marvel Comics: The Untold Story*. New York: Harper-Perennial.

Lin, Patrick, Abney, Keith, and Bekey, George A., eds. 2014. *Robot Ethics: The Ethical and Social Implications of Robotics*. Cambridge: Massachusetts Institute of Technology.

Paur, Joey. 2019. "An AI Bot Writes a Hilarious Episode of *Star Trek: The Next Generation*." Geek Tyrant, November 16, 2019. https://geektyrant.com/news/an-ai-bot -writes-a-hilarious-episode-of-star-trek-the-next-generation.

Pellissier, Hank. 2013. "Robots and Slavery: What Do Humans Want When We Are 'Masters'?" Institute for Ethics and Emerging Technologies, September 13, 2013. https://ieet.org/index.php/IEET2/more/pellissier20130913.

Randall, Terri, dir. 2016. *NOVA*. "Rise of the Robots: Inside the DARPA Robotics Challenge." Aired February 24, 2016, on PBS. DVD.

Roberts-Griffin, Christopher. 2011. "What Is a Good Friend: A Qualitative Analysis of Desired Friendship Qualities." *Penn McNair Research Journal* 3, no. 1 (December 21, 2011). https://repository.upenn.edu/cgi/viewcontent.cgi?article=1019&context =mcnair_scholars.

Robitzski, Dan. 2018. "Artificial Consciousness: How to Give a Robot a Soul." Futurism, June 25, 2018. https://futurism.com/artificial-consciousness.

Rouhiainen, Lasse. 2018. *Artificial Intelligence: 101 Things You Must Know Today about Our Future*. Scotts Valley, CA: CreateSpace.

Spaeth, Dennis. 2018. "From Single-Task Machines to Backflipping Robots: The Evolution of Robots." *Cutting Tool Engineering*, January 15, 2018. https://www.ctemag .com/news/articles/evolution-of-robots.

Staff. 2020. "Shimon: Now a Singing, Songwriting Robot: Marimba-Playing Robot Composes Lyrics and Melodies with Human Collaborators." Georgia Tech Online, February 25, 2020. https://www.news.gatech.edu/2020/02/25/shimon-now-singing -songwriting-robot.

Thagard, Paul. 2017. "Will Robots Ever Have Emotions?" *Psychology Today*, December 14, 2017. https://www.psychologytoday.com/us/blog/hot-thought/201712/will-robots -ever-have-emotions.

Thomas, Roy. 2017. *The Marvel Age of Comics: 1961–1978*. Los Angeles: Taschen.

Wallach, Wendell, and Allen, Colin. 2009. *Moral Machines: Teaching Robots Right from Wrong*. New York: Oxford University Press.

Wright, Bradford W. 2003. *Comic Book Nation: The Transformation of Youth Culture in America*. Baltimore, MD: Johns Hopkins University Press.

Human Torch

Marvel Comics

Due to the massive success of Superman at National/Allied (DC) Comics in 1938, the other comic book publishers of the day were under tremendous pressure to create their own "superheroes" to take advantage of the burgeoning genre. In 1939, publisher Timely Comics bought an idea from writer-artist Carl Burgos (1916–1984) for a superhero called the "Human Torch," which it published in *Marvel Comics #1* (August 1939). Not to be confused with the modern-era superhero of the same name, the original Human Torch, the creation of Dr. Phineas Horton, was an android, a humanoid robot that, when exposed to oxygen, could burst into flames without damaging itself. It also possessed the ability to fly and shoot "balls" or "rays" of flame from his hands (Burgos, "The Human Torch," *Marvel Comics #1*, Timely Comics, August 1939). This hero launched the Timely (and later Marvel) superhero comic books. He was introduced simultaneously with fellow Timely hero the Sub-Mariner (also in *Marvel Comics #1*), and later followed by Captain America (in *Captain America Comics #1*, December, 1940).

Originally designed to be a "Frankenstein"-styled tale, the success of superheroes led to the Human Torch quickly joining their ranks. In his "human" persona, he held the identity of New York City Police Officer Jim Hammond. Once the character received his own self-titled comic book, Human Torch was given a teenage sidekick, Toro. Unlike his partner, Toro was a human boy, mutated by his scientist parents' radiation experiments, giving him the same powers as the android Torch (Burgos, *Human Torch Comics #2*, Timely Comics, October, 1940). Throughout World War II, Human Torch and Toro teamed with the Sub-Mariner, Captain America, and Bucky to fight the fascist threats to the United States. The character was cancelled in the late-1940s as superhero comic sales began to slide. He was briefly revived for six months spanning 1953 and 1954 but failed to catch on. When Stan Lee launched the "Marvel age" of superhero comics in the early 1960s with the Fantastic Four, he utilized the name "Human Torch" and the same powers for a new, human superhero (who, within the pages of the comic, admits to taking his name from the 1940s hero). In 1975, Marvel Comics writer Roy Thomas (b. 1940), a fan of the World War II comics, revived the character, telling tales alongside his wartime partners in the comic book *The Invaders*.

Burgos's original idea for Human Torch was to make a modern-day Frankensteinesque cautionary tale of science gone too far. The decision to, instead, make him a superhero brought about the idea of android heroes, and it still exists today with characters such as Marvel's Vision and *Star Trek*'s Commander Data. Human Torch follows Isaac Asimov's laws of robotics. He is dedicated to protecting humankind. Unlike Vision or Data, however, Hammond/Human Torch does not exhibit melancholy for not being "human." He is comfortable in his own identity and existence. Like Vision and Data, he is recognized as an individual life-form with free will and self-consciousness, and he does not "belong" to a human master. He is the ideal of what AI can be.

Richard A. Hall

See also: Androids, B.A.T.s, Bernard, Bishop, Buffybot, Dolores, Doombots, Fembots, Inspector Gadget, Iron Legion, Jocasta, Lieutenant Commander Data, LMDs, Lore/B4, Marvin the Paranoid Android, Maschinenmensch/Maria, Replicants, Soji and Dahj Asha, Stepford Wives, Ultron, VICI, Vision; *Thematic Essay:* Heroic Robots and Their Impact on Sci-Fi Narratives.

Further Reading

Alkon, Paul K. 1987. *Origins of Futuristic Fiction*. Athens: University of Georgia Press.

Ambrosino, Brandon. 2018. "What Would It Mean for AI to Have a Soul?" BBC, June 17, 2018. https://www.bbc.com/future/article/20180615-can-artificial-intelligence-have-a-soul-and-religion.

Ashby, LeRoy. 2006. *With Amusement for All: A History of American Popular Culture since 1830*. Lexington: University Press of Kentucky.

Byrne, Emma. 2013. "Innovation Isn't Safe: The Future According to Kevin Warwick." *Forbes*, September 30, 2013. https://www.forbes.com/sites/netapp/2013/09/30/kevin -warwick-captain-cyborg/#48b704c13560.

Carper, Steve. 2019. *Robots in American Popular Culture*. Jefferson, NC: McFarland.

Cavaler, Chris. 2015. *On the Origins of Superheroes: From the Big Bang to Action Comics No. 1*. Iowa City: University of Iowa Press.

Chanthadavong, Aimee. 2019. "AI and Ethics: The Debate That Needs to Be Had." ZDnet, September 16, 2019. https://www.zdnet.com/article/ai-and-ethics-the-debate-that -needs-to-be-had/#ftag=CAD-03-10abf5f.

Dinh, Thien-Nam. 2018. *Silicon Minds: The Science, Impact, and Promise of Artificial Intelligence*. Independently Published.

Dowden, Bradley. 1993. *Logical Reasoning*. Belmont, CA: Wadsworth Publishing Company.

Fryer-Biggs, Zachary. 2019. "Coming Soon to a Battlefield: Robots That Can Kill." *The Atlantic*, September 3, 2019. https://www.theatlantic.com/technology/archive/2019 /09/killer-robots-and-new-era-machine-driven-warfare/597130.

Hitchcock, Susan Tyler. 2007. *Frankenstein: A Cultural History*. New York: W. W. Norton.

Howe, Sean. 2012. *Marvel Comics: The Untold Story*. New York: Harper-Perennial.

Krishnan, Armin. 2009. *Killer Robots: Legality and Ethicality of Autonomous Weapons*. London: Routledge.

Lin, Patrick, Abney, Keith, and Bekey, George A., eds. 2014. *Robot Ethics: The Ethical and Social Implications of Robotics*. Cambridge: Massachusetts Institute of Technology.

Mayor, Adrienne. 2018. *Gods and Robots: Myths, Machines, and Ancient Dreams of Technology*. Princeton, NJ: Princeton University Press.

Randall, Terri, dir. 2016. *NOVA*. "Rise of the Robots: Inside the DARPA Robotics Challenge." Aired February 24, 2016, on PBS. DVD.

Robitzski, Dan. 2018. "Artificial Consciousness: How to Give a Robot a Soul." Futurism, June 25, 2018. https://futurism.com/artificial-consciousness.

Rouhiainen, Lasse. 2018. *Artificial Intelligence: 101 Things You Must Know Today about Our Future*. Scotts Valley, CA: CreateSpace.

Thagard, Paul. 2017. "Will Robots Ever Have Emotions?" *Psychology Today*, December 14, 2017. https://www.psychologytoday.com/us/blog/hot-thought/201712/will-robots -ever-have-emotions.

Wallach, Wendell, and Allen, Colin. 2009. *Moral Machines: Teaching Robots Right from Wrong*. New York: Oxford University Press.

Wright, Bradford W. 2003. *Comic Book Nation: The Transformation of Youth Culture in America*. Baltimore, MD: Johns Hopkins University Press.

I

IG-88/IG-11

Star Wars

In 1977, writer-director George Lucas (b. 1944) introduced the world to the sci-fi/fantasy phenomenon *Star Wars* (Lucasfilm/20th Century Fox, 1977), one of the most successful pop culture franchises in American history. His original trilogy of films (1977–1983) focused on the rise of its protagonist, Luke Skywalker (played by Mark Hamill, b. 1951), against the backdrop of a Galactic Civil War against the human forces of the evil Empire, led by its Emperor, Sheev Palpatine (played by Ian McDiarmid, b. 1944), and his henchman, Darth Vader (played by David Prowse, b. 1935; voiced by James Earl Jones, b. 1931). In the second film of the original trilogy, Darth Vader pursues the spaceship *Millennium Falcon*, on which is the leader of the Rebellion, Princess Leia Organa (played by Carrie Fisher, 1956–2016). As the captain of the *Falcon*, Han Solo (played by Harrison Ford, b. 1942) had a bounty on his head due to a debt owed to crime lord Jabba the Hutt, Vader called for help from several bounty hunters, one of which was the droid IG-88 (Lucas, Leigh Brackett, and Lawrence Kasdan, *Star Wars: Episode V—The Empire Strikes Back*, Lucasfilm, 1980).

In the *Star Wars* universe, IG units are "assassin droids," mass-produced in the waning years of the Galactic Republic. With the rise of the Empire, some IG units gained their independence and struck out on their own, primarily as bounty hunters/assassins. Though IG-88 has only a few moments of screen time (and no dialogue), its image was memorable. Roughly a half-meter taller than standard humans, with a skeletal humanoid body structure, a vertically tubular head, and small protruding eye stalks, IG-88 was an imposing figure that sparked the imaginations of fans. Over the years since that first brief appearance, IG units have appeared in various *Star Wars* comics, novels, and television series. Recently, IG-88's status in *Star Wars* was surpassed by the introduction of IG-11.

In the live-action television series *The Mandalorian* (Disney+, 2019–present), the titular bounty hunter meets a new IG hunter, IG-11 (created by showrunner Jon Favreau, b. 1966; voiced by Taika Waititi, b. 1975). Both are searching for a quarry they discover to be an infant child. The Mandalorian immediately feels for the small creature, killing IG-11 to prevent the robot from taking it (Favreau, "Chapter 1: The Mandalorian," *The Mandalorian*, season 1, episode 1, November 12, 2019). Later, as the Mandalorian has taken the task of protecting the child, he once more runs into IG-11, the droid having been found and reprogrammed by the Ugnaut Kuiil (played by Misti Rosas, b. 1973; voiced by Nick Nolte, b. 1941) to be a helpmate (all violent programming having been erased). Kuiil further programs IG-11 to act as a "nurse" to the child, and the droid takes its new caregiver duties deadly seriously. Over the course of two episodes, IG-11 kills two Imperial troops who have absconded with the child (and murdered Kuiil), going on to kill dozens more troops ordered to take the child. Ultimately, IG-11 sacrifices himself to ensure that the Mandalorian and child can safely escape (Favreau, "Chapter 7: The Reckoning"/"Chapter 8: Redemption," *The Mandalorian*, season 1, episodes 7 and 8, December 18 and 27, 2019).

Though intended to be a machine of death, IG-11 proved that droids/robots can go beyond their original intent/programming. As Kuiil points out in "The Reckoning," "Droids are not 'good' or 'bad.' They are neutral reflections of those who imprint on them." Though not entirely accurate—IG-11 did not change by being "imprinted" upon; he was reprogrammed—the quote is a reminder that robots are what their programmers make of them. Assassin or nurse, robots are what we make them to be, and they cannot be "judged" by the same standards as creatures of free will. Just as "bad" humans can seek redemption, "bad" robots can be reprogrammed and placed on a new, more benevolent course.

Richard A. Hall

See also: B.A.T.s, Battle Droids, BB-8, Buffybot, Darth Vader, D-O, Doombots, Echo/CT-1409, ED-209, EV-9D9, Fembots, General Grievous, Iron Legion, K-2SO, L3-37/*Millennium Falcon*, Medical Droids, Metalhead, Metallo, R2-D2, Robots, Sentinels, Terminators, Tin Woodsman, Warlock; *Thematic Essays:* Heroic Robots and Their Impact on Sci-Fi Narratives, Villainous Robots and Their Impact on Sci-Fi Narratives.

Further Reading

Ashby, LeRoy. 2006. *With Amusement for All: A History of American Popular Culture since 1830.* Lexington: University Press of Kentucky.

Bray, Adam, Barr, Tricia, Horton, Cole, and Windham, Ryder. 2019. *Ultimate* Star Wars, *New Edition.* London: DK Publishing.

Bray, Adam, and Horton, Cole. 2017. Star Wars: *Absolutely Everything You Need to Know, Updated and Expanded.* London: DK Children.

Byrne, Emma. 2013. "Innovation Isn't Safe: The Future According to Kevin Warwick." *Forbes*, September 30, 2013. https://www.forbes.com/sites/netapp/2013/09/30/kevin-warwick-captain-cyborg/#48b704c13560.

Chanthadavong, Aimee. 2019. "AI and Ethics: The Debate That Needs to Be Had." ZDnet, September 16, 2019. https://www.zdnet.com/article/ai-and-ethics-the-debate-that-needs-to-be-had/#ftag=CAD-03-10abf5f.

Dinh, Thien-Nam. 2018. *Silicon Minds: The Science, Impact, and Promise of Artificial Intelligence.* Independently Published.

Eberl, Jason T., and Decker, Kevin S., eds. 2015. *The Ultimate* Star Wars *and Philosophy: You Must Unlearn What You Have Learned.* Hoboken, NJ: Wiley-Blackwell.

Ford, Martin. 2015. *Rise of the Robots: Technology and the Threat of a Jobless Future.* New York: Basic.

Fryer-Biggs, Zachary. 2019. "Coming Soon to a Battlefield: Robots That Can Kill." *The Atlantic*, September 3, 2019. https://www.theatlantic.com/technology/archive/2019/09/killer-robots-and-new-era-machine-driven-warfare/597130.

Hampton, Gregory Jerome. 2017. *Imagining Slaves and Robots in Literature, Film, and Popular Culture: Reinventing Yesterday's Slave with Tomorrow's Robot.* Lanham, MD: Lexington.

Hitchcock, Susan Tyler. 2007. *Frankenstein: A Cultural History.* New York: W. W. Norton.

Kaminski, Michael. 2008. *The Secret History of* Star Wars*: The Art of Storytelling and the Making of a Modern Epic*. Kingston, Canada: Legacy.

Krishnan, Armin. 2009. *Killer Robots: Legality and Ethicality of Autonomous Weapons*. London: Routledge.

Lin, Patrick, Abney, Keith, and Bekey, George A., eds. 2014. *Robot Ethics: The Ethical and Social Implications of Robotics*. Cambridge: Massachusetts Institute of Technology.

Longoni, Chiara, Bonezzi, Andrea, and Morewedge, Carey K. 2019. "Resistance to Medical Artificial Intelligence." *Journal of Consumer Research* 46, no. 4 (December): 629–50. https://academic.oup.com/jcr/article-abstract/46/4/629/5485292.

Murdock, Jason. 2019. "Former Google Engineer Warns AI Might Accidentally Start a War: 'These Things Will Start to Behave in Unexpected Ways.'" *Newsweek*, September 16, 2019. https://www.newsweek.com/google-project-maven-artificial-intelligence-laura-nolan-killer-robots-department-defense-1459358.

Randall, Terri, dir. 2016. *NOVA*. "Rise of the Robots: Inside the DARPA Robotics Challenge." Aired February 24, 2016, on PBS. DVD.

Reagin, Nancy R., and Liedl, Janice, eds. 2013. Star Wars *and History*. New York: John Wiley & Sons.

Rouhiainen, Lasse. 2018. *Artificial Intelligence: 101 Things You Must Know Today about Our Future*. Scotts Valley, CA: CreateSpace.

Singer, Peter W. 2009. "Isaac Asimov's Laws of Robotics Are Wrong." Brookings Institute, May 18, 2009. https://www.brookings.edu/opinions/isaac-asimovs-laws-of-robotics-are-wrong.

Slavicsek, Bill. 2000. *A Guide to the* Star Wars *Universe*. San Francisco, CA: LucasBooks.

Spaeth, Dennis. 2018. "From Single-Task Machines to Backflipping Robots: The Evolution of Robots." *Cutting Tool Engineering*, January 15, 2018. https://www.ctemag.com/news/articles/evolution-of-robots.

Sumerak, Marc. 2018. *Star Wars: Droidography*. New York: HarperFestival.

Sunstein, Cass R. 2016. *The World According to* Star Wars. New York: Dey Street.

Sweet, Derek R. 2015. Star Wars *in the Public Square: The Clone Wars as Political Dialogue*. Critical Explorations in Science Fiction and Fantasy series, edited by Donald E. Palumbo and Michael Sullivan. Jefferson, NC: McFarland.

Taylor, Chris. 2015. *How* Star Wars *Conquered the Universe: The Past, Present, and Future of a Multibillion Dollar Franchise*. New York: Basic Books.

Wallace, Daniel. 2006. Star Wars*: The New Essential Guide to Droids*. New York: Del Rey.

Wallach, Wendell, and Allen, Colin. 2009. *Moral Machines: Teaching Robots Right from Wrong*. New York: Oxford University Press.

Inspector Gadget

Inspector Gadget

In 1983, creators Bruno Bianchi (1955–2011), Jean Chalopin (b. 1950), and Andy Heyward (b. 1949) introduced audiences to their cyborg police detective, Inspector Gadget. Nostalgia for the wildly popular animated series of the same name (syndication, 1983–1986) launched several feature films and other animated series in the late 1990s and early 2000s. In the original series, the titular character was voiced by Don Adams (1923–2005), and in the character's feature film debut in 1999, Inspector Gadget was played by Matthew Broderick (b. 1962). The dedicated—though somewhat slow-witted—inspector investigates criminal activity of the organization M.A.D. and its leader, Dr. Claw. Inspector Gadget's successes come almost exclusively from the assistance of his niece, Penny, her dog Brain, and the Inspector's many cybernetically installed "gadgets."

Cyborg Inspector Gadget's human brain and heart remain intact, but the remainder of his body has been enhanced with hundreds (if not thousands) of instruments to assist in his investigative endeavors. The inspector calls forth these gadgets with the command, "Go, go, gadget . . ." (followed by the name of the gadget). For example, one of his more frequently used gadgets is his "Gadget-Copter," releasing helicopter blades from the top of his head along with extended guidance handles the inspector uses to control flight. Unfortunately, his die-hard dedication to his job far surpasses his actual

The classic cartoon character Inspector Gadget at The Musée de l'Homme in Paris, France. (Guillohmz/Dreamstime.com)

mental ability to be successful (leading one to question why his brain was not enhanced along with his body). Though played for cartoonish entertainment, the underlying danger of such massive cybernetic power in the hands of someone clearly incompetent to adequately control said power is evident.

Despite his cybernetic components far outnumbering his remaining organic ones, Inspector Gadget remains far more "robotic human" than "organic robot." His organic brain and heart define him far more than his technological enhancements do. He possesses the human heart and soul of a hero, never giving into his overwhelming mechanical components to become as devoid of love and compassion as a Cyberman or Dalek. Further, unlike the superhero Cyborg, Inspector Gadget never exhibits doubt as to his humanity in the face of his cybernetic implants. He is an idealistic example of what a cyborg could be, with the character's primary purpose clearly being the entertainment and joy of children.

Richard A. Hall

See also: Adam, Alita, Borg, Cybermen, Cyborg, Daleks, Darth Vader, Doctor Octopus, Echo/CT-1409, General Grievous, Jaime Sommers, Nardole, RoboCop, Steve Austin; *Thematic Essays:* Heroic Robots and Their Impact on Sci-Fi Narratives, Cyborgs: Robotic Humans or Organic Robots?

Further Reading

Ashby, LeRoy. 2006. *With Amusement for All: A History of American Popular Culture since 1830.* Lexington: University Press of Kentucky.

Bryant, D'Orsay D., III. 1985. "Spare-Part Surgery: The Ethics of Organ Transplantation." *Journal of the National Medical Association* 77 (2). https://www.ncbi.nlm.nih.gov/pmc/articles/PMC2561842/pdf/jnma00245-0055.pdf.

Byrne, Emma. 2014. "Cybernetic Implants: No Longer Science Fiction." *Forbes*, March 11, 2014. https://www.forbes.com/sites/netapp/2014/03/11/cybernetic-implants-not-sci-fi/#7399c57e77ba.

Calvert, Bronwen. 2017. *Being Bionic: The World of TV Cyborgs.* London: I. B. Tauris.

Carper, Steve. 2019. *Robots in American Popular Culture.* Jefferson, NC: McFarland.

Castleman, Harry, and Podrazik, Walter J. 2016. *Watching TV: Eight Decades of American Television, Third Edition.* Syracuse, NY: Syracuse University Press.

Choi, Charles Q. 2009. "Human Evolution: The Origin of Tool Use." LiveScience, November 11, 2009. https://www.livescience.com/7968-human-evolution-origin-tool.html.

Randall, Terri, dir. 2016. *NOVA.* "Rise of the Robots: Inside the DARPA Robotics Challenge." Aired February 24, 2016, on PBS. DVD.

Iron Giant

The Iron Giant/The Iron Man

In 1968, British author Ted Hughes published the children's book *The Iron Man*, which was published in the United States as *The Iron Giant*. In 1999, America screenwriter Tim McCanlies (b. 1953) adapted the book for film based on a story treatment by director Brad Bird (b. 1957), producing a film under the original's American title as an animated feature for Warner Brothers. The story, set at the height of the U.S.-Soviet Cold War (1947–1992), the "Iron Giant" is an alien robot that lands in the United States. As the U.S. government investigates the robot's purpose (presuming a threat of some kind), the robot is befriended by a local boy. When conventional American military force proves no match for the robot, a sinister government agent orders a nuclear strike (which would, of course, kill everyone in the town). Inspired by *Superman* comics read to him by his young human friend, the Iron Giant intercepts the nuclear missile, sacrificing his own metallic life in the process (Hughes, *The Iron Man/The Iron Giant*, 1968).

The "Iron Giant" is a robot in the very basic sense of the word. It is a mechanical device programmed for a specific purpose (in this case, likely, observation of the inhabitants of planet Earth). It does possess powerful defense mechanisms that are meant to protect it from potential threat. However, in the case of the Iron Giant, the robot is capable of learning human traits such as protectiveness, kindness, and self-sacrifice. It is ironic that the human inspiration for the robot's heroics is Superman, known also as the "Man of Steel"; and the robot does, indeed, respond to a nuclear threat in the same fashion as the comic book hero has in his own adventures. In the end, the Iron Giant shows us that robots, even after initial programming, learn from what they observe. Whether such automatons become "good" or "bad" is based, to a degree, on how they interpret the information they take in.

Richard A. Hall

See also: Baymax, BB-8, Buffybot, C-3PO, D-O, HERBIE, Iron Man, Johnny 5/S.A.I.N.T. Number 5, K-2SO, K9, Mechagodzilla, Muffit, R2-D2, Robby the Robot, Robot, Rosie, Tin Woodsman, Transformers, Voltron, WALL-E, Warlock; *Thematic Essay:* Heroic Robots and Their Impact on Sci-Fi Narratives.

Further Reading

Albert, Robert S., and Brigante, Thomas R. 1962. "The Psychology of Friendship Relations: Social Factors." *Journal of Social Psychology* 56 (1). https://www.tandfon line.com/doi/abs/10.1080/00224545.1962.9919371.

Amati, Viviana, Meggiolaro, Silvia, Rivellini, Giulia, and Zaccarin, Susanna. 2018. "Social Relations and Life Satisfaction: The Role of Friends." *Genus* 74, no. 1 (May 4). https://www.ncbi.nlm.nih.gov/pmc/articles/PMC5937874.

Ambrosino, Brandon. 2018. "What Would It Mean for AI to Have a Soul?" BBC, June 17, 2018. https://www.bbc.com/future/article/20180615-can-artificial-intelligence-have -a-soul-and-religion.

Ashby, LeRoy. 2006. *With Amusement for All: A History of American Popular Culture since 1830.* Lexington: University Press of Kentucky.

Carper, Steve. 2019. *Robots in American Popular Culture.* Jefferson, NC: McFarland.

Daniels, Les. 2004. *Superman: The Complete History—The Life and Times of the Man of Steel.* New York: DC Comics.

Dinh, Thien-Nam. 2018. *Silicon Minds: The Science, Impact, and Promise of Artificial Intelligence.* Independently Published.

Fryer-Biggs, Zachary. 2019. "Coming Soon to a Battlefield: Robots That Can Kill." *The Atlantic*, September 3, 2019. https://www.theatlantic.com/technology/archive/2019 /09/killer-robots-and-new-era-machine-driven-warfare/597130.

Krishnan, Armin. 2009. *Killer Robots: Legality and Ethicality of Autonomous Weapons.* London: Routledge.

Lin, Patrick, Abney, Keith, and Bekey, George A., eds. 2014. *Robot Ethics: The Ethical and Social Implications of Robotics.* Cambridge: Massachusetts Institute of Technology.

Roberts-Griffin, Christopher. 2011. "What Is a Good Friend: A Qualitative Analysis of Desired Friendship Qualities." *Penn McNair Research Journal* 3, no. 1 (December 21). https://repository.upenn.edu/cgi/viewcontent.cgi?article=1019&context=mcnair _scholars.

Robitzski, Dan. 2018. "Artificial Consciousness: How to Give a Robot a Soul." Futurism, June 25, 2018. https://futurism.com/artificial-consciousness.

Rouhiainen, Lasse. 2018. *Artificial Intelligence: 101 Things You Must Know Today about Our Future.* Scotts Valley, CA: CreateSpace.

Seed, David. 1999. *American Science Fiction and the Cold War: Literature and Film.* Abingdon, UK: Routledge.

Starck, Kathleen, ed. 2010. *Between Fear and Freedom: Cultural Representations of the Cold War.* Newcastle, UK: Cambridge Scholars.

Wallach, Wendell, and Allen, Colin. 2009. *Moral Machines: Teaching Robots Right from Wrong.* New York: Oxford University Press.

Iron Legion

Marvel Cinematic Universe

In 2008, Marvel Comics launched its new Marvel Studios with the debut of the "Marvel Cinematic Universe" with the film *Iron Man*, starring Robert Downey Jr. (b. 1965). Downey played classic Marvel Comics billionaire engineer and weapons manufacturer Tony Stark. After being taken hostage by terrorists in Afghanistan and forced to produce weapons for them, Stark, whose heart was damaged during capture, instead manufactures a weaponized suit of armor powered by the device he has built to keep the metal fragments near his heart from killing him. Escaping and returning home to the United States, Stark decides to perfect his suit of armor idea and fight wrong as Iron Man (Jon Favreau, dir., *Iron Man*, Marvel Studios, 2008). Constantly upgrading his armor, Stark eventually comes to own numerous versions of his Iron Man suits (some with very specified abilities). In *Iron Man 3* (2013), facing a particularly dangerous threat, Stark utilizes the JARVIS AI program to call forth all of his armors as the "Iron Legion," an army of automatons with each suit carefully manipulated by JARVIS.

By the second Avengers film, *Avengers: Age of Ultron* (2015), the Iron Legion has been modified to act as first responders and peacekeepers to assist with civilian casualties and/or unrest during and after Avengers battles. Stark sees this as an opportunity to "build a suit of armor around the world," with the new "Ultron" AI as the command-and-control of the overall program. When Ultron achieves sentience and sees Stark and all humanity as the greatest threat facing the world, the AI takes over the Iron Legion, turning its members on the Avengers and utilizing them as his own private army in his scheme to wipe out humanity and evolve himself (Joss Whedon, *Avengers: Age of Ultron*, Marvel/Disney, 2015). At their core, the Iron Legion is a team of robots, completely at the behest of whoever possesses control over them. They can be utilized as a force for good or evil, making them—on the whole—too dangerous to continue.

With the Iron Legion, audiences are made aware of the massive dangers of unintended consequences. Stark began with the premise that he could control his AI and gave no real consideration to the idea that someone—or something—smarter than he might gain control of what could be perverted into a major force for evil. These robots are the Frankenstein's monster motif multiplied by a factor of hundreds. If, in the not-too-distant future, humanity allows hubris to override the need for intense safety protocols, the promise of the Iron Legion could very easily—and quickly—turn into the threat posed by Terminators (see "Terminators").

Richard A. Hall

See also: B.A.T.s, Battle Droids, Cylons, Doombots, Fembots, Inspector Gadget, Iron Man, JARVIS/Friday, LMDs, Replicants, Robots, Sentinels, Terminators, Transformers, Ultron, Vision, Warlock; *Thematic Essays:* Robots and Slavery, AI and the Apocalypse: Science Fiction Meeting Science Fact.

Further Reading

Bastani, Aaron. 2019. *Fully Automated Luxury Communism*. New York: Verso.

Byrne, Emma. 2013. "Innovation Isn't Safe: The Future According to Kevin Warwick." *Forbes*, September 30, 2013. https://www.forbes.com/sites/netapp/2013/09/30/kevin-warwick-captain-cyborg/#48b704c13560.

Choi, Charles Q. 2009. "Human Evolution: The Origin of Tool Use." LiveScience, November 11, 2009. https://www.livescience.com/7968-human-evolution-origin-tool.html.

Fryer-Biggs, Zachary. 2019. "Coming Soon to a Battlefield: Robots That Can Kill." *The Atlantic*, September 3, 2019. https://www.theatlantic.com/technology/archive/2019/09/killer-robots-and-new-era-machine-driven-warfare/597130.

Hitchcock, Susan Tyler. 2007. *Frankenstein: A Cultural History*. New York: W. W. Norton.

Howe, Sean. 2012. *Marvel Comics: The Untold Story*. New York: Harper-Perennial.

Murdock, Jason. 2019. "Former Google Engineer Warns AI Might Accidentally Start a War: 'These Things Will Start to Behave in Unexpected Ways.'" *Newsweek*, September 16, 2019. https://www.newsweek.com/google-project-maven-artificial-intelligence-laura-nolan-killer-robots-department-defense-1459358.

Randall, Terri, dir. 2016. *NOVA*. "Rise of the Robots: Inside the DARPA Robotics Challenge." Aired February 24, 2016, on PBS. DVD.

Spaeth, Dennis. 2018. "From Single-Task Machines to Backflipping Robots: The Evolution of Robots." *Cutting Tool Engineering*, January 15, 2018. https://www.ctemag.com/news/articles/evolution-of-robots.

Tucker, Reed. 2017. *Slugfest: Inside the Epic 50-Year Battle Between Marvel and DC*. New York: Da Capo Press.

Virk, Rizwan. 2019. *The Simulation Hypothesis: An MIT Computer Scientist Shows Why AI, Quantum Physics, and Eastern Mystics All Agree We Are in a Video Game*. Milwaukee, WI: Bayview.

White, Mark D., ed. 2010. *Iron Man and Philosophy: Facing the Stark Reality*. Hoboken, NJ: Wiley.

Iron Man

Marvel Comics

In 1961, comic book legend Stan Lee (1922–2018) launched the "Marvel age" of comics. With the launching of the Fantastic Four—soon to be followed by the Hulk, Spider-Man, Iron Man, and many, many more—comic book superheroes took a more dramatic turn, with storylines focused on the private lives of those blessed (or burdened) with superpowers. In 1963, Lee, along with the creative team of Larry Lieber (b. 1931), Don Heck (1929–1995), and Jack Kirby (1917–1994), introduced Iron Man. Iron Man was billionaire engineer and military industrialist Tony Stark. While visiting U.S. troops in Vietnam, Stark is captured and forced to make weapons for the communist North Vietnamese. Instead, Stark builds a weaponized suit of armor powered by the device he engineered to keep him alive after his heart was damaged during capture. Making his way back to the United States, Stark decides to fight evil as the superhero Iron Man (Lee, Lieber, Heck, and Kirby, *Tales of Suspense #39*, Marvel Comics, March 1963). In 2008, Marvel Comics launched their new Marvel Studios with the debut of the "Marvel Cinematic Universe" with the film *Iron Man*, starring Robert Downey Jr. (b. 1965).

The Iron Man armor possesses a vast array of weaponry, with each new version of the armor more advanced than the last (and some with very specialized gadgets). Rocket thrusters in the boots and gauntlets allow Iron Man to fly. The gauntlet thrusters can also be utilized to fire bolts at enemies. It can launch minimissiles, project force fields, and, at one point in the comics, even produce "roller skates" in

The official Madame Tussaud's wax sculpture of the comic book hero Iron Man, cocreated by Stan Lee, Larry Lieber, Don Heck, and Jack Kirby. (Mihail Ivanov/Dreamstime.com)

the boots. In the modern comics and feature films, Stark connects his armor to an AI program (which, in the films, are JARVIS and Friday). Other than the magnetically powered reactor in his chest (originally designed to keep the metal fragments that threatened his heart at bay), the armor is not "connected" to Stark. It is a suit of armor that can be worn/deployed or discarded at Stark's whim.

In his comic book origins, Iron Man is very much a product of the Cold War. In the early years of his comic book adventures, Iron Man faces off against a Soviet version of his armored suit: Crimson Dynamo. In the overall zeitgeist of American pop culture, however, Iron Man is much more commonly associated with the feature film version brought to life by Downey. In those films, Iron Man/Tony Stark is much more an allegory for a modern-day Dr. Frankenstein. While some of his inventions—the Iron Man and War Machine armors (including the "Hulk-Buster" armor affectionately called "Veronica," the AI programs JARVIS and Friday, and the synthetic AI life-form "Vision")—become forces for good, others, such as the Iron Legion and Ultron, are warped into forces for evil.

Iron Man is not a "robot," nor is it a "cyborg" or an "AI." It is a highly advanced weaponized suit of armor, making Tony Stark a modern-day "knight." Despite this distinction, no volume of robotics in pop culture would be complete without a discussion of Iron Man. The Iron Man armor is the culmination of what robotics/AI holds for the future of humankind. However, as is seen from the examples given above, what Stan Lee so brilliantly stated in his Spider-Man comics most definitely applies to Tony Stark/Iron Man as well: "With great power, there must also come . . . great responsibility." As technology works its way closer and closer to the ideal presented by Iron Man, keeping this quote in mind would be wise.

Richard A. Hall

See also: Arnim Zola, Cerebro/Cerebra, Doctor Octopus, Doombots, Human Torch, Inspector Gadget, Iron Legion, JARVIS/Friday, Jocasta, LMDs, Sentinels, Ultron, Vision, Warlock; *Thematic Essay:* Heroic Robots and Their Impact on Sci-Fi Narratives.

Further Reading

Allan, Kathryn, ed. 2013. *Disability in Science Fiction.* New York: Palgrave Macmillan.

Ashby, LeRoy. 2006. *With Amusement for All: A History of American Popular Culture since 1830.* Lexington: University Press of Kentucky.

Bastani, Aaron. 2019. *Fully Automated Luxury Communism.* New York: Verso.

Carper, Steve. 2019. *Robots in American Popular Culture.* Jefferson, NC: McFarland.

Cavaler, Chris. 2015. *On the Origins of Superheroes: From the Big Bang to Action Comics No. 1.* Iowa City: University of Iowa Press.

Costello, Matthew J. 2009. *Secret Identity Crisis: Comic Books & the Unmasking of Cold War America.* New York: Continuum.

Dinh, Thien-Nam. 2018. *Silicon Minds: The Science, Impact, and Promise of Artificial Intelligence.* Independently Published.

Ford, Martin. 2015. *Rise of the Robots: Technology and the Threat of a Jobless Future.* New York: Basic.

Howe, Sean. 2012. *Marvel Comics: The Untold Story.* New York: Harper-Perennial.

Lin, Patrick, Abney, Keith, and Bekey, George A., eds. 2014. *Robot Ethics: The Ethical and Social Implications of Robotics.* Cambridge: Massachusetts Institute of Technology.

Lin, Patrick, Jenkins, Ryan, and Abney, Keith, eds. 2017. *Robot Ethics 2.0: From Autonomous Cars to Artificial Intelligence.* New York: Oxford University Press.

Randall, Terri, dir. 2016. *NOVA.* "Rise of the Robots: Inside the DARPA Robotics Challenge." Aired February 24, 2016, on PBS. DVD.

Seed, David. 1999. *American Science Fiction and the Cold War: Literature and Film.* Abingdon, UK: Routledge.

Starck, Kathleen, ed. 2010. *Between Fear and Freedom: Cultural Representations of the Cold War.* Newcastle, UK: Cambridge Scholars.

Thomas, Roy. 2017. *The Marvel Age of Comics: 1961–1978.* Los Angeles: Taschen.

Tucker, Reed. 2017. *Slugfest: Inside the Epic 50-Year Battle between Marvel and DC.* New York: Da Capo Press.

White, Mark D., ed. 2010. *Iron Man and Philosophy: Facing the Stark Reality.* Hoboken, NJ: Wiley.

Wright, Bradford W. 2003. *Comic Book Nation: The Transformation of Youth Culture in America.* Baltimore, MD: Johns Hopkins University Press.

J

Jaime Sommers

The Bionic Woman

In *The Bionic Woman* (ABC 1976–1977, NBC 1977–1978), Jaime Sommers (played by Lindsay Wagner, b. 1949) is a human with bionic body parts who works as an undercover agent for the OSI (Office of Scientific Intelligence), where she saves the world from evildoers time and again. When the character, created by Kenneth Johnson, originally appeared in *The Six Million Dollar Man* (ABC, 1974–1978) in back-to-back episodes (Johnson, "The Bionic Woman, Parts 1 and 2," *The Six Million Dollar Man*, season 2, episodes 19 and 20, March 16 and 23, 1975), it was supposed to be the last anyone saw of her. Sommers was introduced as a tennis pro who reunites with high school sweetheart Steve Austin (played by Lee Majors, b. 1939), the titular hero of *The Six Million Dollar Man*. They fall in love before a skydiving accident leaves Jaime badly injured. To save her life, Steve asks his OSI handler Oscar Goldman (played by Richard Anderson, 1926–2017) to replace her injured legs, right arm, and right ear with cybernetic implants similar to the ones he has. Goldman assigns Dr. Rudy Wells (played by Alan Oppenheimer, b. 1930) to perform the surgery, which appears to be successful at first, but Jaime's body soon rejects the parts, and she dies.

Fans demanded the return of the popular character, so in a later episode (Johnson, "The Return of the Bionic Woman: Part 1," *The Six Million Dollar Man*, season 3, episode 1, September 14, 1975), it is revealed that Jaime's body had been placed in "cryogenic suspension" after she presumably died. She is revived by Dr. Wells (now played by Martin E. Brooks, 1925–2015), who is now able to safely remove the cerebral clot that killed her and repair her body. Some of her memories are lost in the process, including those of her love for Steve. Because trying to remember causes her great physical pain, Steve, not wanting to cause her harm, accepts that they will move forward as friends.

In *The Bionic Woman*, Jaime works as a teacher by day and lives in an apartment on the ranch owned by Steve's mother and stepfather. She takes on many missions in which she goes undercover, posing in such jobs as a nun, a police officer, a college student, and even a professional wrestler. Storylines often involved Jaime being drugged and kidnapped, forcing her to rely on her bionic powers to escape dangerous situations. Although Jaime is very powerful, she does have limits. In "Kill Oscar" (Arthur Rowe and Oliver Crawford, *The Bionic Woman*, season 2, episode 5, October 27, 1976) she jumps from a building that is too tall, resulting in a severe injury to her bionic legs. All of her implants, except the one encased in her ear, are susceptible to extreme cold and malfunction in such conditions. Among the most memorable enemies that Jaime faces are the

Jarvik-7

In 1982, the Jarvik-7 was the first successful transplant of an artificial heart into a human in world history. Named for design team leader Dr. Robert Jarvik (b. 1946), the Jarvik-7 represented a revolution in health care. Not to be confused with ventricular assist devices, machines that assist failing hearts, or the cardiopulmonary bypass machine utilized to temporarily act as heart and lung during surgery, the Jarvik-7 artificial heart was a fully functioning mechanical device meant to permanently replace the original human organ. The first successful artificial heart transplant was done in the Soviet Union in 1937, when Dr. Vladimir Demikhov (1916–1998) transplanted an artificial heart into a dog. The Jarvik-5 had experienced some success, keeping two calves alive for a while (one for six months, the other for eight). The Jarvik-7 was implanted into Barney Clark (1921–1983). He lived for just over three-and-a-half months (Clark's condition required that he also be connected to a pneumatic compressor). The second recipient of the Jarvik-7 was William Schroeder (1932–1986) in 1984; he lived for just under twenty-one months. The latest incarnation of the Jarvik-7 is known as the SynCardia and is currently used as a temporary device, allowing patients to live more comfortably while awaiting an organic heart transplant. While great strides have been made in the area of artificial limbs, permanent artificial organs are not yet feasible. In 2001, the company Abiomed received FDA approval for its titanium–based artificial heart, the Abiocor; but its design parameters make it only feasible for roughly 50 percent of patients, and its potential life span is from one to two years maximum. In 2017, the Swiss research facility ETH Zurich announced it was working on a silicone–based artificial heart, with a goal of making it functional for fifteen years.

Richard A. Hall

"Fembots," an army of female robots controlled by an evil scientist ("Kill Oscar"), one of a few crossover storylines with *The Six Million Dollar Man*. The Fembots reappear in "Fembots in Las Vegas, Parts 1 and 2" (Rowe, *The Bionic Woman*, season 3, episodes 3 and 4, September 24 and October 1, 1977). In another *Six Million Dollar Man* crossover, the bionic friends face off with Sasquatch (Rowe, "The Return of Bigfoot: Part 2," *The Bionic Woman*, season 2, episode 1, September 22, 1976).

In the third-season opener, Jaime gains a canine partner when Max, a German shepherd with bionic parts, is introduced (James D. Parriott and Harve Bennett, "The Bionic Dog, Parts 1 and 2," *The Bionic Woman*, season 3, episodes 1 and 2, September 10 and 17, 1977). Max, who had been used to test early bionic prosthetics, appears to be rejecting his bionic parts, but before he is put down, Jaime uncovers that the real cause for his condition is psychological and saves his life. A spin-off starring Max was rejected by the network, so he stayed with Jaime for the remainder of *The Bionic Woman*'s run. In the last episode of the series (Steven E. deSouza, "On the Run," *The Bionic Woman*, season 3, episode 22, May 13, 1978), Jaime wants to retire from OSI, but the organization is concerned about her leaving, claiming that her bionic parts make her the property of the company. When higher-ups at OSI try to place her where they can keep an eye on her, she goes on the run at Oscar's urging. She eventually comes to accept her bionic side and returns, agreeing to stay with OSI on the condition of being assigned fewer missions and having more free time.

The first of three TV movies, *The Return of the Six Million Dollar Man and the Bionic Woman* (NBC, 1987), reunites Jaime and Steve after nearly a decade. Upon recovering from a concussion, Jaime's memory is fully restored, and she must come to terms with her feelings for Steve. The second film, *Bionic Showdown: The Six Million Dollar Man and the Bionic Woman* (NBC, 1989), introduced Kate Mason (played by Sandra Bullock, b. 1964) as a paraplegic who is given bionic parts. Jaime helps train the new human cyborg. In the final reunion film, *Bionic Ever After?* (CBS, 1994), a computer virus corrupts Jaime's bionics. She receives a major upgrade that increases her powers and provides her with night vision, after which she and Steve finally marry. A remake of the series, *Bionic Woman* (NBC), was produced in 2007. Its darker tone failed to match the more hopeful tone of the original, and only eight episodes were produced and aired.

When Jaime Sommers came along in 1975, the sexual revolution in the United States was in full swing. For young girls who were ready to exercise their newly gained freedom, she provided a strong, smart female role model, transcending the sometimes one-dimensional roles women were often relegated to in film and television. Part of her appeal was that she did not rely solely on her bionic powers to complete missions, rarely using them against human opponents and never with deadly force. She leaned as much on her wits and other feminine strengths as on her physical capabilities. With her classic beauty and superhuman strength, she easily could have fallen into the archetype of "goddess in human form" or the "divine feminine," but the character remained down to earth, a bionic girl-next-door who struggled with her place in the world as a half human/half cyborg, offering a glimpse into a future where this fictitious phenomenon could become reality.

Lisa C. Bailey

See also: Adam, Alita, Borg, Cybermen, Cyborg, Daleks, Darth Vader, Doctor Octopus, Echo/CT-1409, General Grievous, Inspector Gadget, Nardole, RoboCop, Steve Austin; *Thematic Essay*: Cyborgs: Robotic Humans or Organic Robots?

Further Reading

Binns, Donna. 2013. "The Bionic Woman: Machine or Human?" In *Disability in Science Fiction*, edited by Kathryn Allan, 89–101. New York: Palgrave Macmillan.

Branson-Trent, Gregory M. 2009. The Bionic Woman: *Complete Episode Guide*. Scotts Valley, CA: CreateSpace.

Bryant, D'Orsay D., III. 1985. "Spare-Part Surgery: The Ethics of Organ Transplantation." *Journal of the National Medical Association* 77 (2). https://www.ncbi.nlm.nih.gov/pmc/articles/PMC2561842/pdf/jnma00245-0055.pdf.

Calvert, Bronwen. 2017. *Being Bionic: The World of TV Cyborgs*. London: I. B. Tauris.

McAdams, Taylor. 2019. "Fans Were Never Meant to Know What Happened Off Camera in *Bionic Woman*." Brain-Sharper, October 6, 2019. https://www.brain-sharper.com/entertainment/bionic-woman-fb.

Means, Sean P. 2019. "At FanX, 'Bionic' Stars Lindsay Wagner and Lee Majors Recall Their TV Glory, and Lots of Running." *Salt Lake Tribune*, September 6, 2019. https://www.sltrib.com/artsliving/2019/09/06/fanx-bionic-stars.

Pilato, Herbie J. 2007. *The Bionic Book:* The Six Million Dollar Man *and the* Bionic Woman *Reconstructed*. Albany, GA: BearManor.

Pilato, Herbie J. 2016. "A 40th Anniversary Tribute to the Bionic Woman and Wonder Woman Part I: *The Bionic Woman*." Emmys, December 19, 2016. https://www

.emmys.com/news/online-originals/40th-anniversary-tribute-bionic-woman-and
-wonder-woman-part-1-bionic-woman.

Tramel, Jimmie. 2019. "Bionic and Iconic: Lindsay Wagner, TV's Bionic Woman, Shares
Memories before Tulsa Pop Culture Expo." *Tulsa World*, October 30, 2019. https://
www.tulsaworld.com/entertainment/television/bionic-and-iconic-lindsay-wagner
-tv-s-bionic-woman-shares/article_a702a343-6afb-5127-9916-827077db598a.html.

Janet

The Good Place

In 2016, television producer Michael Schur (b. 1975) introduced audiences to
one of the most unique television sitcoms in the genre's history: *The Good Place*.
Set in a generic "afterlife," the series examines the human philosophy of "good"
and "evil" through the afterlives of four flawed humans. Rather than a "heaven" or
"hell," the world of the sitcom places souls in the "Good Place" or the "Bad Place"
(each possessing the overall ideologies of heaven or hell, respectively) based on a
"point system" that humans accumulate during their mortal lives. A demon named
Michael (played by Ted Danson, b. 1947), disguised as an angel, places four
human souls bound for the "Bad Place" into an imaginary "Good Place," their
eternal torture being their realization that they do not deserve eternal reward.
However, through an exploration of human philosophy, the four, led by Eleanor
(played by Kristen Bell, b. 1980), discover the ruse and—with the help of
Michael—become "better people," deserving of an eternity in the "Good Place."
Michael's primary assistant throughout this character study is the otherworldly AI
program "Janet" (played by D'Arcy Carden, b. 1980).

Though the nature of Janet's existence remains unclear throughout the series,
she does identify herself as a "program." She manifests as a human female (her
"personality" and garb depending on the nature of the particular "Janet" of the
moment, good or bad). Janet possesses the ability to appear on command and to
provide an infinite number of individual requests. When not performing duties,
Janet exists in her "Void," a seemingly boundless area of white nothingness.
Though it would be presumed that the "Janet" that is serving Michael in his ruse
against the human souls should be a "bad" Janet, the primary Janet of the series
shows all the characteristics of a "good" Janet. Though she admits that, as an AI
program, she is incapable of "feeling," she does develop romantic feelings for the
human soul Jason (played by Manny Jacinto, b. 1987). She also develops a deep
sense of loyalty and affection for Michael and the four humans. Though frequently
referred to as a servant to the various "living" denizens of the afterlife, Janet plays
a vital role in helping Michael and the humans force the afterlife to reexamine
their methods of judging human souls.

The character of Janet, at her core, is a typical AI program, bound by the rules
and specifications of her programming. She does, however, evolve to the point of
being able to manipulate and reinterpret her programming to be more in line with
her own growing sense of "right" and "wrong." Through the fantastic writing of
the series and the masterful performance of Carden, Janet emerges as a fully
rounded and well-developed character in her own right. Though a program, Janet

becomes a life of her own, and a model around which humans can build their own ambitions.

Richard A. Hall

See also: AI/Ziggy, Batcomputer, Brainiac, Cerebro/Cerebra, Control, Doctor/EMH, The Great Intelligence, HAL 9000, HERBIE, JARVIS/Friday, Landru, The Matrix/Agent Smith, OASIS, Oz, Rehoboam, Skynet, Starfleet Computer, TARDIS, V-GER, WOPR; *Thematic Essays:* Robots and Slavery, Heroic Robots and Their Impact on Sci-Fi Narratives, AI and the Apocalypse: Science Fiction Meeting Science Fact.

Further Reading

Albert, Robert S., and Brigante, Thomas R. 1962. "The Psychology of Friendship Relations: Social Factors." *Journal of Social Psychology* 56 (1). https://www.tandfon line.com/doi/abs/10.1080/00224545.1962.9919371.

Allen, Arthur. 2019. "There's a Reason We Don't Know Much about AI." Politico, September 16, 2019. https://www.politico.com/agenda/story/2019/09/16/artficial-intel ligence-study-data-000956?cid=apn.

Ambrosino, Brandon. 2018. "What Would It Mean for AI to Have a Soul?" BBC, June 17, 2018. https://www.bbc.com/future/article/20180615-can-artificial-intelligence-have -a-soul-and-religion.

Bastani, Aaron. 2019. *Fully Automated Luxury Communism*. New York: Verso.

Chandler, Simon. 2019. "Artificial Intelligence Has Become a Tool for Classifying and Ranking People." *Forbes*, October 1, 2019. https://www.forbes.com/sites/simon chandler/2019/10/01/artificial-intelligence-has-become-a-tool-for-classifying-and -ranking-people/#7431657f1d7c.

Chanthadavong, Aimee. 2019. "AI and Ethics: The Debate That Needs to Be Had." ZDnet, September 16, 2019. https://www.zdnet.com/article/ai-and-ethics-the-debate-that -needs-to-be-had/#ftag=CAD-03-10abf5f.

Conrad, Dean. 2018. *Space Sirens, Scientists and Princesses: The Portrayal of Women in Science Fiction Cinema*. Jefferson, NC: McFarland.

Dinh, Thien-Nam. 2018. *Silicon Minds: The Science, Impact, and Promise of Artificial Intelligence*. Independently Published.

Dowden, Bradley. 1993. *Logical Reasoning*. Belmont, CA: Wadsworth Publishing Company.

Ford, Martin. 2015. *Rise of the Robots: Technology and the Threat of a Jobless Future*. New York: Basic.

Hampton, Gregory Jerome. 2017. *Imagining Slaves and Robots in Literature, Film, and Popular Culture: Reinventing Yesterday's Slave with Tomorrow's Robot*. Lanham, MD: Lexington.

Leetaru, Kalev. 2019. "Automatic Image Captioning and Why Not Every AI Problem Can Be Solved Through More Data." *Forbes*, July 7, 2019. https://www.forbes.com /sites/kalevleetaru/2019/07/07/automatic-image-captioning-and-why-not-every -ai-problem-can-be-solved-through-more-data/#20b943476997.

Marks, Robert J. 2020. "*2084* vs. *1984*: The Difference AI Could Make to Big Brother." Podcast interview with author John Lennox. Mind Matters, July 3, 2020. https:// mindmatters.ai/2020/07/2084-vs-1984-the-difference-ai-could-make-to-big -brother.

Mayor, Adrienne. 2018. *Gods and Robots: Myths, Machines, and Ancient Dreams of Technology*. Princeton, NJ: Princeton University Press.

Pellissier, Hank. 2013. "Robots and Slavery: What Do Humans Want When We Are 'Masters'?" Institute for Ethics and Emerging Technologies, September 13, 2013. https://ieet.org/index.php/IEET2/more/pellissier20130913.

Roberts-Griffin, Christopher. 2011. "What Is a Good Friend: A Qualitative Analysis of Desired Friendship Qualities." *Penn McNair Research Journal* 3, no. 1 (December 21). https://repository.upenn.edu/cgi/viewcontent.cgi?article=1019&context=mcnair_scholars.

Robitzski, Dan. 2018. "Artificial Consciousness: How to Give a Robot a Soul." Futurism, June 25, 2018. https://futurism.com/artificial-consciousness.

Rouhiainen, Lasse. 2018. *Artificial Intelligence: 101 Things You Must Know Today about Our Future*. Scotts Valley, CA: CreateSpace.

Schroeder, Stan. 2020. "Samsung Just Launched an 'Artificial Human' Called Neon, and Wait, What?" Mashable, January 7, 2020. https://mashable.com/article/samsung-star-labs-neon-ces.

Thagard, Paul. 2017. "Will Robots Ever Have Emotions?" *Psychology Today*, December 14, 2017. https://www.psychologytoday.com/us/blog/hot-thought/201712/will-robots-ever-have-emotions.

Virk, Rizwan. 2019. *The Simulation Hypothesis: An MIT Computer Scientist Shows Why AI, Quantum Physics, and Eastern Mystics All Agree We Are in a Video Game*. Milwaukee, WI: Bayview.

Wallach, Wendell, and Allen, Colin. 2009. *Moral Machines: Teaching Robots Right from Wrong*. New York: Oxford University Press.

JARVIS/Friday

Marvel Cinematic Universe

In 2008, Marvel Comics launched their new Marvel Studios with the debut of the "Marvel Cinematic Universe" with the film *Iron Man*, starring Robert Downey Jr. (b. 1965). Downey played classic Marvel Comics billionaire engineer and weapons manufacturer Tony Stark. After being taken hostage by terrorists in Afghanistan and forced to produce weapons for them, Stark, whose heart was damaged during capture, instead manufactures a weaponized suit of armor powered by the device he has built to keep the metal fragments near his heart from killing him. Escaping and returning home to the United States, Stark decides to perfect his suit of armor idea and fight wrong as Iron Man (Jon Favreau, dir., *Iron Man*, Marvel Studios, 2008). Stark's primary ally is his AI assistant, JARVIS (Just A Rather Very Intelligent System). In the *Iron Man* and *Avengers* films, JARVIS is voiced by Paul Bettany (b. 1971).

The concept of the JARVIS AI in the Marvel films is based on the human butler (Edwin Jarvis) who works for Tony Stark and the Avengers in the Marvel Comics. In the films, the system is named after Tony's father's butler, Edwin Jarvis (played by James D'Arcy, b. 1975, in the television series *Agent Carter*, ABC, 2015–2016). At the beginning of the film *Iron Man*, JARVIS is the AI embedded in Stark's home, handling standard functions (opening blinds, regulating temperature, turning devices on or off, home security, providing weather updates, schedule reminders, alarm clock, etc.). More than a generic AI along the lines of Siri or Alexa, JARVIS has a distinct personality, complete with sarcasm and the ability, as if he was human, to completely interact with Stark. JARVIS is also Stark's personal assistant in his laboratory, running computer simulations and searching files and/or the internet, among other duties. When Stark decides to become "Iron Man," he

Siri and Alexa

In 2008, moviegoers around the world were introduced to JARVIS, the super advanced AI program personal assistant to Tony Stark in that year's *Iron Man*. Not long after, such AI assistants began popping up, first in iPhones and later in other household units around the world. In 2011, Apple launched "Siri" (available on all Apple iOS systems, but primarily associated with the iPhone). Rather than looking up various bits of information on their phones or performing basic functions by themselves, users can simply say, for example, "Siri, where is the closest pizza place?" or "Siri, set alarm for six a.m." Users can allow Siri access to everything within their device, allowing Siri to call up photos, videos, songs, and more. Popularly recognized with a female voice speaking American English, Siri can be programmed to sound like any gender, in various languages, and with accents. As an amusing plus, designers even programmed Siri with the comedy routine "Who's on First?" and causing her response to that specific question to be, "That is correct, Who is on first." Along very similar lines, Amazon launched "Alexa" in 2014, first used as part of Amazon's Echo device. Alexa possesses similar abilities to Siri and is capable of looking up nearby restaurants, providing weather updates, placing phone calls, and so forth. Alexa also keeps track of a user's internet search history and online purchasing history to better advise and "serve" the user. As both AI programs, at their core, are required to keep intricate records of their owner's behaviors, however, the companies that operate these systems also have access to the most intimate details of a user's private life. This has raised considerable concern among consumers; but, to date, the luxury provided by these virtual assistants has outweighed any negative long-term effects.

Richard A. Hall

networks the suit of armor with JARVIS, who can monitor the suit's systems, warn Stark of malfunctions, and—to a degree—take over the suit should Stark lose consciousness. At the end of *Iron Man 3* (2013), JARVIS controls the army of unmanned Iron Man suits, creating the "Iron Legion."

In the film *Avengers: Age of Ultron* (2015), JARVIS is seemingly "killed" by the superior AI program "Ultron." However, calculating Ultron's likely victory over his systems, JARVIS scatters his consciousness, working through various systems to keep a semblance of himself intact. When the Avengers capture the synthetic body meant to be Ultron's next incarnation, Stark utilizes the synthetic body, the Infinity Stone known as the "Mind Stone," and the remaining remnants of JARVIS to create—with the help of lightning provided by the Norse god of thunder, Thor—a new AI, synthetic life-form: Vision (also played by Bettany). Without JARVIS as his assistant for the Iron Man armor, Stark implements a new AI: Friday (Female Replacement Intelligent Digital Assistant Youth). Friday (voiced by Kerry Condon, b. 1983) receives her name from the term "girl Friday," as in the 1940 film *His Girl Friday*, referring to a woman assistant (Joss Whedon, *Avengers: Age of Ultron*, Marvel/Disney, 2015).

JARVIS/Friday represent the ideal in interactive AI assistance. As much as Americans have come to rely on AI such as Siri and Alexa, most would agree that they would prefer an AI that can interact more personally. With Tony Stark being such a highly intelligent and sarcastic person, it is logical that he would model an AI program that could keep up with his conversations on an equal basis. The implementation of the AI's ability to seemingly "care" makes JARVIS/Friday a character equal to the humans in the narrative; and the "death" of JARVIS in *Age of Ultron* hit audiences to the same degree that a human character's death would. A real-world JARVIS/Friday, however, would bring with it the potential for it to evolve beyond its original programming (as the Ultron AI did in the film). This potentiality makes the concept as dangerous as it is appealing.

Richard A. Hall

See also: AI/Ziggy, Arnim Zola, Batcomputer, Brainiac, Cerebro/Cerebra, Control, Doctor/EMH, Dr. Theopolis and Twiki, The Great Intelligence, HAL 9000, HERBIE, Iron Legion, Iron Man, Janet, The Matrix/Agent Smith, OASIS, Rehoboam, Skynet, Starfleet Computer, TARDIS, Ultron, V-GER, Vision, Warlock, WOPR, Zordon/Alpha-5; *Thematic Essays:* Robots and Slavery, AI and the Apocalypse: Science Fiction Meeting Science Fact.

Further Reading

Albert, Robert S., and Brigante, Thomas R. 1962. "The Psychology of Friendship Relations: Social Factors." *The Journal of Social Psychology* 56 (1). *Taylor & Francis.* https://www.tandfonline.com/doi/abs/10.1080/00224545.1962.9919371.

Amati, Viviana, Meggiolaro, Silvia, Rivellini, Giulia, and Zaccarin, Susanna. 2018. "Social Relations and Life Satisfaction: The Role of Friends." *Genus* 74, no. 1 (May 4). https://www.ncbi.nlm.nih.gov/pmc/articles/PMC5937874.

Ambrosino, Brandon. 2018. "What Would It Mean for AI to Have a Soul?" BBC, June 17, 2018. https://www.bbc.com/future/article/20180615-can-artificial-intelligence-have-a-soul-and-religion.

Bastani, Aaron. 2019. *Fully Automated Luxury Communism*. New York: Verso.

Byrne, Emma. 2013. "Innovation Isn't Safe: The Future According to Kevin Warwick." *Forbes*, September 30, 2013. https://www.forbes.com/sites/netapp/2013/09/30/kevin-warwick-captain-cyborg/#48b704c13560.

Chandler, Simon. 2019. "Artificial Intelligence Has Become a Tool for Classifying and Ranking People." *Forbes*, October 1, 2019. https://www.forbes.com/sites/simonchandler/2019/10/01/artificial-intelligence-has-become-a-tool-for-classifying-and-ranking-people/#7431657f1d7c.

Chanthadavong, Aimee. 2019. "AI and Ethics: The Debate That Needs to Be Had." ZDnet, September 16, 2019. https://www.zdnet.com/article/ai-and-ethics-the-debate-that-needs-to-be-had/#ftag=CAD-03-10abf5f.

Ford, Martin. 2015. *Rise of the Robots: Technology and the Threat of a Jobless Future.* New York: Basic.

Howe, Sean. 2012. *Marvel Comics: The Untold Story.* New York: Harper-Perennial.

Lin, Patrick, Jenkins, Ryan, and Abney, Keith, eds. 2017. *Robot Ethics 2.0: From Autonomous Cars to Artificial Intelligence.* New York: Oxford University Press.

Linder, Courtney. 2020. "This AI Robot Just Nabbed the Lead Role in a Sci-Fi Movie: Meet Erica, the Movie Star Who May Put Human Actors Out of Work." *Popular Mechanics*, June 25, 2020. https://www.popularmechanics.com/technology/robots/a32968811/artificial-intelligence-robot-movie-star-erica.

Marks, Robert J. 2020. "*2084* vs. *1984*: The Difference AI Could Make to Big Brother." Podcast interview with author John Lennox. Mind Matters, July 3, 2020. https://mindmatters.ai/2020/07/2084-vs-1984-the-difference-ai-could-make-to-big-brother.

Murdock, Jason. 2019. "Former Google Engineer Warns AI Might Accidentally Start a War: 'These Things Will Start to Behave in Unexpected Ways.'" *Newsweek*, September

16, 2019. https://www.newsweek.com/google-project-maven-artificial-intelligence-laura-nolan-killer-robots-department-defense-1459358.

Paur, Joey. 2019. "An AI Bot Writes a Hilarious Episode of *Star Trek: The Next Generation*." Geek Tyrant, November 16, 2019. https://geektyrant.com/news/an-ai-bot-writes-a-hilarious-episode-of-star-trek-the-next-generation.

Tucker, Reed. 2017. *Slugfest: Inside the Epic 50-Year Battle Between Marvel and DC*. New York: Da Capo Press.

Virk, Rizwan. 2019. *The Simulation Hypothesis: An MIT Computer Scientist Shows Why AI, Quantum Physics, and Eastern Mystics All Agree We Are in a Video Game*. Milwaukee, WI: Bayview.

White, Mark D., ed. 2010. *Iron Man and Philosophy: Facing the Stark Reality*. Hoboken, NJ: Wiley.

Jocasta

Marvel Comics

In 1961, comic book legend Stan Lee (1922–2018) launched the "Marvel age" of comics. With the launching of the Fantastic Four—soon to be followed by the Hulk, Spider-Man, Iron Man, and many, many more—comic book superheroes took a more dramatic turn, with storylines focused on the private lives of those blessed (or burdened) with superpowers. One of Marvel's early successes was the superhero team *The Avengers* (1963). Over the decades, the Avengers have had a continuously revolving roster of members. One such member was the robot, Jocasta, the creation of one of the Avengers' greatest foes, the robot Ultron. In the comics, Ultron was the creation of Avenger Hank Pym. When Ultron gained sentience, it turned on its master like a modern-day Frankenstein's monster. Like that creature from classic literature, Ultron also sought a "bride." In a vengeful act against his creator, Ultron–based Jocasta's brain waves on those of Pym's own love interest, Janet Van Dyne (a.k.a. the "Wasp"), even forcing Pym to utilize Van Dyne's own life energy to bring his creation to life. Aware that this would kill the human woman, Jocasta resisted and betrayed her robot love (Jim Shooter and George Pérez, *The Avengers #162*, August 1977).

In another literary connection, Jocasta was named after the wife/mother character from the classic Greek play *Oedipus Rex* (c. 429 BCE); the AI character was both wife to Ultron and a copy of Ultron's "father's" wife. While Jocasta was initially programmed to "love" Ultron, she was repeatedly able to overcome that programming to do what was "right." Like her creator, Jocasta possesses enhanced strength, eye lasers, the ability to interact with computers, and the ability to create protective force fields. Unlike Ultron, Jocasta also possesses holographic projectors that allow her to manipulate her appearance. In yet another comparison, this time to the classic film *The Bride of Frankenstein* (1935), while Ultron's appearance is that of a nightmarish skeletal creature, Jocasta possesses the bodily appearance of a beautiful human female. Unlike her cinematic forebear, however, Jocasta does "love" Ultron (at least as much as her programming allows for such emotion).

The creation of Jim Shooter (b. 1951) and George Pérez (b. 1954), Jocasta is much more a modern-day take on women of classic literature and film than a traditional cautionary tale of the future of robotics and AI. Though initially created to be a de facto "love slave," she emerges beyond her primary programming to be a hero in every sense of the word. She is not defined by her original purpose. She charts her own course and becomes a "person." Though likely tortured by the logical inconsistencies between her programmed "love" and her sense of "right," Jocasta triumphs as a modern symbol of the 1970s women's liberation movement. She is not a mindless automaton. She is her own woman.

Richard A. Hall

See also: Buffybot, Dolores, Doombots, Fembots, Iron Man, Janet, L3-37/*Millennium Falcon*, LMDs, Maschinenmensch/Maria, Soji and Dahj Asha, Stepford Wives, Ultron, Vision, Warlock; *Thematic Essays:* Robots and Slavery, Heroic Robots and Their Impact on Sci-Fi Narratives.

Further Reading

Albert, Robert S., and Brigante, Thomas R. 1962. "The Psychology of Friendship Relations: Social Factors." *Journal of Social Psychology* 56 (1). https://www.tandfon line.com/doi/abs/10.1080/00224545.1962.9919371.

Amati, Viviana, Meggiolaro, Silvia, Rivellini, Giulia, and Zaccarin, Susanna. 2018. "Social Relations and Life Satisfaction: The Role of Friends." *Genus* 74, no. 1 (May 4). https://www.ncbi.nlm.nih.gov/pmc/articles/PMC5937874.

Ambrosino, Brandon. 2018. "What Would It Mean for AI to Have a Soul?" BBC, June 17, 2018. https://www.bbc.com/future/article/20180615-can-artificial-intelligence-have -a-soul-and-religion.

Bastani, Aaron. 2019. *Fully Automated Luxury Communism*. New York: Verso.

Blair, Anthony. 2018. "Sex Robots That Feel LOVE and Suffer PAIN When Dumped Coming Soon Claims Expert." *Daily Star*, December 6, 2018. https://www.dailys tar.co.uk/news/latest-news/sex-robot-feel-love-pain-16824160.

Cavaler, Chris. 2015. *On the Origins of Superheroes: From the Big Bang to Action Comics No. 1*. Iowa City: University of Iowa Press.

Chanthadavong, Aimee. 2019. "AI and Ethics: The Debate that Needs to Be Had." ZDnet, September 16, 2019. https://www.zdnet.com/article/ai-and-ethics-the-debate-that -needs-to-be-had/#ftag=CAD-03-10abf5f.

Costello, Matthew J. 2009. *Secret Identity Crisis: Comic Books & the Unmasking of Cold War America*. New York: Continuum.

Dinh, Thien-Nam. 2018. *Silicon Minds: The Science, Impact, and Promise of Artificial Intelligence*. Independently Published.

Hitchcock, Susan Tyler. 2007. *Frankenstein: A Cultural History*. New York: W. W. Norton.

Howe, Sean. 2012. *Marvel Comics: The Untold Story*. New York: Harper-Perennial.

Lin, Patrick, Abney, Keith, and Bekey, George A., eds. 2014. *Robot Ethics: The Ethical and Social Implications of Robotics*. Cambridge: Massachusetts Institute of Technology.

Pellissier, Hank. 2013. "Robots and Slavery: What Do Humans Want When We Are 'Masters'?" Institute for Ethics and Emerging Technologies, September 13, 2013. https://ieet.org/index.php/IEET2/more/pellissier20130913.

Randall, Terri, dir. 2016. *NOVA*. "Rise of the Robots: Inside the DARPA Robotics Challenge." Aired February 24, 2016, on PBS. DVD.

Roberts-Griffin, Christopher. 2011. "What Is a Good Friend: A Qualitative Analysis of Desired Friendship Qualities." *Penn McNair Research Journal* 3, no. 1 (December 21).

https://repository.upenn.edu/cgi/viewcontent.cgi?article=1019&context=mcnair
_scholars.

Robitzski, Dan. 2018. "Artificial Consciousness: How to Give a Robot a Soul." Futurism, June 25, 2018. https://futurism.com/artificial-consciousness.

Rouhiainen, Lasse. 2018. *Artificial Intelligence: 101 Things You Must Know Today about Our Future*. Scotts Valley, CA: CreateSpace.

Schroeder, Stan. 2020. "Samsung Just Launched an 'Artificial Human' Called Neon, and Wait, What?" Mashable, January 7, 2020. https://mashable.com/article/samsung
-star-labs-neon-ces.

Seed, David. 1999. *American Science Fiction and the Cold War: Literature and Film*. Abingdon, UK: Routledge.

Spaeth, Dennis. 2018. "From Single-Task Machines to Backflipping Robots: The Evolution of Robots." *Cutting Tool Engineering*, January 15, 2018. https://www.ctemag
.com/news/articles/evolution-of-robots.

Thomas, Roy. 2017. *The Marvel Age of Comics: 1961–1978*. Los Angeles: Taschen.

Wallach, Wendell, and Allen, Colin. 2009. *Moral Machines: Teaching Robots Right from Wrong*. New York: Oxford University Press.

Johnny 5/S.A.I.N.T. Number 5

Short Circuit

In the 1980s, the U.S.-Soviet Cold War was in one of its most heated periods, and American films of the 1980s reflected this growing tension. One such film was *Short Circuit* (TriStar, 1986). In this film, two robotics engineers develop a series of robots for use by the U.S. military against the Soviet Union. The series of robots was called S.A.I.N.T. (Strategic Artificially Intelligent Nuclear Transport). One of the units—"Number 5"—is struck by lightning, which gives the robot "sentience"; at that point, Number 5 escapes. Naturally, the laboratory wants its valuable property returned. Number 5 meets a sympathetic young woman and, through interactions with her, "learns" about humanity, life, and death. After successfully evading the lab's security (and some of his fellow S.A.I.N.T.s), and faking his "destruction," Number 5 escapes to the country with one of his initial creators and his female friend, giving himself the name "Johnny" (S. S. Wilson and Brent Maddock, *Short Circuit*, TriStar Pictures, 1986).

Johnny 5 closely resembles modern-day bomb-disposal robots, and—to a lesser degree—NASA rovers. He has a torso with two arms, a dual-camera head (resembling complex binoculars); and dual front treads and rear wheels for motion (with a hydraulic "life" allowing his torso to raise or lower). Being a transportation robot, he is not equipped with any weaponry or defense mechanisms. He is a simple roving robot. The film, like many in the '80s, is a cautionary tale against nuclear war, conveying the message that if the robot Johnny can discover the importance of life, then so, too, should humans. Johnny 5 was a popular pop culture icon and definitive of sci-fi of that decade.

Richard A. Hall

See also: AWESOM-O 4000, Baymax, BB-8, Cambot/Gypsy/Tom Servo/Crow, D-O, HERBIE, Iron Giant, K9, Muffit, R2-D2, VICI, WOPR; *Thematic Essays:* Robots and Slavery, Heroic Robots and Their Impact on Sci-Fi Narratives.

A replica of Johnny 5, from the sci-fi film *Short Circuit* (1986). (Jose Gil/Dreamstime.com)

Mars Rovers

Mars rovers are small, wheeled vehicles mounted with scientific equipment and a camera and powered by solar panels on the vehicles' "back." NASA and the Jet Propulsion Laboratory have, to date, had four successful Mars rover landings since the late 1990s. In 1997, *Sojourner* (a.k.a. *Pathfinder*) successfully landed and began transmissions to Earth, but communication was lost eleven weeks later. *Spirit* landed in January 2004 and covered almost five miles before becoming trapped in sand; it continued to communicate with NASA for another six years before losing signal (but remained trapped). *Opportunity* landed just a few weeks after *Spirit* but was much more successful. Beating all expectations, *Opportunity* remained in contact for fifteen years, covering almost thirty miles of territory. The most advanced rover to date, *Curiosity* (a.k.a. *Mars Science Laboratory*) landed in 2012 and, at the time of this writing, remains in contact with NASA and the Jet Propulsion Laboratory. In July 2020, NASA launched the most recent rover, *Perseverance*, just a week after China's first rover mission, *Tianwen-1*. Through the rover missions, NASA and the United States hope to gather enough data about Mars to pave the way for future human-staffed missions to the "Red Planet." The rovers conduct research and exploration missions examining Mars's climate and geology and are also attempting to determine whether "life" (as we understand it) ever existed on the planet and, if so, what happened to that life.

Richard A. Hall

Further Reading

Albert, Robert S., and Brigante, Thomas R. 1962. "The Psychology of Friendship Relations: Social Factors." *Journal of Social Psychology* 56 (1). https://www.tandfon line.com/doi/abs/10.1080/00224545.1962.9919371.

Amati, Viviana, Meggiolaro, Silvia, Rivellini, Giulia, and Zaccarin, Susanna. 2018. "Social Relations and Life Satisfaction: The Role of Friends." *Genus* 74, no. 1 (May 4). https://www.ncbi.nlm.nih.gov/pmc/articles/PMC5937874.

Ambrosino, Brandon. 2018. "What Would It Mean for AI to Have a Soul?" BBC, June 17, 2018. https://www.bbc.com/future/article/20180615-can-artificial-intelligence-have-a-soul-and-religion.

Ashby, LeRoy. 2006. *With Amusement for All: A History of American Popular Culture since 1830.* Lexington: University Press of Kentucky.

Bastani, Aaron. 2019. *Fully Automated Luxury Communism.* New York: Verso.

Carper, Steve. 2019. *Robots in American Popular Culture.* Jefferson, NC: McFarland.

Lin, Patrick, Abney, Keith, and Bekey, George A., eds. 2014. *Robot Ethics: The Ethical and Social Implications of Robotics.* Cambridge: Massachusetts Institute of Technology.

Randall, Terri, dir. 2016. *NOVA.* "Rise of the Robots: Inside the DARPA Robotics Challenge." Aired February 24 on PBS. DVD.

Resnick, Brian. 2020. "NASA's Latest Rover Is Our Best Chance Yet to Find Life on Mars." Vox, July 30, 2020. https://www.vox.com/science-and-health/2020/7/29/21340464/nasa-perseverance-ingenuity-launch-live-stream-how-to-watch-science-life-on-mars.

Seed, David. 1999. *American Science Fiction and the Cold War: Literature and Film.* Abingdon, UK: Routledge.

Spaeth, Dennis. 2018. "From Single-Task Machines to Backflipping Robots: The Evolution of Robots." *Cutting Tool Engineering*, January 15, 2018. https://www.ctemag.com/news/articles/evolution-of-robots.

Starck, Kathleen, ed. 2010. *Between Fear and Freedom: Cultural Representations of the Cold War.* Newcastle, UK: Cambridge Scholars.

Wallach, Wendell, and Allen, Colin. 2009. *Moral Machines: Teaching Robots Right from Wrong.* New York: Oxford University Press.

K

K-2SO

Star Wars

In 1977, writer-director George Lucas (b. 1944) introduced the world to the sci-fi/fantasy phenomenon *Star Wars* (Lucasfilm/20th Century Fox, 1977), one of the most successful pop culture franchises in American history. His original trilogy of films (1977–1983) focused on the rise of its protagonist, Luke Skywalker (played by Mark Hamill, b. 1951), against the backdrop of a Galactic Civil War against the human forces of the evil Empire, led by its Emperor, Sheev Palpatine (played by Ian McDiarmid, b. 1944), and his henchman, Darth Vader (played by David Prowse, b. 1935; voiced by James Earl Jones, b. 1931). In 2012, Walt Disney Corporation announced that it had purchased Lucasfilm and the rights to *Star Wars*. In 2016, the second *Star Wars* film of the Disney era was released: *Rogue One: A Star Wars Story*—an immediate "prequel" to the original film—which told the tale of how the Rebellion achieved the stolen plans to the Empire's Death Star battle station just prior to the opening of the original *Star Wars*. One member of the team of heroes who sacrificed their lives to retrieve the stolen plans was the reprogrammed Imperial droid K-2SO (voiced by Alan Tudyk, b. 1971).

The brainchild of *Rogue One* writer-producer John Knoll (b. 1962), K-2SO began its existence as an Imperial security droid during the Galactic Civil War between the ruling Empire and the Rebel Alliance. Captured and reprogrammed by the Alliance, K-2SO becomes the partner/property of Alliance spy Cassian Andor (played by Diego Luna, b. 1979). Freed from its emotionless original programming, K-2SO takes on a distinct personality, laced with sarcasm. Cassian refuses to allow the droid to carry a weapon, doubtless due to the fear that its original programming could reemerge at any time. When the two capture criminal Jyn Erso (played by Felicity Jones, b. 1983), she, too, distrusts the Imperial droid. Over the course of their mission to steal the Death Star plans, K-2SO gains Erso's trust; leading her to give the droid a weapon in order to defend itself as it guides her and Andor to the plans. Ultimately, K-2SO sacrifices itself in order to protect his human friends and their mission (Chris Weitz and Tony Gilroy, *Rogue One: A Star Wars Story*, Lucasfilm/Disney, 2016).

As an Imperial droid, K-2SO is a valuable asset to the Alliance. He can mingle in Imperial areas without raising suspicion. He can interact with Imperial computers, and he possesses an encyclopedic knowledge of Imperial procedures and regulations. He proves to be a true hero in every sense of the word but is—as are most droids are in the *Star Wars* franchise—a slave. K-2SO is "property." Even after being "rescued" by the freedom-loving Rebels, he remains the property of one of their agents. Though now possessing "free will," he is limited as to what degree he can act on it. This is the case for most "heroic" droids in the saga: they are property despite having distinct personalities, a clear sense of self-awareness, and the ability to think and act upon free will. This raises the question as to how "good" the "good guys" are in *Star Wars*. They are clearly willing to sacrifice their lives for the freedom of all organic life; but, conversely, they are not willing to bestow such freedom on the inorganic life-forms on whom they so desperately depend for their mission to succeed.

Richard A. Hall

See also: B.A.T.s, Battle Droids, BB-8, C-3PO, Darth Vader, D-O, EV-9D9, General Grievous, IG-88/IG-11, L3-37/*Millennium Falcon*, Medical Droids, R2-D2, Tin Woodsman, Warlock; *Thematic Essays*: Robots and Slavery, Heroic Robots and Their Impact on Sci-Fi Narratives.

Further Reading

Albert, Robert S., and Brigante, Thomas R. 1962. "The Psychology of Friendship Relations: Social Factors." *Journal of Social Psychology* 56 (1). https://www.tandfon line.com/doi/abs/10.1080/00224545.1962.9919371.

Amati, Viviana, Meggiolaro, Silvia, Rivellini, Giulia, and Zaccarin, Susanna. 2018. "Social Relations and Life Satisfaction: The Role of Friends." *Genus* 74, no. 1 (May 4). https://www.ncbi.nlm.nih.gov/pmc/articles/PMC5937874/.

Bray, Adam, and Horton, Cole. 2017. Star Wars*: Absolutely Everything You Need to Know, Updated and Expanded.* London, England, UK: DK Children.

Carper, Steve. 2019. *Robots in American Popular Culture.* Jefferson, NC: McFarland.

Dinh, Thien-Nam. 2018. *Silicon Minds: The Science, Impact, and Promise of Artificial Intelligence.* Independently Published.

Eberl, Jason T., and Decker, Kevin S., eds. 2015. *The Ultimate* Star Wars *and Philosophy: You Must Unlearn What You Have Learned.* Hoboken, NJ: Wiley-Blackwell.

Fryer-Biggs, Zachary. 2019. "Coming Soon to a Battlefield: Robots That Can Kill." *The Atlantic*, September 3, 2019. https://www.theatlantic.com/technology/archive/2019 /09/killer-robots-and-new-era-machine-driven-warfare/597130.

Hampton, Gregory Jerome. 2017. *Imagining Slaves and Robots in Literature, Film, and Popular Culture: Reinventing Yesterday's Slave with Tomorrow's Robot.* Lanham, MD: Lexington.

Krishnan, Armin. 2009. *Killer Robots: Legality and Ethicality of Autonomous Weapons.* London: Routledge.

Lin, Patrick, Abney, Keith, and Bekey, George A., eds. 2014. *Robot Ethics: The Ethical and Social Implications of Robotics.* Cambridge: Massachusetts Institute of Technology.

Pellissier, Hank. 2013. "Robots and Slavery: What Do Humans Want When We Are 'Masters'?" Institute for Ethics and Emerging Technologies, September 13, 2013. https://ieet.org/index.php/IEET2/more/pellissier20130913.

Reagin, Nancy R., and Liedl, Janice, eds. 2013. Star Wars *and History.* New York: John Wiley & Sons.

Sumerak, Marc. 2018. Star Wars*: Droidography.* New York: HarperFestival.

Sunstein, Cass R. 2016. *The World According to* Star Wars. New York: Dey Street.

Taylor, Chris. 2015. *How* Star Wars *Conquered the Universe: The Past, Present, and Future of a Multibillion Dollar Franchise.* New York: Basic Books.

Wallach, Wendell, and Allen, Colin. 2009. *Moral Machines: Teaching Robots Right from Wrong.* New York: Oxford University Press.

K9

Doctor Who

In 1963, BBC executive Sydney Newman (1917–1997) and producer Verity Lambert (1935–2007) created the British television series *Doctor Who* (BBC, 1963–1989; FOX, 1996; BBC, 2005–present), the longest running science fiction television series in history. The series centers on the character of the "Doctor," an alien "Time Lord" possessing the power to "regenerate," or gain a new body and personality, when his or her (depending on the incarnation) physical body dies. The Doctor travels through time and space in a device known as a TARDIS (Time and Relative Dimension in Space). Though TARDISes are designed to change their outward appearance in order to blend in with the environment, the Doctor's TARDIS is "stuck" in the constant form of a 1960s British police call box. The fourth incarnation of the Doctor (played by Tom Baker, b. 1934) encountered a small robot dog named K9 (sometimes written as "K-9"), who became a companion of the Doctor's for a while afterward.

K9 (first voiced by John Leeson, b. 1943) was a small, trapezoid-shaped robot with a dog-shaped "head" and antenna "tail." His head had two radar disk protrusions for "ears." He referred to the Doctor as "Master," and his female companion(s) as "Mistress." Created in the year 5000 by Professor Marius (played by Frederick Jaeger, 1928–2004), K9 assists the Doctor against the villain Nucleus (also voiced by Leeson). Marius then gives K9 to the Doctor, and the small robot leaves for new adventures aboard the TARDIS (Bob Baker and Dave Martin, "The Invisible Enemy, Parts 1–4," *Doctor Who*, original series, season 15, episodes 5–9, October 1, 8, 15, 22, 1977). K9 was a frequently appearing character on the series for the next three seasons and made a brief appearance in the twentieth-anniversary special, *Doctor Who: The Five Doctors* (1983). He also appeared in three spin-off series from BBC: *K-9 and Company* (1981); *The Sarah Jane Adventures* (2007–2011); and *K-9* (2009–2010).

Though clearly intelligent and resourceful, the robot dog K9 clearly sees itself as being the "property" of living beings. It is difficult to ascertain whether this is because he is a robot or because he identifies as a dog (referencing the Doctor as "Master" could be explained either way). Like Twiki on *Buck Rogers in the 25th Century* and Muffit on *Battlestar Galactica*, K9 is clearly a response to the global success of *Star Wars* and the massive popularity of that film's own small robot helper, R2-D2. What all of these robots have in common, other than their repeated usefulness to their respective human "owners," is their appeal to young viewers. No other incarnation of the Doctor traveled with K9, the robot having been (mysteriously) given to the fourth Doctor's prior human companion, Sarah Jane Smith (who was no longer traveling with the Doctor by the time he encountered K9). Despite his brief time on the program, K9 remains a popular aspect of the long-running sci-fi classic.

Richard A. Hall

See also: Baymax, BB-8, C-3PO, Cybermen, Daleks, D-O, Dr. Theopolis and Twiki, The Great Intelligence, HERBIE, K-2SO, Muffit, Nardole, R2-D2, Robot, Rosie, TARDIS, Teselecta, WALL-E; *Thematic Essays:* Robots and Slavery, Heroic Robots and Their Impact on Sci-Fi Narratives.

Further Reading

Allen, Arthur. 2019. "There's a Reason We Don't Know Much about AI." Politico, September 16, 2019. https://www.politico.com/agenda/story/2019/09/16/artficial-intelligence-study-data-000956?cid=apn.

Ambrosino, Brandon. 2018. "What Would It Mean for AI to Have a Soul?" BBC, June 17, 2018. https://www.bbc.com/future/article/20180615-can-artificial-intelligence-have-a-soul-and-religion.

Bastani, Aaron. 2019. *Fully Automated Luxury Communism*. New York: Verso.

Campbell, Mark. 2010. Doctor Who: *The Complete Guide*. London: Running Press.

Dinh, Thien-Nam. 2018. *Silicon Minds: The Science, Impact, and Promise of Artificial Intelligence*. Independently Published.

Dowden, Bradley. 1993. *Logical Reasoning*. Belmont, CA: Wadsworth Publishing Company.

Hills, Matt, ed. 2013. *New Dimensions of* Doctor Who: *Adventures in Space, Time and Television*. London: I. B. Tauris.

Kistler, Alan. 2013. Doctor Who: *Celebrating Fifty Years, A History*. Guilford, CT: Lyons.

Lewis, Courtland, and Smithka, Paula, eds. 2010. Doctor Who *and Philosophy: Bigger on the Inside*. Chicago: Open Court.

Lewis, Courtland, and Smithka, Paula, eds. 2015: *More* Doctor Who *and Philosophy: Regeneration Time*. Chicago: Open Court.

Lin, Patrick, Jenkins, Ryan, and Abney, Keith, eds. 2017. *Robot Ethics 2.0: From Autonomous Cars to Artificial Intelligence*. New York: Oxford University Press.

Muir, John Kenneth. 2007. *A Critical History of* Doctor Who *on Television*. Jefferson, NC: McFarland.

Nahin, Paul J. 1999. *Time Machines: Time Travel in Physics, Metaphysics, and Science*. 2nd ed. Woodbury, NY: AIP Press; New York: Springer.

Rouhiainen, Lasse. 2018. *Artificial Intelligence: 101 Things You Must Know Today about Our Future*. Scotts Valley, CA: CreateSpace.

Virk, Rizwan. 2019. *The Simulation Hypothesis: An MIT Computer Scientist Shows Why AI, Quantum Physics, and Eastern Mystics All Agree We Are in a Video Game*. Milwaukee, WI: Bayview.

Wasserman, Ryan. 2018. *Paradoxes of Time Travel*. Oxford: Oxford University Press.

KITT

Knight Rider

In 1982, legendary television creator-producer Glen A. Larson (1937–2014) introduced audiences around the world to his series *Knight Rider* (NBC, 1982–1986). In the series, billionaire philanthropist Wilton Knight orders the rescue/abduction of badly injured policeman Michael Long, giving him reconstructive facial surgery and a new identity: Michael Knight. Michael (played by David Hasselhoff, b. 1952) is then recruited by Wilson's private organization, the Foundation for Law and Government (FLAG) to act as a detective seeking justice around the United States. Michael is also given control of KITT (Knight Industries Two Thousand): an AI supercomputer housed in a black 1982 Pontiac Trans Am (the AI voiced by William Daniels, b. 1927). Decades before the Marvel AI JARVIS, KITT built a bond of trust and friendship with Michael Knight.

HERBIE ("The Love Bug")

In 1968, Walt Disney Studios released the live-action film *The Love Bug*, based on the 1961 book *Car, Boy, Girl* by Gordon Buford (life years unknown). The film featured Dean Jones (1931–2015), Michele Lee (b. 1942), and Buddy Hackett (1924–2003). The true star of the film, however, was the 1963 Volkswagen two-door Beetle named HERBIE. The car was painted off-white with off-white hubcaps and interior. It had one red, two white, and two blue racing stripes running along the center of the vehicle, with a black "53" in a white circle on the hood and doors. Its California license plate (with yellow letters on a black field) read "OFP 857." From 1968 to 2005, HERBIE appeared in six feature films. What distinguishes HERBIE from other race cars in the films is the fact that HERBIE is—inexplicably— "alive." Though he does not "speak" like the robot car KITT (or have any advanced technology to speak of), HERBIE can communicate through the honking of his horn and the opening of his doors, hood, and trunk. It is difficult to properly identify HERBIE. He possesses no technological gimmickry that would qualify him as a "robot." He possesses no artificial intelligence that can be ascertained or identified. He is best identified, then, as a "living machine." HERBIE exhibits such emotions as frustration, disappointment, sadness, fear, and happiness. What adds to the comedic construct of the franchise is the fact that a Volkswagen Beetle, which is not noted for its speed or aerodynamics, could be a "race car." He did precede KITT, however, by more than fifteen years, laying the groundwork for the believability of a "living" car.

Richard A. Hall

Aside from boundless intellect and interactivity with a vast array of computer systems and networks (long before the internet went public), KITT also had bullet- and missile-proof casing, "turbo boost" for accelerated travel, an optical/X-ray/ infrared visual device known as an "anamorphic equalizer" (more commonly known as the roving red light on the front of the hood), multiple scanners, oil slick, smoke screen, flamethrower, communication jamming technology, and a tear gas launcher (just to name his primary components). Through his interactions with Michael, KITT is able to further understand the nuances of "humanity." Early on in the series, it is discovered that prior to KITT, there was a prototype named KARR (Knight Automated Roving Robot). Originally identical to KITT (the only visible difference being KARR's greenish onscreen dashboard voice modulator, while KITT's was red), KARR was programmed with its prime directive being "self-preservation." This turned out to make KARR potentially "evil," as there were multiple possibilities where following his prime directive could endanger innocent human life.

Though relatively short-lived as a series, *Knight Rider* was a pop culture phenomenon; and many scholars have linked its global popularity to being a factor in the end of the Cold War (bootleg copies of the series proved wildly popular in communist Eastern Europe, inspiring many to rise up against their communist governments). Due to the enduring popularity of the series (and KITT in particular), NBC attempted an updated series in 2008. The new *Knight Rider* (NBC, 2008–2009) saw the next generation of FLAG's crusade against crime, with Michael Knight's son (also named Michael and played by Justin Bruening, b. 1979) receiving an updated car, the Knight Industries Three Thousand (also called KITT, and voiced by Val Kilmer, b. 1959). The new KITT was a 2008 Ford Shelby GT500KR and naturally possessed updated tech. The original Michael Knight and KITT appeared in the series' premiere. Decades after its initial cancellation, *Knight Rider* and KITT live on in the American pop culture zeitgeist.

KITT was the ultimate AI partner, always having the back of his human counterpart and exhibiting all the attributes of friendship commonly associated with human friends. To this day, he remains among the most iconic cars in television history. As advanced as automobiles have become in the twenty-first century, they remain light-years behind the abilities of the original KITT. Though technically the "property" of FLAG, the only restrictions that the human organization places on KITT have to do with response to program malfunctions. Otherwise, KITT is free to "live" a normal "life" with his best friend. As such, KITT acts as partner, "pet," and "gadget." Though a seemingly odd combination of roles, it does reflect current attitudes toward robotics. The ethics of such attitudes will someday soon be a matter of debate.

Richard A. Hall

See also: Batcomputer, HAL 9000, Iron Legion, Iron Man, L3-37/*Millennium Falcon*, R2-D2, Speed Buggy, TARDIS, Transformers; *Thematic Essay:* Heroic Robots and Their Impact on Sci-Fi Narratives.

Further Reading

Albert, Robert S., and Brigante, Thomas R. 1962. "The Psychology of Friendship Relations: Social Factors." *Journal of Social Psychology* 56 (1). https://www.tandfon line.com/doi/abs/10.1080/00224545.1962.9919371.

Amati, Viviana, Meggiolaro, Silvia, Rivellini, Giulia, and Zaccarin, Susanna. 2018. "Social Relations and Life Satisfaction: The Role of Friends." *Genus* 74, no. 1 (May 4). https://www.ncbi.nlm.nih.gov/pmc/articles/PMC5937874.

Ashby, LeRoy. 2006. *With Amusement for All: A History of American Popular Culture since 1830.* Lexington: University Press of Kentucky.

Carper, Steve. 2019. *Robots in American Popular Culture.* Jefferson, NC: McFarland.

Castleman, Harry, and Podrazik, Walter J. 2016. *Watching TV: Eight Decades of American Television, Third Edition.* Syracuse, NY: Syracuse University Press.

Chanthadavong, Aimee. 2019. "AI and Ethics: The Debate That Needs to Be Had." ZDnet, September 16, 2019. https://www.zdnet.com/article/ai-and-ethics-the-debate-that -needs-to-be-had/#ftag=CAD-03-10abf5f.

Dinh, Thien-Nam. 2018. *Silicon Minds: The Science, Impact, and Promise of Artificial Intelligence.* Independently Published.

Krishnan, Armin. 2009. *Killer Robots: Legality and Ethicality of Autonomous Weapons.* London: Routledge.

Lin, Patrick, Jenkins, Ryan, and Abney, Keith, eds. 2017. *Robot Ethics 2.0: From Autonomous Cars to Artificial Intelligence.* New York: Oxford University Press.

Randall, Terri, dir. 2016. *NOVA.* "Rise of the Robots: Inside the DARPA Robotics Challenge." Aired February 24, 2016, on PBS. DVD.

Robitzski, Dan. 2018. "Artificial Consciousness: How to Give a Robot a Soul." Futurism, June 25, 2018. https://futurism.com/artificial-consciousness.

Rouhiainen, Lasse. 2018. *Artificial Intelligence: 101 Things You Must Know Today about Our Future.* Scotts Valley, CA: CreateSpace.

Said, Carolyn. 2019. "Robot Cars Are Getting Better—But Will True Self-Driving Ever Arrive?" *San Francisco Chronicle*, February 14, 2019. https://www.sfchronicle.com/business/article/Robot-cars-are-getting-better-but-will-true-13614280.php.

Seed, David. 1999. *American Science Fiction and the Cold War: Literature and Film.* Abingdon, UK: Routledge.

Starck, Kathleen, ed. 2010. *Between Fear and Freedom: Cultural Representations of the Cold War.* Newcastle, UK: Cambridge Scholars.

Wallach, Wendell, and Allen, Colin. 2009. *Moral Machines: Teaching Robots Right from Wrong.* New York: Oxford University Press.

L

L3-37/Millennium Falcon

Star Wars

In 1977, writer-director George Lucas (b. 1944) introduced the world to the sci-fi/fantasy phenomenon *Star Wars* (Lucasfilm/20th Century Fox, 1977), one of the most successful pop culture franchises in American history. His original trilogy of films (1977–1983) focused on the rise of its protagonist, Luke Skywalker (played by Mark Hamill, b. 1951), against the backdrop of a Galactic Civil War against the human forces of the evil Empire, led by its Emperor, Sheev Palpatine (played by Ian McDiarmid, b. 1944), and his henchman, Darth Vader (played by David Prowse, b. 1935; voiced by James Earl Jones, b. 1931). In 2012, Lucas sold his production company, Lucasfilm, to Walt Disney Studios. Disney then embarked on producing new *Star Wars* films, beginning with *Star Wars, Episode VII: The Force Awakens* (Lucasfilm/Disney, 2015), which began a new trilogy that would also include *Episode VIII: The Last Jedi* (Lucasfilm/Disney, 2017) and *Episode IX: The Rise of Skywalker* (Lucasfilm/Disney, 2019). The most iconic spacecraft of the original and sequel trilogies (as well as one of the most popular spaceships in all sci-fi history) is the smuggler vessel, *Millennium Falcon*. In the second stand-alone film of the Disney *Star Wars* era, *Solo: A Star Wars Story* (2018), audiences learned much more about the history of "the fastest hunk of junk in the galaxy," including its odd connection to a peculiar droid named L3-37.

In the original *Star Wars* trilogy, it is established that the *Millennium Falcon* is the property of Han Solo (played by Harrison Ford, b. 1942) and that he won the ship in a card game with his longtime friend, Lando Calrissian (played by Billy Dee Williams, b. 1937). Unlike the other spaceships in the film franchise, however, the *Falcon* has always seemed to have a distinct "personality" (in *The Empire Strikes Back*, Solo even enlists the protocol droid C-3PO to "talk" to the *Falcon* to find out what is wrong with it). This ethereal personality is explained in the film *Solo: A Star Wars Story*. In this origin story, the *Falcon* was under the ownership of Calrissian (this time played by Donald Glover, b. 1983) and had a droid copilot named L3-37 (voiced by Phoebe Waller-Bridge, b. 1985). Unlike most droids in the *Star Wars* galaxy, L3 was a self-built droid (meaning that she was not the creation of an organic droid maker but, rather, she constructed her own body, though the creation of her programming/personality remains a mystery). As such, she has no "owner." She is not a "slave," as is standard for droids in *Star Wars*, and frequently makes trouble by attempting to convince other droids to rise up and demand their rights. Another interesting aspect to the character of L3-37 is that a romantic/physical relationship between the droid and the human Calrissian is directly implied. This is the first time in the history of *Star Wars* that a sexual identity (specifically that of pansexualism) is applied to a droid character.

With skeletal limbs extending from a barrel-shaped torso and a horizontal/circular "head," L3 does look unique among *Star Wars* droids. On the planet Kessel, during a heist undertaken by her captain, Calrissian, a young Han Solo (this time played by Alden Ehrenreich, b. 1989) and Chewbacca (played by Joonas Suotamo, b. 1986), L3 does manage to start a droid riot, removing the restraining bolts from a roomful of droids, giving them freedom for the first time in their collective existences. As the heroes attempt to escape Kessel, however, L3 is severely damaged, to a point beyond repair. What made L3 such a valuable copilot was that she was programmed with the most advanced navigational system in the galaxy. Unable to assist in this capacity any longer, the human crew attempt to download her nav system into the *Falcon*; ultimately downloading most of her overall personality (though not her sense of identity). As a result, L3 becomes permanently incorporated into the overall essence of the vessel (hence the later references to the ship having a personality). Though many fans complained about this storyline, it did successfully explain why this seemingly junky craft had such a superior navigational system (Jonathan Kasdan and Lawrence Kasdan, *Solo: A Star Wars Story*, Lucasfilm/Disney, 2018).

L3-37 is a fascinating addition to the *Star Wars* mythos. For the first time in the franchise's history, the fact that droids were viewed as "property" and, in essence, "slaves" (even among heroic life-forms) was addressed. Since L3 had no "creator" to speak of, her essence and identity were her own. She stands as a voice for the oppressed and a reminder that, in the *Star Wars* galaxy, even heroes possess some unattractive attributes (i.e., their continuance of the concept of slavery toward nonorganic beings). Her incorporation into the *Millennium Falcon* explains the unique nature of this ship among *Star Wars* vehicles. The fact, however, that her "death" resulted in her being absorbed into an object that *is* the property of someone else does add a sense of melancholy to her overall story of heroism and her quest for freedom.

Richard A. Hall

See also: Alita, Battle Droids, BB-8, Bender, C-3PO, Darth Vader, D-O, Dolores, EV-9D9, Fembots, HAL 9000, IG-88/IG-11, JARVIS/FRIDAY, K-2SO, KITT, Medical Droids, R2-D2, Rosie, Tin Woodsman, Warlock; *Thematic Essays:* Robots and Slavery, Heroic Robots and Their Impact on Sci-Fi Narratives.

Further Reading
Carper, Steve. 2019. *Robots in American Popular Culture*. Jefferson, NC: McFarland.

Eberl, Jason T., and Decker, Kevin S., eds. 2015. *The Ultimate* Star Wars *and Philosophy: You Must Unlearn What You Have Learned*. Hoboken, NJ: Wiley-Blackwell.

Hampton, Gregory Jerome. 2017. *Imagining Slaves and Robots in Literature, Film, and Popular Culture: Reinventing Yesterday's Slave with Tomorrow's Robot*. Lanham, MD: Lexington.

Kaminski, Michael. 2008. *The Secret History of* Star Wars*: The Art of Storytelling and the Making of a Modern Epic*. Kingston, Canada: Legacy Books.

Lin, Patrick, Abney, Keith, and Bekey, George A., eds. 2014. *Robot Ethics: The Ethical and Social Implications of Robotics*. Cambridge: Massachusetts Institute of Technology.

Older, Daniel Jose. 2018. Star Wars: *Last Shot*. New York: Del Rey.

Reagin, Nancy R., and Liedl, Janice, eds. 2013. Star Wars *and History*. New York: John Wiley & Sons.

Spaeth, Dennis. 2018. "From Single-Task Machines to Backflipping Robots: The Evolution of Robots." *Cutting Tool Engineering*, January 15, 2018. https://www.ctemag.com/news/articles/evolution-of-robots.

Stanford University. n.d. "Robotics: A Brief History." Stanford University. https://cs.stanford.edu/people/eroberts/courses/soco/projects/1998-99/robotics/history.html.

Sunstein, Cass R. 2016. *The World According to* Star Wars. New York: Dey Street.

Taylor, Chris. 2015. *How* Star Wars *Conquered the Universe: The Past, Present, and Future of a Multibillion Dollar Franchise*. New York: Basic Books.

Wallach, Wendell, and Allen, Colin. 2009. *Moral Machines: Teaching Robots Right from Wrong*. New York: Oxford University Press.

Landru

Star Trek

In 1966, creator-writer-producer Gene Roddenberry (1921–1991) launched his iconic science fiction television series *Star Trek* (NBC, 1966–1969). Set in the twenty-third century, the series centers on the adventures of the starship *USS Enterprise*, its commander, Captain James T. Kirk (played by William Shatner, b. 1931), and crew, headed by the alien first officer, Mr. Spock (played by Leonard Nimoy, 1931–2015), representing the United Federation of Planets. Though short-lived in its initial run, the series has become legendary, not only for its ground-breaking special effects but also for the sophistication of its stories, many of which were grounded in real-world issues of the day, such as racism and a fear of the Cold War. Its success in syndicated reruns revived the franchise, leading to an animated series, thirteen feature films, and five live-action "sequel" series (at the time of this writing). In the original series' twenty-first episode, the crew of the *Enterprise* encountered one of the most dangerous villains of the franchise's history: the omnipotent AI program "Landru."

Visiting the planet Beta III in an attempt to discover the fate of the starship *Archon* that disappeared from the planet over a century prior, Captain Kirk and his crew discover a simple society similar to Earth's early twentieth century. The people appear kind and gentle until a clock in the town square chimes and they become crazed and violent. Seeking shelter in a nearby building and discovering some of the town's elders, Kirk and company discover they have arrived during "Festival," an annual event during which all of the people's pent-up desires and frustrations can be freely expressed. Quickly exposed as foreign interlopers, Kirk discovers that the ruler of the people of this planet is an enigmatic entity known only as "Landru." Landru is all-seeing and all-powerful, controlling the minds of every person on Beta III. Eventually, Kirk and Spock uncover the fact that Landru is an AI program run by computers, the original, human Landru having died thousands of years earlier and having left this program to ensure a stable and peaceful society. Kirk convinces the program that its denial of free will makes the program itself "evil." On this realization, Landru sees no option other than self-destruction, thereby freeing the people to seek out their own destiny (Gene Roddenberry and Boris Sobelman, "Return of the Archons," *Star Trek*, season 1, episode 21, February 9, 1967).

The story of Landru exhibits all the potential unintended consequences of the traditional Frankenstein's monster narrative. Intended as a means of ensuring peace and prosperity, the program instead created a society of mindless drones confined to a static existence devoid of emotion, ingenuity, or growth. Landru shows the limitations of an AI program's reliance on original programming and logic without the ability to apply nuance and creative thinking. While the program did provide a society free of crime and violence 99 percent of the time, the constrained and pent-up physical need of organic beings to express their darker natures (even if only in their minds) resulted in one day a year of unrestrained verbal and physical assault, rape, and even murder on a societal scale. Similar in some ways to the twenty-first-century *The Purge* franchise, the *Star Trek* narrative adds the equally nightmarish scenario of a mind-controlling computer program essentially enslaving an entire civilization.

Richard A. Hall

See also: AI/Ziggy, Androids, Arnim Zola, Batcomputer, Borg, Cerebro/Cerebra, Control, Doctor/EMH, The Great Intelligence, HAL 9000, Janet, JARVIS/Friday, Lieutenant Commander Data, Lore/B4, The Matrix/Agent Smith, Max Headroom, OASIS, Oz, Rehoboam, Soji and Dahj Asha, Starfleet Computer, V-GER, WOPR, Zordon/Alpha-5; *Thematic Essays:* Villainous Robots and Their Impact on Sci-Fi Narratives, AI and the Apocalypse: Science Fiction Meeting Science Fact.

Further Reading

Allen, Arthur. 2019. "There's a Reason We Don't Know Much about AI." Politico, September 16, 2019. https://www.politico.com/agenda/story/2019/09/16/artficial-intelligence-study-data-000956?cid=apn.

Ashby, LeRoy. 2006. *With Amusement for All: A History of American Popular Culture since 1830.* Lexington: University Press of Kentucky.

Bastani, Aaron. 2019. *Fully Automated Luxury Communism.* New York: Verso.

Byrne, Emma. 2013. "Innovation Isn't Safe: The Future According to Kevin Warwick." *Forbes*, September 30, 2013. https://www.forbes.com/sites/netapp/2013/09/30/kevin-warwick-captain-cyborg/#48b704c13560.

Carper, Steve. 2019. *Robots in American Popular Culture.* Jefferson, NC: McFarland.

Castleman, Harry, and Podrazik, Walter J. 2016. *Watching TV: Eight Decades of American Television, Third Edition.* Syracuse, NY: Syracuse University Press.

Chandler, Simon. 2019. "Artificial Intelligence Has Become a Tool for Classifying and Ranking People." *Forbes*, October 1, 2019. https://www.forbes.com/sites/simonchandler/2019/10/01/artificial-intelligence-has-become-a-tool-for-classifying-and-ranking-people/#7431657f1d7c.

Chanthadavong, Aimee. 2019. "AI and Ethics: The Debate That Needs to Be Had." ZDnet, September 16, 2019. https://www.zdnet.com/article/ai-and-ethics-the-debate-that-needs-to-be-had/#ftag=CAD-03-10abf5f.

Decker, Kevin S., and Eberl, Jason T., eds. 2016. *The Ultimate* Star Trek *and Philosophy: The Search for Socrates.* Hoboken, NJ: Wiley-Blackwell.

Dinh, Thien-Nam. 2018. *Silicon Minds: The Science, Impact, and Promise of Artificial Intelligence.* Independently Published.

Gross, Edward, and Altman, Mark A. 2016. *The Fifty-Year Mission, The First 25 Years: The Complete, Uncensored, Unauthorized Oral History of* Star Trek. New York: St. Martin's.

Hitchcock, Susan Tyler. 2007. *Frankenstein: A Cultural History.* New York: W. W. Norton.

Lin, Patrick, Jenkins, Ryan, and Abney, Keith, eds. 2017. *Robot Ethics 2.0: From Autonomous Cars to Artificial Intelligence.* New York: Oxford University Press.

Murdock, Jason. 2019. "Former Google Engineer Warns AI Might Accidentally Start a War: 'These Things Will Start to Behave in Unexpected Ways.'" *Newsweek*, September 16, 2019. https://www.newsweek.com/google-project-maven-artificial-intelligence-laura-nolan-killer-robots-department-defense-1459358.

Paur, Joey. 2019. "An AI Bot Writes a Hilarious Episode of *Star Trek: The Next Generation*." Geek Tyrant, November 16, 2019. https://geektyrant.com/news/an-ai-bot-writes-a-hilarious-episode-of-star-trek-the-next-generation.

Reagin, Nancy R., and Liedl, Janice, eds. 2013. Star Trek *and History.* Hoboken, NJ: John Wiley & Sons.

Rouhiainen, Lasse. 2018. *Artificial Intelligence: 101 Things You Must Know Today about Our Future.* Scotts Valley, CA: CreateSpace.

Starck, Kathleen, ed. 2010. *Between Fear and Freedom: Cultural Representations of the Cold War.* Newcastle, UK: Cambridge Scholars.

Stark, Steven D. 1997. *Glued to the Set: The 60 Television Shows and Events that Made Us Who We Are Today.* New York: Delta Trade Paperbacks.

Wallach, Wendell, and Allen, Colin. 2009. *Moral Machines: Teaching Robots Right from Wrong.* New York: Oxford University Press.

Lieutenant Commander Data

Star Trek: The Next Generation

Set in Earth's twenty-fourth century, *Star Trek: The Next Generation* (syndication, 1987–1994) follows the adventures of the *USS Enterprise-D,* a "Starfleet" starship of the "United Federation of Planets," in its mission to explore the Milky Way galaxy. Lieutenant Commander Data (played by Brent Spiner, b. 1949), a self-aware and anatomically fully functional android, first appeared in "Encounter at Farpoint" (written by D. C. Fontana and Gene Roddenberry, season 1, September 28, 1987), the first episode of *The Next Generation.* With the exception of the few other synthetic life-forms created by Dr. Noonien Soong (also played by Brent Spiner), Data is the only sentient android known to exist in the galaxy. Gene Roddenberry, creator and producer of both *Star Trek: The Original Series* (NBC, 1966–1969) and *The Next Generation*, created the character. Data serves as the ship's second officer and chief operations officer. He is treated as an equal in every respect by the crew of the *Enterprise,* which includes Captain Jean-Luc Picard (played by Patrick Stewart, b. 1940).

Dr. Soong (also played by Spiner) designed and built Data in his own likeness. The android was the sole survivor of a science colony on "Omicron Theta" after it was attacked by the "Crystalline Entity," a creature that feeds on organic life-forms for energy. Found by Starfleet personnel and reactivated in 2338, Data attended Starfleet Academy from 2341 to 2345, after which he served aboard the *USS Trieste* prior to being assigned to the *Enterprise* in 2364 (Maurice Hurley,

---

Jean-Luc Picard

Captain (later Admiral) Jean-Luc Picard was the primary human protagonist in the hit sci-fi television series *Star Trek: The Next Generation* (syndication, 1987–1994). As Captain of the *USS Enterprise-D* through seven seasons of television and four feature films, Picard (played by Sir Patrick Stewart, b. 1940) was established as one of the most popular heroes both in the genre of science fiction in general and the *Star Trek* franchise more specifically. Throughout the series and films, one of Picard's most trusted crew members was the android science officer, Lieutenant Commander Data (played by Brent Spiner, b. 1949), who sacrificed his "life" to save Picard in the group's final feature film, *Star Trek: Nemesis* (Paramount Pictures, 2002). The enduring popularity of the Picard character led to a spin-off television series launched on the CBS All Access streaming service in January 2020: *Star Trek: Picard*. In the final episode of the series' first season, Picard—suffering from a brain malady—succumbs to his illness and dies. However, having discovered a planet of highly advanced androids created by the son of Data's own creator, Picard's consciousness is "uploaded" into an android body built to look identical to his own, aged human form. Though now having the potential for immortality, Picard is informed that his new body has been programmed to "age" accordingly and to "die" at the end of a normal human life span (Michael Chabon, "Et in Arcadia Ego, Part 2," *Star Trek: Picard*, season 1, episode 10, March 26, 2020). As such, from this point forward, the character of Admiral Picard will continue as an android with a human "soul."

Richard A. Hall

Robert Lewin, and Gene Roddenberry, "Datalore," *Star Trek: The Next Generation*, season 1, episode 12, January 16, 1988). He has a "positronic brain" that gives him exceptional computational abilities—his memory capacity is eight hundred quadrillion bits, and his total linear computational speed is sixty trillion operations per second (Melinda M. Snodgrass, "The Measure of a Man," *Star Trek: The Next Generation*, season 2, episode 9, February 11, 1989). However, he is unable to feel emotion or understand much human behavior. His drive to develop his own humanity is a frequent focus of the show's storylines and a source of humor throughout the series.

Data can be shut down by remote control or through use of his off switch, which is located on his lower back (Hurley, Lewin, and Roddenberry, "Datalore"). His upper spinal support is "a poly-alloy, designed to withstand extreme stress," and his skull "is composed of 'cortenide' and 'duranium'" (Joe Menosky and Ronald D. Moore, "The Chase," *Star Trek: The Next Generation*, season 6, episode 20, April 24, 1993). Because Data's construction allows him to be dismantled and reassembled, Chief Engineer Geordi La Forge (played by LeVar Burton, b. 1957) rather than the ship's doctor often treats his mechanical or cognitive function failures (Joe Menosky and Michael Piller, "Time's Arrow," *Star Trek: The Next Generation*, season 5, episode 26, June 13, 1992).

During the first season, the *Enterprise* returns to Omicron Theta, where Data learns he has a "brother," Lore (also played by Brent Spiner), who was created before him (Hurley, Lewin, and Roddenberry, "Datalore"). Lore, who consistently displays a lack of morals, attempts to betray the *Enterprise* to the Crystalline Entity, but his plan is foiled by Wesley Crusher (played by Wil Wheaton), who

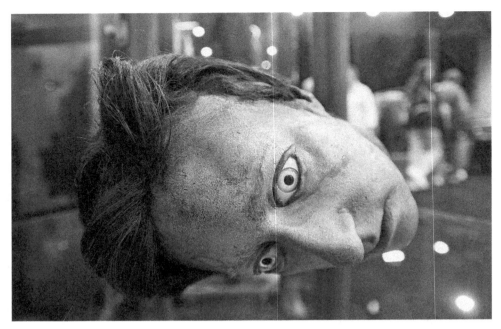

A prop from the television series *Star Trek: The Next Generation* (1987–1994) featuring the "head" of the android Data, played in the series by Brent Spiner. (Chrisharvey/Dreamstime.com)

"beams" him into space. In "The Measure of a Man" (Snodgrass), the question of whether Data is a piece of property or an individual with the same rights as other sentient beings is the focus of a trial in which Picard represents the android. Picard argues that Data is both intelligent and self-aware and challenges the court to present a means of measuring consciousness, after which the judge rules that Data is not the property of Starfleet and has defined rights.

Data creates Lal (played by Hallie Todd, b. 1962), a robot he refers to as his daughter, in the third season (René Echevarria, "The Offspring," *Star Trek: The Next Generation*, season 3, episode 16, March 10, 1989). Using current advances in Federation cybernetics, he bases Lal on his own structural design, but she perishes by the end of the episode due to a "cascade failure" in her positronic brain. In the fourth season of *The Next Generation*, Data and his "father," Dr. Soong, are reunited (Rick Berman, "Brothers," *Star Trek: The Next Generation*, season 4, episode 3, October 6, 1990). The doctor had created an emotion chip meant for Data, but Lore steals it and then fatally wounds Soong. Data's pet cat, Spot, is introduced during this season (Harold Apter and Ronald D. Moore, "Data's Day," *Star Trek: The Next Generation*, season 4, episode 11, January 5, 1991). Spot, an orange tabby, appears in several episodes as well as in two feature films.

Lore returns for a third time at the end of the show's sixth season and beginning of the seventh and final season (Ronald D. Moore and Jeri Taylor, "Descent," *Star Trek: The Next Generation*, season 6, episode 26, June 19, 1993; and René Echevarria, "Descent, Part II," *Star Trek: The Next Generation*, season 7, episode 1, September 18, 1993). He uses the emotion chip to control Data, forcing his brother to aid him in an attempt to turn the "Borg," cybernetic organisms linked in a hive mind, into entirely artificial life-forms. After eventually deactivating Lore, Data decides against installing the damaged emotion chip. Later Data reunites with Dr. Soong's former wife, Juliana (played by Fionnula Flanagan, b. 1941), Data's "mother" (Dan Koeppel and René Echevarria, "Inheritance," *Star Trek: The Next Generation*, season 7, episode 10, November 20, 1993). She is unaware that she is an android duplicate built by Soong after the real Juliana's death during the Crystalline Entity attack. Data chooses to keep this from her, allowing her to continue with her life as she knows it. In the final episode of *The Next Generation* (Ronald D. Moore and Brannon Braga, "All Good Things," *Star Trek: The Next Generation*, season 7, episode 25, May 23, 1994), Captain Picard time hops among past, present, and future. In the future setting, Data is now a luminary physicist and holder of the Lucasian Chair at Cambridge University.

In *Star Trek Generations* (Paramount Pictures, 1994), the first of four films featuring the *Next Generation* crew, Data installs his emotion chip, but the full range of emotions he experiences proves difficult to control. By the next film, *Star Trek: First Contact* (Paramount Pictures, 1996), he has mastered the emotion chip, which includes deactivating it at times. In this film, he betrays a Borg queen, saving the *Enterprise* and thwarting a Borg plot to conquer Earth. In the third film, *Star Trek: Insurrection* (Paramount Pictures, 1998), Data is sent undercover to observe the Ba'ku people. He experiences a malfunction, which leads him and Picard to uncover a nefarious plan to remove the Ba'ku to another planet.

In the final film featuring the *Next Generation* crew, *Star Trek: Nemesis* (Paramount Pictures, 2002), Data beams Picard off an enemy ship before destroying it, thereby sacrificing himself and saving the crew of the *Enterprise*. With the begrudging help of La Forge, he had copied his core memories into B-4 (also played by Brent Spiner), a lost brother who is introduced in the movie, before the enemy ship's destruction. Brent Spiner has reprised his role as Data in dream sequences in the series *Star Trek: Picard* (CBS All Access, 2020–present). Data now has two "daughters," Dahj and Soji Asha (both played by Isa Briones, b. 1999), who were created by Dr. Bruce Maddox (played by Brian Brophy, b. 1959) through fractal neuronic cloning from one of Data's positronic neurons that Maddox retrieved from B-4 (Akiva Goldsman, James Duff, Michael Chabon, Kirsten Beyer, and Alex Kurtzman, "Remembrance," *Star Trek: Picard*, season 1, episode 1, January 23, 2020).

Like Spock (played by Leonard Nimoy, 1931–2015), the Vulcan first officer from *Star Trek: The Original Series,* Data provides an outsider's view on humanity. He offers a nonjudgmental and innocent window through which the audience may observe the behavior of his fellow crew members and others. Unlike Spock, who chooses to suppress his human side, Data aspires to increase his humanity and experience human emotion. This is a consistent theme throughout the series and often provides comic relief as well as pathos. Although his goal to experience more humanity often seems just out of reach, Data subtly projects wisdom, sensitivity, and curiosity throughout the series and the films, garnering the respect and admiration of his colleagues and of his many fans. But perhaps the most outstanding accomplishment of Data's role in television and film history lies in bringing to light the point that sentience in artificial intelligence should allow them the same inalienable rights as other sentient beings. This question, which reached beyond pop culture and out into the fields of science and ethics, is one that only grows in prevalence as technology continues to "boldly go where no one has gone before."

Lisa C. Bailey

See also: Androids, B.A.T.s, Bernard, Bishop, Borg, Buffybot, Doctor/EMH, Dolores, Fembots, Human Torch, LMDs, Lore/B4, Marvin the Paranoid Android, Replicants, Soji and Dahj Asha, Starfleet Computer, Stepford Wives, VICI, Vision, Warlock; *Thematic Essays:* Robots and Slavery, Heroic Robots and Their Impact on Sci-Fi Narratives.

Further Reading

Asher-Perrin, Emmet. 2014. "*Star Trek*, Why Was This a Good Idea Again?—Data's Human Assimilation." Tor, January 29, 2014. https://www.tor.com/2014/01/29/star -trek-why-was-this-a-good-idea-again-datas-human-assimilation.

Baird, Scott. 2018. "*Star Trek*: 20 Strange Details about Data's Body." ScreenRant, August 25, 2018. https://screenrant.com/star-trek-data-body-abilities-hidden-trivia.

Delpozo, Brian. 2016. "Trek at 50: Enduring Archetypes, Part Two: The Alien Other." Comicsverse, August 16, 2016. https://comicsverse.com/trek-50-enduring-archetypes -part-two-alien.

Gross, Edward, and Altman, Mark A. 2016. *The Fifty-Year Mission: The Next 25 Years.* New York: Thomas Dunne.

Harris, Molly. 2020. "Data: All His Best Ever *Star Trek: The Next Generation* Moments." Film Daily, April 10, 2020. https://filmdaily.co/news/best-data-star-trek-tng-moments.

Kelly, Andy. 2020. "*Star Trek: Picard*: How Data Died, and His Appearance in *Picard* Explained." TechRadar, January 24, 2020. https://www.techradar.com/news/star -trek-picard-data-death-explained.

Reeves-Stevens, Garfield, and Reeves-Stevens, Judith. 1997. Star Trek: The Next Generation: *The Continuing Mission.* New York: Pocket Books.

Shapiro, Allen. 2013. *Star Trek and the Android Data.* Seattle, WA: Amazon.com Services.

Vary, Adam B. 2007. "*Star Trek: The Next Generation*: An Oral History." *Entertainment Weekly*, September 25, 2007. https://ew.com/article/2007/09/25/star-trek-tng-oral -history.

LMDs (Life Model Decoys)

Marvel Comics

In 1961, comic book legend Stan Lee (1922–2018) launched the "Marvel age" of comics. With the launching of the Fantastic Four—soon to be followed by The Hulk, Spider-Man, Iron Man, and many, many more—comic book superheroes took a more dramatic turn, with storylines focused on the private lives of those blessed (or burdened) with superpowers. Early in the burgeoning Marvel comics universe, Lee and artist Jack Kirby (1917–1994) introduced audiences to the international spy agency S.H.I.E.L.D. (originally standing for "Supreme Headquarters, International Espionage and Law-Enforcement Division," though that has changed a few times over the decades). S.H.I.E.L.D. is commanded by master spy Colonel Nick Fury and staffed by thousands of human agents. In addition to their human legions, however, S.H.I.E.L.D. also makes use of "Life Model Decoys" (LMDs), robots made to look human (and often used to create robot copies of actual people). Over the decades, LMDs have been used in a multitude of capacities (Lee and Kirby, *Strange Tales #135*, Marvel Comics, August 1965).

As robots, LMDs possess increased strength, speed, and agility over their human counterparts. They are often used as stand-ins for potential political targets and have, from time to time, stood in for actual heroes (from Black Widow to Captain America, and even for Deadpool). The most famous (or infamous) use of LMDs, however, has been by Nick Fury himself. Fury frequently utilizes LMDs to stand in for him on dangerous missions, leading those who witness Fury's "death" to be surprised when he later returns, alive and well. In the 2014 crossover event, "Original Sin," the Marvel heroes discover that the slow-aging Fury they have known for years is actually one of an army of Fury LMDs that assist a very elderly Fury as watchdog over all of Earth (Jason Aaron and Mike Deodato, *Original Sin*, Marvel Comics, August-September, 2014). They have also played a major role in the television series *Agents of S.H.I.E.L.D.* (ABC, 2013–2020).

Like B.A.T.s, Battle Droids, and Doombots, LMDs provide a literal army of killer robots. Unlike those other examples, however, LMDs are just as frequently utilized for espionage purposes, taking the place of living people (for good or ill). They are, however, the property of S.H.I.E.L.D. In one storyline, Fury discovers that a large group of "sentient" LMDs (known as "Deltites") have infiltrated both S.H.I.E.L.D. and their enemy organization, Hydra. Their ability to perfectly mimic anyone makes them a particularly dangerous weapon in S.H.I.E.L.D.'s arsenal—and a potential liability if control is lost.

Richard A. Hall

See also: Androids, Arnim Zola, B.A.T.s, Battle Droids, Buffybot, Cylons, Doombots, Fembots, HERBIE, Iron Legion, Iron Man, Nardole, Robots, Sentinels, Soji and Dahj Asha, Stepford Wives, Ultron, VICI, Vision, Warlock; *Thematic Essay:* Robots and Slavery.

Further Reading

Ashby, LeRoy. 2006. *With Amusement for All: A History of American Popular Culture since 1830*. Lexington: University Press of Kentucky.

Bastani, Aaron. 2019. *Fully Automated Luxury Communism*. New York: Verso.

Byrne, Emma. 2013. "Innovation Isn't Safe: The Future According to Kevin Warwick." *Forbes*, September 30, 2013. https://www.forbes.com/sites/netapp/2013/09/30/kevin -warwick-captain-cyborg/#48b704c13560.

Costello, Matthew J. 2009. *Secret Identity Crisis: Comic Books & the Unmasking of Cold War America*. New York: Continuum.

Hampton, Gregory Jerome. 2017. *Imagining Slaves and Robots in Literature, Film, and Popular Culture: Reinventing Yesterday's Slave with Tomorrow's Robot*. Lanham, MD: Lexington.

Howe, Sean. 2012. *Marvel Comics: The Untold Story*. New York: Harper-Perennial.

Lin, Patrick, Abney, Keith, and Bekey, George A., eds. 2014. *Robot Ethics: The Ethical and Social Implications of Robotics*. Cambridge: Massachusetts Institute of Technology.

Murdock, Jason. 2019. "Former Google Engineer Warns AI Might Accidentally Start a War: 'These Things Will Start to Behave in Unexpected Ways.'" *Newsweek*, September 16, 2019. https://www.newsweek.com/google-project-maven-artificial-intel ligence-laura-nolan-killer-robots-department-defense-1459358.

Pellissier, Hank. 2013. "Robots and Slavery: What Do Humans Want When We Are 'Masters'?" Institute for Ethics and Emerging Technologies, September 13, 2013. https://ieet.org/index.php/IEET2/more/pellissier20130913.

Randall, Terri, dir. 2016. *NOVA*. "Rise of the Robots: Inside the DARPA Robotics Challenge." Aired February 24, 2016, on PBS. DVD.

Spaeth, Dennis. 2018. "From Single-Task Machines to Backflipping Robots: The Evolution of Robots." *Cutting Tool Engineering*, January 15, 2018. https://www.ctemag.com/news/articles/evolution-of-robots.

Thomas, Roy. 2017. *The Marvel Age of Comics: 1961–1978*. Los Angeles: Taschen.

Wallach, Wendell, and Allen, Colin. 2009. *Moral Machines: Teaching Robots Right from Wrong*. New York: Oxford University Press.

Wright, Bradford W. 2003. *Comic Book Nation: The Transformation of Youth Culture in America*. Baltimore, MD: Johns Hopkins University Press.

Lore/B4

Star Trek

In 1966, creator-writer-producer Gene Roddenberry (1921–1991) launched his iconic science fiction television series *Star Trek* (NBC, 1966–1969). Set in the twenty-third century, the series centers on the adventures of the starship *USS Enterprise*, its commander, Captain James T. Kirk (played by William Shatner, b. 1931), and crew, headed by the alien first officer, Mr. Spock (played by Leonard Nimoy, 1931–2015), representing the United Federation of Planets. The series' success in syndicated reruns revived the franchise, leading to an animated series, thirteen feature films, and six live-action "sequel" series (at the time of this writing). The first of these sequel series was *Star Trek: The Next Generation* (a.k.a. *TNG*; Syndication, 1987–1994), set a century after the original, with a new *Enterprise*, under the command of Captain Jean-Luc Picard (played by Patrick Stewart, b. 1940). One of Picard's most valuable crew members is the android officer, Lieutenant Commander Data (played by Brent Spiner, b. 1949). Though Data is originally to be unique, in the first season of *TNG*, he and Picard encounter his predecessor, Lore.

Data was the creation of Dr. Noonien Soong (also eventually played by Spiner). When the new *Enterprise* visits the planet Omicron Theta (where Data was originally found), the crew discovers the disassembled body of a duplicate android. Once reassembled, the android identifies itself as "Lore" (also played by Spiner), an improved model of the original Data (as evidenced by his ability to use contractions in his speech). However, it is soon discovered that Lore was actually the precursor and that he had been disassembled by Soong when the fellow colonists exhibited fear of him. Soong's second model, Data, would then be programmed to be more "robotic" and less "human" (Robert Lewin and Maurice Hurley, "Datalore," *Star Trek: The Next Generation*, season 1, episode 12, January 18, 1988). Lore would return later to trick his presumed-dead creator to give him an "emotion chip" designed for Data, in order to continue as the religious leader of a group of emotion-infected Borg. Lore was, in essence, the traditional "evil twin" for the Data storyline.

Artificial Intelligence

The term "artificial intelligence" refers to machines given the ability to move beyond simple programming to "think," that is, to learn from experience to adjust programming and find solutions to problems not easily arrived at through traditional calculation. The concept of human–made intelligent mechanisms dates back to ancient times, primarily as storytelling devices. The most famous modern incarnation of the idea came with publication of *Frankenstein* in 1818 by Mary Wollstonecraft Shelley (1797–1851), which gave birth to the modern ethical question of humans "playing God," or creating intelligent beings. In 1940, nuclear physicist Edward Condon (1902–1974) presented the Nimatron, a computer that could successfully play the mathematical game Nim. In 1951, the University of Manchester presented the Ferranti Mark 1, a computer programmed to play checkers and chess. Ten years later, General Motors tested the Unimate, an assembly-line-labor robot. In 1969, Shakey the Robot was unveiled by the Stanford Research Institute, the first mobile robot with the ability to "reason" with regard to its actions. In 1980, Stanford University hosted the first meeting of the American Association for Artificial Intelligence. In recent decades, popular culture has both glamorized and provided cautionary tales of the advancing of artificial intelligence. While most people would love to have an interactive automobile such as KITT or a personal assistant such as JARVIS, the flip-side nightmare scenarios of *The Terminator* and *Westworld* give pause to many as to the unintended consequences of artificial intelligence. Additionally, when artificial intelligence does advance to the next level, at what point will machines become "sentient," and, therefore, earn—and demand—basic rights?

Richard A. Hall

In the 2002 film, *Star Trek: Nemesis*, Picard and company discover yet another Data ancestor, "B4" (yet again played by Spiner). Believed to be the original prototype to both Lore and Data, B4's programming is so deteriorated that he exists as little more than a robotic shell. Data hopes to remedy this by sharing some programming with this new "brother." After Data is destroyed, Picard takes the slowly responding B4 back to Starfleet Command. In the spin-off series *Picard* (CBS All Access, 2020–present), audiences discover that although B4 could not be saved, what was left of him was invaluable to Starfleet's eventually cracking the technology for creating an entire race of androids that was soon to be doomed to an existence of de facto slavery, leading to an apparent revolt that causes Starfleet to outlaw artificial life-forms.

Like Data before, and Soji and Dahj Asha afterward, Lore and B4 are the ultimate examples of what artificial life can become (for good or ill). Like Data, Lore possesses phenomenal abilities designed to assist humans in the continuing quest for knowledge throughout the universe. Unlike Data, however, Lore possessed a clear understanding of his superiority to organic life and, with that knowledge, a strong desire to hold some role fit to his superior status. B4, by contrast, shows the early steps toward the Lore/Data model, providing the audience with an understanding that this work was a progression rather than immediate success. These three original sentient *Star Trek* androids—B4, Lore, and Data—set the groundwork for the future of *Star Trek*'s exploration of the possibilities as well as the dangers of artificial intelligence.

Richard A. Hall

See also: Androids, B.A.T.s, Battle Droids, Borg, Cylons, Doctor/EMH, Doombots, Fembots, Landru, Lieutenant Commander Data, LMDs, Nardole, Replicants, Robots, Sentinels, Soji and Dahj Asha, Starfleet Computer, Terminators, VICI, Vision, Warlock; *Thematic Essay:* Villainous Robots and Their Impact on Sci-Fi Narratives.

Further Reading

Allen, Arthur. 2019. "There's a Reason We Don't Know Much about AI." Politico, September 16, 2019. https://www.politico.com/agenda/story/2019/09/16/artficial-intelligence-study-data-000956?cid=apn.

Ashby, LeRoy. 2006. *With Amusement for All: A History of American Popular Culture since 1830.* Lexington: University Press of Kentucky.

Asher-Perrin, Emmet. 2014. "*Star Trek*, Why Was This a Good Idea Again?—Data's Human Assimilation." Tor, January 29, 2014. https://www.tor.com/2014/01/29/star-trek-why-was-this-a-good-idea-again-datas-human-assimilation.

Baird, Scott. 2018. "*Star Trek*: 20 Strange Details about Data's Body." ScreenRant, August 25, 2018. https://screenrant.com/star-trek-data-body-abilities-hidden-trivia.

Bastani, Aaron. 2019. *Fully Automated Luxury Communism.* New York: Verso.

Byrne, Emma. 2013. "Innovation Isn't Safe: The Future According to Kevin Warwick." *Forbes*, September 30, 2013. https://www.forbes.com/sites/netapp/2013/09/30/kevin-warwick-captain-cyborg/#48b704c13560.

Carper, Steve. 2019. *Robots in American Popular Culture.* Jefferson, NC: McFarland.

Castleman, Harry, and Podrazik, Walter J. 2016. *Watching TV: Eight Decades of American Television, Third Edition.* Syracuse, NY: Syracuse University Press.

Chandler, Simon. 2019. "Artificial Intelligence Has Become a Tool for Classifying and Ranking People." *Forbes*, October 1, 2019. https://www.forbes.com/sites/simonchandler/2019/10/01/artificial-intelligence-has-become-a-tool-for-classifying-and-ranking-people/#7431657f1d7c.

Chanthadavong, Aimee. 2019. "AI and Ethics: The Debate That Needs to Be Had." ZDnet, September 16, 2019. https://www.zdnet.com/article/ai-and-ethics-the-debate-that-needs-to-be-had/#ftag=CAD-03-10abf5f.

Decker, Kevin S., and Eberl, Jason T. eds. 2016. *The Ultimate* Star Trek *and Philosophy: The Search for Socrates.* Hoboken, NJ: Wiley-Blackwell.

Dinh, Thien-Nam. 2018. *Silicon Minds: The Science, Impact, and Promise of Artificial Intelligence.* Independently Published.

Gross, Edward, and Altman, Mark A. 2016. *The Fifty-Year Mission, The Next 25 Years: The Complete, Uncensored, Unauthorized Oral History of* Star Trek. New York: St. Martin's.

Hampton, Gregory Jerome. 2017. *Imagining Slaves and Robots in Literature, Film, and Popular Culture: Reinventing Yesterday's Slave with Tomorrow's Robot.* Lanham, MD: Lexington.

Hitchcock, Susan Tyler. 2007. *Frankenstein: A Cultural History.* New York: W. W. Norton.

Kelly, Andy. 2020. "*Star Trek: Picard*: How Data Died, and His Appearance in *Picard* Explained." TechRadar, January 24, 2020. https://www.techradar.com/news/star-trek-picard-data-death-explained.

Lin, Patrick, Abney, Keith, and Bekey, George A., eds. 2014. *Robot Ethics: The Ethical and Social Implications of Robotics.* Cambridge: Massachusetts Institute of Technology.

Murdock, Jason. 2019. "Former Google Engineer Warns AI Might Accidentally Start a War: 'These Things Will Start to Behave in Unexpected Ways.'" *Newsweek*,

September 16, 2019. https://www.newsweek.com/google-project-maven-artificial -intelligence-laura-nolan-killer-robots-department-defense-1459358.

Paur, Joey. 2019. "An AI Bot Writes a Hilarious Episode of *Star Trek: The Next Generation*." Geek Tyrant, November 16, 2019. https://geektyrant.com/news/an-ai-bot -writes-a-hilarious-episode-of-star-trek-the-next-generation.

Reagin, Nancy R., and Liedl, Janice, eds. 2013. Star Trek *and History*. Hoboken, NJ: John Wiley & Sons.

Rouhiainen, Lasse. 2018. *Artificial Intelligence: 101 Things You Must Know Today about Our Future*. Scotts Valley, CA: CreateSpace.

Shapiro, Allen. 2013. Star Trek *and the Android Data*. Seattle, Washington: Amazon.com Services.

Spaeth, Dennis. 2018. "From Single-Task Machines to Backflipping Robots: The Evolution of Robots." *Cutting Tool Engineering*, January 15, 2018. https://www.ctemag .com/news/articles/evolution-of-robots.

Wallach, Wendell, and Allen, Colin. 2009. *Moral Machines: Teaching Robots Right from Wrong*. New York: Oxford University Press.

M

Marvin the Paranoid Android

The Hitchhiker's Guide to the Galaxy

In 1978, author Douglas Adams (1952–2001) brought his science fiction series, *The Hitchhiker's Guide to the Galaxy* to BBC Radio (later to be translated into novels, a short–lived television series, and eventually a feature film in 2005). The story (altered some from format to format) revolves around a human from Earth named Arthur Dent, who is rescued just prior to Earth's destruction by a researcher named Ford Prefect, who is writing the titular galaxy guide. The two are soon picked up by the president of the Galaxy, Zaphod Beeblebrox, whose spaceship, *Heart of Gold*, is crewed by a depressed and paranoid robot named Marvin (Adams, *The Hitchhiker's Guide to the Galaxy*, book 1, 1979). The first visual appearance of Marvin is from the 1981 television series, where he is presented as essentially a simple robot form—squared torso, jointed arms with pincer hands, and a face with a saddened expression—whereas the 2005 film version gives Marvin a more rounded torso, squat legs, joined arms with more humanlike "hands," and a large, spherical head.

The reason Marvin is so depressed, bored, and paranoid is that his mental capacity, as compared to the human brain, is roughly the size of a planet. With such massive intelligence, there is no task that can remotely begin to challenge him. Likewise, because of his superior intelligence, he understands the potential threat that he represents to mere humans; if humans came to similar conclusions, his very existence could become

A representation of Marvin the Paranoid Android from the radio program, *The Hitchhiker's Guide to the Galaxy* (1978). (Wirestock/Dreamstime.com)

threatened. In the *Hitchhiker's* franchise, he essentially represents basic comic relief. He also represents an example of human-made technology far surpassing the abilities of its creator(s). While possession of a "Marvin" would prove invaluable to anyone, its existence also presents two unattractive possibilities: (1) it is able to take on all human tasks, which could lead humanity to eventually "devolve" in both ambition and intelligence; and (2) it holds the potential for overtaking its human "masters" to rule in an apocalyptic "A.I. overlord" capacity.

Richard A. Hall

See also: Androids, B.A.T.s, Battle Droids, Buffybot, Cylons, Doombots, Fembots, Lieutenant Commander Data, LMDs, Lore/B4, Maschinenmensch/Maria, Replicants, Robots, Sentinels, Soji and Dahj Asha, Terminators; *Thematic Essay:* Heroic Robots and Their Impact on Sci-Fi Narratives.

Further Reading

Allen, Arthur. 2019. "There's a Reason We Don't Know Much about AI." Politico, September 16, 2019. https://www.politico.com/agenda/story/2019/09/16/artificial-intel ligence-study-data-000956?cid=apn.

Ashby, LeRoy. 2006. *With Amusement for All: A History of American Popular Culture since 1830*. Lexington: University Press of Kentucky.

Bastani, Aaron. 2019. *Fully Automated Luxury Communism*. New York: Verso.

Byrne, Emma. 2013. "Innovation Isn't Safe: The Future According to Kevin Warwick." *Forbes*, September 30, 2013. https://www.forbes.com/sites/netapp/2013/09/30/kevin -warwick-captain-cyborg/#48b704c13560.

Carper, Steve. 2019. *Robots in American Popular Culture*. Jefferson, NC: McFarland.

Chandler, Simon. 2019. "Artificial Intelligence Has Become a Tool for Classifying and Ranking People." *Forbes*, October 1, 2019. https://www.forbes.com/sites/simon chandler/2019/10/01/artificial-intelligence-has-become-a-tool-for-classifying-and -ranking-people/?sh=23b55c091d7c.

Chanthadavong, Aimee. 2019. "AI and Ethics: The Debate That Needs to Be Had." ZDnet, September 16, 2019. https://www.zdnet.com/article/ai-and-ethics-the-debate-that -needs-to-be-had/#ftag=CAD-03-10abf5f.

Dinh, Thien-Nam. 2018. *Silicon Minds: The Science, Impact, and Promise of Artificial Intelligence*. Independently Published.

Hampton, Gregory Jerome. 2017. *Imagining Slaves and Robots in Literature, Film, and Popular Culture: Reinventing Yesterday's Slave with Tomorrow's Robot*. Lanham, MD: Lexington.

Lin, Patrick, Abney, Keith, and Bekey, George A., eds. 2014. *Robot Ethics: The Ethical and Social Implications of Robotics*. Cambridge: Massachusetts Institute of Technology.

Paur, Joey. 2019. "An AI Bot Writes a Hilarious Episode of *Star Trek: The Next Generation*." Geek Tyrant, November 16, 2019. https://geektyrant.com/news/an-ai-bot -writes-a-hilarious-episode-of-star-trek-the-next-generation.

Rouhiainen, Lasse. 2018. *Artificial Intelligence: 101 Things You Must Know Today about Our Future*. Scotts Valley, CA: CreateSpace.

Spaeth, Dennis. 2018. "From Single-Task Machines to Backflipping Robots: The Evolution of Robots." *Cutting Tool Engineering*, January 15, 2018. https://www.ctemag .com/news/articles/evolution-of-robots.

Wallach, Wendell, and Allen, Colin. 2009. *Moral Machines: Teaching Robots Right from Wrong*. New York: Oxford University Press.

Maschinenmensch/Maria

Metropolis

In 1927, groundbreaking German director Fritz Lang (1890–1976) released his masterpiece silent film, *Metropolis*. The film is set in the futuristic world of 2030, where the upper classes live a life of luxury while the working classes struggle below the earth, keeping the gleaming city of Metropolis powered. A young man named Freder (played by Gustav Fröhlich, 1902–1987), the son of the city's "master," falls in love with the beautiful Maria (Brigitte Helm, 1908–1996), a social crusader seeking to negotiate a better life for the workers. When Freder meets an engineer named Rotwang (played by Rudolf Klein-Rogge, 1885–1955), he is shown the scientist's secret invention: a golden robot in the form of a woman, based primarily on the love of Rotwang's life, Freder's deceased mother. Freder's father convinces Rotwang to give the robot Maria's appearance in order to damage her reputation with the workers. The robot Maria does lead the workers to nearly destroy themselves and ultimately leave their children in danger. When the workers fear their children dead, they turn on "Maria," only to discover then that she was a robot. In the end, Freder and the real Maria do succeed in bringing the workers and master together (Lang and Thea von Harbou, *Metropolis*, Parufamet, 1927).

Maschinenmensch/Maria was one of the first robots to appear in a video medium of science fiction. In 1977, her iconic form would be an influence on artist Ralph McQuarrie (1929–2012) and his design of the *Star Wars* droid, C-3PO. In her basic robotic form, Maschinenmensch appears to be a golden woman, with feminine hips, breasts, and facial features. She is the property of Rotwang, created from his obsession with his deceased lost love (though it is unclear what the robot's actual purpose was meant to be). Disguised as "Maria," she has a fully human appearance, capable of passing as a human in order to create devastation at the behest of her master. She was one of the first sci-fi examples of a robot slave, with a very subtle Frankenstein's monster subtext.

Richard A. Hall

See also: Androids, B.A.T.s, Battle Droids, Buffybot, C-3PO, Cylons, Dolores, Doombots, Fembots, Lieutenant Commander Data, LMDs, Lore/B4, Marvin the Paranoid Android, Replicants, Robots, Sentinels, Soji and Dahj Asha, Stepford Wives, Terminators, Tin Woodsman, VICI; *Thematic Essays:* Robots and Slavery, Villainous Robots and Their Impact on Sci-Fi Narratives.

Further Reading

Allen, Arthur. 2019. "There's a Reason We Don't Know Much about AI." Politico, September 16, 2019. https://www.politico.com/agenda/story/2019/09/16/artificial-intelligence-study-data-000956?cid=apn.

Ashby, LeRoy. 2006. *With Amusement for All: A History of American Popular Culture since 1830*. Lexington: University Press of Kentucky.

Bastani, Aaron. 2019. *Fully Automated Luxury Communism*. New York: Verso.

Byrne, Emma. 2013. "Innovation Isn't Safe: The Future According to Kevin Warwick." *Forbes*, September 30, 2013. https://www.forbes.com/sites/netapp/2013/09/30/kevin-warwick-captain-cyborg/#48b704c13560.

Carper, Steve. 2019. *Robots in American Popular Culture*. Jefferson, NC: McFarland.

Chandler, Simon. 2019. "Artificial Intelligence Has Become a Tool for Classifying and Ranking People." *Forbes*, October 1, 2019. https://www.forbes.com/sites/simon chandler/2019/10/01/artificial-intelligence-has-become-a-tool-for-classifying-and -ranking-people/#7431657f1d7c.

Chanthadavong, Aimee. 2019. "AI and Ethics: The Debate That Needs to Be Had." ZDnet, September 16, 2019. https://www.zdnet.com/article/ai-and-ethics-the-debate-that -needs-to-be-had/#ftag=CAD-03-10abf5f.

Dinh, Thien-Nam. 2018. *Silicon Minds: The Science, Impact, and Promise of Artificial Intelligence*. Independently Published.

Hampton, Gregory Jerome. 2017. *Imagining Slaves and Robots in Literature, Film, and Popular Culture: Reinventing Yesterday's Slave with Tomorrow's Robot*. Lanham, MD: Lexington.

Lin, Patrick, Abney, Keith, and Bekey, George A., eds. 2014. *Robot Ethics: The Ethical and Social Implications of Robotics*. Cambridge: Massachusetts Institute of Technology.

Paur, Joey. 2019. "An AI Bot Writes a Hilarious Episode of *Star Trek: The Next Generation*." Geek Tyrant, November 16, 2019. https://geektyrant.com/news/an-ai-bot -writes-a-hilarious-episode-of-star-trek-the-next-generation.

Rouhiainen, Lasse. 2018. *Artificial Intelligence: 101 Things You Must Know Today about Our Future*. Scotts Valley, CA: CreateSpace.

Spaeth, Dennis. 2018. "From Single-Task Machines to Backflipping Robots: The Evolution of Robots." *Cutting Tool Engineering*, January 15, 2018. https://www.ctemag .com/news/articles/evolution-of-robots.

Wallach, Wendell, and Allen, Colin. 2009. *Moral Machines: Teaching Robots Right from Wrong*. New York: Oxford University Press.

The Matrix/Agent Smith

The Matrix

In 1999, the Wachowski sisters, Lana and Lilly (b. 1965 and 1967, respectively) introduced audiences around the world to their sci-fi masterpiece: *The Matrix*. The film—and its two sequels—tells the story of "Neo" (played by Keanu Reeves, b. 1964), a computer hacker in late twentieth-century America. Neo seeks out and meets the enigmatic "Morpheus" (played by Laurence Fishburne, b. 1961), who offers Neo a choice of two pills: a red one that will reveal the truth or a blue one that will return Neo to his "normal" life. Neo chooses the red and awakens in a pod, covered in goo, and he observes that his is one of millions of such pods connected to a massive machine. Later, on Morpheus's ship, Neo is told that "reality" as he has known it does not exist. What he thought of as twentieth-century reality was, in fact, far into the future. He is also told that in the early twenty-first century, humanity had gone to war with highly advanced machines. When humans attempted to cut off the machines from their power supply, the machines discovered that the natural bioelectric energy of humans also served as power. The machines, then, began harvesting humans and plugging their minds into the "Matrix," a virtual reality program allowing humans to live out "normal" lives in the mid-1990s. Neo then joins Morpheus and his crew to free humanity from the Matrix (The Wachowskis, *The Matrix*, Warner Brothers, 1999).

Virtual Reality

The term "virtual reality" refers to computer-based imagery experienced through headsets for the eyes and ears (and, in advanced cases, the nose) and, often, gloves that immerse the wearer in a fabricated environment that is designed to look, sound, feel, and (occasionally) smell like a real environment. These environments can provide the appearance of the real world, a fully fictional world, or some mixture of the two. VR has multiple uses but is primarily used for entertainment (i.e., video games) and military training (immersing soldiers in real-world situations). Primary research into VR first occurred at NASA's Jet Propulsion Laboratory in the late 1970s/early 1980s. Two of the most popular expressions of VR in American pop culture were presented in the television series *Star Trek: The Next Generation* (syndication, 1987–1994) and the movie franchise *The Matrix* (Warner Brothers, 1999–2003). In *Star Trek*, the starship *USS Enterprise-D* possessed a "holodeck," where crew members could immerse themselves in any virtual environment, real or imagined. Additionally, they could turn on "safety protocols" that would prevent the user from any physical harm (turning off this feature could open the user to physical harm or even fatality). On the spin-off series, *Star Trek: Deep Space Nine* (syndication, 1993–1999), the Starfleet space station *DS9* contained a similar space but it was called a "holosuite." *The Matrix* franchise put forth the idea that all of reality as we know it is, in fact, a virtual program (see "The Matrix/Agent Smith"). In recent years, VR technology in video games has achieved incredible degrees of realism.

Richard A. Hall

The Matrix is the virtual reality program in which all of humanity is trapped. Sensory input allows people to believe that they see, smell, taste, hear, and feel their "real" world. Only a handful of humans are aware of the truth, and they are hunted by virtual "agents" (clad as the legendary "men in black" of UFO conspiracy fame) led by "Agent Smith" (Hugo Weaving, b. 1960). The concept within *The Matrix* (the film) is a conundrum: on the one hand, humans are content in their virtual existence, and "escape" from it would expose them to a toxic world of intelligent machines, a world in which the humans would not be able to develop food (or possibly even breathe without assistance); on the other hand, their virtual existence is fake, and their individual lives playing out however the program decides they should. In order to maintain their power system, the machines need for humanity to continue in the Matrix, with the VR "Agent Smith" leading his VR agents to maintain stability within the program.

The Matrix opens a discussion on artificial intelligence and virtual reality. As both AI and VR technology advances, the time will soon come when humans will be able to immerse themselves in a virtual reality of their choosing, experiencing every sensation of "life" within a controlled environment that can, potentially, guarantee happiness and pleasure. If the narrative of *The Matrix* were to become reality, it would unlikely be a situation where humanity was placed within it oblivious of it having been done to them; a much more likely scenario would be that humans would place themselves in such a system of their own volition. How much would humanity sacrifice for a life of peace and happiness? What is more important to a person: free will or controlled contentment? The answer could ultimately prove more frightening than the film.

Richard A. Hall

See also: AI/Ziggy, Arnim Zola, Bernard, Cerebro/Cerebra, Control, The Great Intelligence, HAL 9000, HERBIE, Janet, JARVIS/Friday, Landru, OASIS, Oz, Rehoboam, Skynet, TARDIS, V-GER, WOPR; *Thematic Essay:* AI and the Apocalypse: Science Fiction Meeting Science Fact.

Further Reading

Allen, Arthur. 2019. "There's a Reason We Don't Know Much about AI." Politico, September 16, 2019. https://www.politico.com/agenda/story/2019/09/16/artificial-intelligence-study-data-000956?cid=apn.

Ashby, LeRoy. 2006. *With Amusement for All: A History of American Popular Culture since 1830*. Lexington: University Press of Kentucky.

Bastani, Aaron. 2019. *Fully Automated Luxury Communism*. New York: Verso.

Carper, Steve. 2019. *Robots in American Popular Culture*. Jefferson, NC: McFarland.

Chandler, Simon. 2019. "Artificial Intelligence Has Become a Tool for Classifying and Ranking People." *Forbes*, October 1, 2019. https://www.forbes.com/sites/simonchandler/2019/10/01/artificial-intelligence-has-become-a-tool-for-classifying-and-ranking-people/#7431657f1d7c.

Chanthadavong, Aimee. 2019. "AI and Ethics: The Debate That Needs to Be Had." ZDnet, September 16, 2019. https://www.zdnet.com/article/ai-and-ethics-the-debate-that-needs-to-be-had/#ftag=CAD-03-10abf5f.

Dinh, Thien-Nam. 2018. *Silicon Minds: The Science, Impact, and Promise of Artificial Intelligence*. Independently Published.

Dowden, Bradley. 1993. *Logical Reasoning*. Belmont, CA: Wadsworth Publishing Company.

Grau, Christopher, ed. 2005. *Philosophers Explore* The Matrix. Oxford: Oxford University Press.

Irwin, William, ed. 2002. The Matrix *and Philosophy: Welcome to the Desert of the Real*. Chicago: Open Court.

Leetaru, Kalev. 2019. "Automatic Image Captioning and Why Not Every AI Problem Can Be Solved through More Data." *Forbes*, July 7, 2019. https://www.forbes.com/sites/kalevleetaru/2019/07/07/automatic-image-captioning-and-why-not-every-ai-problem-can-be-solved-through-more-data/#20b943476997.

Lin, Patrick, Jenkins, Ryan, and Abney, Keith, eds. 2017. *Robot Ethics 2.0: From Autonomous Cars to Artificial Intelligence*. New York: Oxford University Press.

Linder, Courtney. 2020. "This AI Robot Just Nabbed the Lead Role in a Sci-Fi Movie: Meet Erica, the Movie Star Who May Put Human Actors Out of Work." *Popular Mechanics*, June 25, 2020. https://www.popularmechanics.com/technology/robots/a32968811/artificial-intelligence-robot-movie-star-erica.

Marks, Robert J. 2020. "*2084* vs. *1984*: The Difference AI Could Make to Big Brother." Podcast interview with author John Lennox. Mind Matters, July 3, 2020. https://mindmatters.ai/2020/07/2084-vs-1984-the-difference-ai-could-make-to-big-brother.

Murdock, Jason. 2019. "Former Google Engineer Warns AI Might Accidentally Start a War: 'These Things Will Start to Behave in Unexpected Ways.'" *Newsweek*, September 16, 2019. https://www.newsweek.com/google-project-maven-artificial-intelligence-laura-nolan-killer-robots-department-defense-1459358.

Paur, Joey. 2019. "An AI Bot Writes a Hilarious Episode of *Star Trek: The Next Generation*." Geek Tyrant, November 16, 2019. https://geektyrant.com/news/an-ai-bot -writes-a-hilarious-episode-of-star-trek-the-next-generation.

Rouhiainen, Lasse. 2018. *Artificial Intelligence: 101 Things You Must Know Today about Our Future*. Scotts Valley, CA: CreateSpace.

Singer, Peter W. 2009. "Isaac Asimov's Laws of Robotics Are Wrong." Brookings Institute, May 18, 2009. https://www.brookings.edu/opinions/isaac-asimovs-laws-of -robotics-are-wrong.

Virk, Rizwan. 2019. *The Simulation Hypothesis: An MIT Computer Scientist Shows Why AI, Quantum Physics, and Eastern Mystics All Agree We Are in a Video Game*. Milwaukee, WI: Bayview.

Yeffeth, Glenn, ed. 2003. *Taking the Red Pill: Science, Philosophy and the Religion of The Matrix*. Smart Pop series. Dallas, TX: BenBella.

Max Headroom

Max Headroom: 20 Minutes into the Future

The 1980s had successfully normalized the concept of the "talking head." From twenty-four-hour news pundits to MTV VJs, American television audiences had grown accustomed to seeing their screens filled with various human faces mounted solely on shoulders talking to them directly, breaking the long-touted "fourth wall." In 1985, British director Rocky Morton (b. 1955) devised the idea for television's first CGI/AI onscreen personality: Max Headroom. Max was the very rudimentary CGI human head of a blond-haired white male with ghostly blue eyes, his upper torso showing a black suit jacket, white shirt, and black necktie. The randomly changing background consisted of stereotypically "'80s" computer graphics. Max's voice (and basic facial design) was that of actor Matt Frewer (b. 1958).

Max Headroom was introduced to the world in the 1985 BBC4 telefilm *Max Headroom: 20 Minutes into the Future*. In the film, Frewer plays Edison Carter, a human journalist investigating a television network's experiments into injecting subliminal messaging into their advertising. Discovered while snooping around the station, Edison is knocked unconscious by a low hanging sign that reads "Max. Headroom." The network's engineers decide to use Edison to create a CGI/AI character for television, but the program proves eccentric, frequently stuttering his name: "Max-max-max-Max Headroom." The discarded AI is discovered by another broadcaster who makes Max a star, while Edison escapes and exposes the crimes of the other network (Steve Roberts, *Max Headroom: 20 Minutes into the Future*, BBC4, 1985).

The massive success of Max Headroom led to the character becoming a VJ (or "video jockey") for music videos on BBC; and an American spin-off of the original film aired briefly: *Max Headroom* (ABC, 1987–1988). Today, Max Headroom is most remembered as a commercial spokesperson for "New Coke," during the infamous "cola wars" of the late 1980s, and for his appearance in the music video *Paranoimia* by the group Art of Noise (1986). Max Headroom introduced the world to the concept of a CGI/AI character. Though, technically, more of a CGI "cartoon" than an actual "AI," the illusion proved impressive for its day. Historically, he remains an iconic symbol of the 1980s' MTV-obsessed culture.

Richard A. Hall

See also: Al/Ziggy, Cambot/Gypsy/Tom Servo/Crow, Johnny 5/S.A.I.N.T. Number 5, KITT, Oz, WOPR; *Thematic Essay:* AI and the Apocalypse: Science Fiction Meeting Science Fact.

Further Reading

Allen, Arthur. 2019. "There's a Reason We Don't Know Much about AI." Politico, September 16, 2019. https://www.politico.com/agenda/story/2019/09/16/artificial-intelligence-study-data-000956?cid=apn.

Ashby, LeRoy. 2006. *With Amusement for All: A History of American Popular Culture since 1830.* Lexington: University Press of Kentucky.

Bastani, Aaron. 2019. *Fully Automated Luxury Communism.* New York: Verso.

Carper, Steve. 2019. *Robots in American Popular Culture.* Jefferson, NC: McFarland.

Castleman, Harry, and Podrazik, Walter J. 2016. *Watching TV: Eight Decades of American Television, Third Edition.* Syracuse, NY: Syracuse University Press.

Chandler, Simon. 2019. "Artificial Intelligence Has Become a Tool for Classifying and Ranking People." *Forbes,* October 1, 2019. https://www.forbes.com/sites/simonchandler/2019/10/01/artificial-intelligence-has-become-a-tool-for-classifying-and-ranking-people/#7431657f1d7c.

Chanthadavong, Aimee. 2019. "AI and Ethics: The Debate That Needs to Be Had." ZDnet, September 16, 2019. https://www.zdnet.com/article/ai-and-ethics-the-debate-that-needs-to-be-had/#ftag=CAD-03-10abf5f.

Dinh, Thien-Nam. 2018. *Silicon Minds: The Science, Impact, and Promise of Artificial Intelligence.* Independently Published.

Leetaru, Kalev. 2019. "Automatic Image Captioning and Why Not Every AI Problem Can Be Solved through More Data." *Forbes,* July 7, 2019. https://www.forbes.com/sites/kalevleetaru/2019/07/07/automatic-image-captioning-and-why-not-every-ai-problem-can-be-solved-through-more-data/#20b943476997.

Lin, Patrick, Jenkins, Ryan, and Abney, Keith, eds. 2017. *Robot Ethics 2.0: From Autonomous Cars to Artificial Intelligence.* New York: Oxford University Press.

Linder, Courtney. 2020. "This AI Robot Just Nabbed the Lead Role in a Sci-Fi Movie: Meet Erica, the Movie Star Who May Put Human Actors Out of Work." *Popular Mechanics,* June 25, 2020. https://www.popularmechanics.com/technology/robots/a32968811/artificial-intelligence-robot-movie-star-erica.

Paur, Joey. 2019. "An AI Bot Writes a Hilarious Episode of *Star Trek: The Next Generation.*" Geek Tyrant, November 16, 2019. https://geektyrant.com/news/an-ai-bot-writes-a-hilarious-episode-of-star-trek-the-next-generation.

Rouhiainen, Lasse. 2018. *Artificial Intelligence: 101 Things You Must Know Today about Our Future.* Scotts Valley, CA: CreateSpace.

Schroeder, Stan. 2020. "Samsung Just Launched an 'Artificial Human' Called Neon, and Wait, What?" Mashable, January 7, 2020. https://mashable.com/article/samsung-star-labs-neon-ces.

Mechagodzilla

Godzilla

In 1954, Japanese director Ishirō Honda (1911–1993) released his iconic monster film, *Godzilla.* To succeed in American markets, the film was reworked and dubbed, inserting new footage with American actor Raymond Burr (1917–1993),

Bruce

"Bruce" was the name given by the film crew of the feature film *Jaws* (Universal Pictures, 1975) to the robot shark used in filming most of the movie's major scenes. The name was derived from Bruce Raynor (life years unknown), attorney to the film's director Steven Spielberg (b. 1946). Though real shark footage was utilized for many of the underwater scenes, close-up interactions with the actors required a shark that could be more easily controlled, leading to the creation of Bruce. The robot was built to resemble a large, twenty-five-foot great white shark. It had a hydraulically controlled lower jaw and could fit an adult human in its mouth. The robot was mounted on tracks that were supposed to allow the shark to rise above and sink below the surface of the water. However, as the robot was designed and tested in warm swimming pool water in California, the icy salt water of the North Atlantic (where the film was shot) proved problematic, resulting in numerous malfunctions. The time delays caused by these malfunctions forced director Spielberg to become more creative in filming the interactive sequences of the film (actually creating a more suspenseful atmosphere as a result). Bruce did, however, work adequately for the climax scenes of the film, where the shark flops onto the deck of the ship that was pursuing it and devours the ship's captain, and where the remaining hero shoots an oxygen tank lodged in Bruce's mouth. As a result of the legend of the Bruce robot, fans of the *Jaws* franchise have unofficially dubbed the shark antagonist in the film by the same name.

Richard A. Hall

and released in the United States as *Godzilla, King of the Monsters!* (1956). Both versions tell the same story: a lizard, irradiated by constant atomic testing in the Pacific Ocean, grew to a height of over 160 feet and had the ability to shoot fire from its mouth. This monster lizard made its way to shore and attacked Tokyo, Japan (Honda and Takeo Murata, *Godzilla*, Toho Pictures, 1954). The film was a massive success, making the titular monster one of the most iconic in the history of film cinema. As a demand for sequels became clear, writers were pressed to come up with repeated reasons why Godzilla might keep attacking. The most logical conclusion was to make Godzilla the "hero," fighting against other monstrous threats to save humanity. Of the many giant monsters fought by Godzilla over the years, one of the most popular has been "Mechagodzilla."

After facing organic monsters such as Mothra and King Ghidorah, Godzilla then faced off against the robot threat of Mechagodzilla. Equal in height to the now-hero lizard (his height would increase in later versions, as would Godzilla's), Mechagodzilla was sent to Earth long ago from outer space. Along with equal height and strength to his lizard foe, the mechanical menace could also fire missiles from his hands, feet, and knee joints. To rival Godzilla's fire breath, Mechagodzilla could fire energy beams from his eyes. In his first film appearance, Mechagodzilla is part of a massive underground bunker of extraterrestrial metal. Once awakened, he poses a threat to Japan. At first, the creature appears to be the real Godzilla, until his metallic structure is exposed. While the people still believe the creature to be Godzilla, the real lizard appears in order to battle its doppelganger. During the battle, Mechagodzilla's true self is revealed. Meanwhile, a priestess releases "King Caesar," a metaphysical monster that assists Godzilla in battle (Hiroyasu Yamamura and Jun Fukuda, *Godzilla vs. Mechagodzilla*, Toho,

1974). The huge success of the film led to an immediate sequel, *Terror of Mechagodzilla*, 1974. In the film *Godzilla vs. SpaceGodzilla* (1994), the new mechanical threat is discovered to have been created from the remains of Mechagodzilla.

Mechagodzilla is the ultimate killer robot from outer space. Although he is primarily pure science fiction escapism, the concept of a giant killer robot is one that comes closer to reality with every passing year. The downside, of course, is that, should a giant killer robot ever become a real-world threat, the world does not have a Godzilla to rely on for help. As offensive technology increases, therefore, it is vitally important that *defensive* technology increase at a commensurate rate. If Godzilla is a cautionary tale of the dangers of atomic experimentation, Mechagodzilla is a cautionary tale in the vein of Frankenstein's monster, on a massive scale.

Richard A. Hall

See also: Alita, Brainiac, Borg, Cybermen, Cylons, Daleks, The Great Intelligence, Iron Giant, Terminators, Transformers, Voltron, Zordon/Alpha-5; *Thematic Essay:* Villainous Robots and Their Impact on Sci-Fi Narratives.

Further Reading

Ashby, LeRoy. 2006. *With Amusement for All: A History of American Popular Culture since 1830.* Lexington: University Press of Kentucky.

Byrne, Emma. 2013. "Innovation Isn't Safe: The Future According to Kevin Warwick." *Forbes*, September 30, 2013. https://www.forbes.com/sites/netapp/2013/09/30/kevin-warwick-captain-cyborg/#48b704c13560.

Carper, Steve. 2019. *Robots in American Popular Culture.* Jefferson, NC: McFarland.

Fryer-Biggs, Zachary. 2019. "Coming Soon to a Battlefield: Robots That Can Kill." *The Atlantic.* https://www.theatlantic.com/technology/archive/2019/09/killer-robots-and-new-era-machine-driven-warfare/597130.

Hendershot, Cyndy. 1999. *Paranoia, the Bomb, and 1950s Science Fiction Films.* Bowling Green, KY: Bowling Green State University Popular Press.

Hoberman, J. 2011. *An Army of Phantoms: American Movies and the Making of the Cold War.* New York: New Press.

Kirshner, Jonathan. 2012. *Hollywood's Last Golden Age: Politics, Society, and the Seventies Film in America.* Ithaca, NY: Cornell University Press.

Krishnan, Armin. 2009. *Killer Robots: Legality and Ethicality of Autonomous Weapons.* London: Routledge.

Randall, Terri, dir. 2016. *NOVA.* "Rise of the Robots: Inside the DARPA Robotics Challenge." Aired February 24, 2016, on PBS. DVD.

Ryfle, Steve. 1998. *Japan's Favorite Mon-Star: The Unauthorized Biography of "The Big G."* Toronto, Canada: ECW.

Seed, David. 1999. *American Science Fiction and the Cold War: Literature and Film.* Abingdon, UK: Routledge.

Singer, Peter W. 2009. "Isaac Asimov's Laws of Robotics Are Wrong." Brookings Institute, May 18, 2009. https://www.brookings.edu/opinions/isaac-asimovs-laws-of-robotics-are-wrong.

Solomon, Brian. 2017. *Godzilla FAQ: All That's Left to Know About the King of the Monsters.* Framingham, MA: Applause.

Starck, Kathleen, ed. 2010. *Between Fear and Freedom: Cultural Representations of the Cold War.* Newcastle, UK: Cambridge Scholars.

Medical Droids

Star Wars

In 1977, writer-director George Lucas (b. 1944) introduced the world to the sci-fi/fantasy phenomenon *Star Wars* (Lucasfilm/20th Century Fox, 1977), one of the most successful pop culture franchises in American history. His original trilogy of films (1977–1983) focused on the rise of its protagonist, Luke Skywalker (played by Mark Hamill, b. 1951), against the backdrop of a Galactic Civil War against the human forces of the evil Empire, led by its Emperor, Sheev Palpatine (played by Ian McDiarmid, b. 1944), and his henchman, Darth Vader (played by David Prowse, b. 1935; voiced by James Earl Jones, b. 1931). In 1999, Lucas returned with a "prequel" trilogy, telling the tale of the fall of Anakin Skywalker to Darth Vader, beginning with *Star Wars: Episode I—The Phantom Menace*. In 2012, Lucasfilm was bought by Walt Disney Studios, and a sequel trilogy to the original was immediately put into production, launched with *Star Wars: Episode VII—The Force Awakens* (2015). "Living" in the *Star Wars* universe—for both good guys and bad—is the slave labor class of robotic "droids." Whereas most droids are utilized for basic labor, one class of droids is used specifically for medical procedures.

Going in chronological order of film release, the first medical droid seen by audiences is the surgical droid 2-1B and his nurse/assistant, FX-7. They nurse Luke Skywalker back to health after he is attacked by the wild Wampa beast and exposed to the icy night of the planet Hoth. 2-1B has a skeletal humanoid body with three pincer "fingers" on one hand and an injection gun on the other. FX-7 is a cylindrical droid with a rotating flat, horizontal circular head with multiple retractable appendages emerging from its barrel-shaped body. Later in the same film, 2-1B returns to replace Luke Skywalker's lost right hand with a mechanical duplicate (Lucas, Lawrence Kasdan, and Leigh Brackett, *Star Wars: Episode V—The Empire Strikes Back*, Lucasfilm, 1980). Precursors to 2-1B and FX-7 also appear as part of the surgical team assigned with saving the life of Anakin Skywalker/Darth Vader (played by Hayden Christensen, b. 1981) and implanting mechanical replacements for his left arm, both legs, and lungs (Lucas, *Star Wars: Episode III—Revenge of the Sith*, Lucasfilm, 2005).

In the prequel era, two types of medical droids are introduced. In the last of the prequel films, the droid GH-7 is assigned with saving the life of the pregnant senator Padme Amidala (played by Natalie Portman, b. 1981) and delivering her twin children. GH-7 "floats" in air thanks to minithrusters in the lower part of its legless torso body. It has two humanoid arms with three pincer "fingers" on each. Attached to the front of his torso is a "tray" on which the droid can carry whatever medical implements are needed (Lucas, *Revenge of the Sith*, 2005). Three years after the last prequel film, Lucasfilm brought *Star Wars* to the small screen with the CGI animated series *Star Wars: The Clone Wars* (Cartoon Network, 2008–2013; Netflix, 2014; Disney+, 2020). In this series, the clone soldiers of the Galactic Republic are treated by the medical droid IM-6. This droid is similar in appearance to GH-7 except that it has a white finish, whereas GH-7 has a metallic gray finish.

In the Disney-era *Star Wars* films, a medical droid appears in the second film of the sequel trilogy. When First Order leader Kylo Ren (played by Adam Driver, b. 1983) was injured in light-saber combat, the slash to his face is repaired by IT-S00.2. This surgical droid is a floating spherical ball with three pointed appendages (Rian Johnson, *Star Wars: Episode VIII—The Last Jedi*, Lucasfilm/Disney, 2017). This droid is very similar in appearance to the ball-shaped "torture" droid that appeared in the original *Star Wars* (1977). Unlike other sci-fi franchises such as *Star Trek* and *Battlestar Galactica*, the *Star Wars* universe never shows a human/humanoid doctor. All medical needs appear to be fulfilled by droids.

Medical droids are perhaps the closest to reality in all of sci-fi. For decades, doctors and nurses have relied on computers for diagnoses and monitoring of patients. Since the late 1990s, robotic "arms" have been useful in microsurgeries around the world. While robots in popular culture are usually used as either dangerous enemies or comedic sidekicks, the medical droids of *Star Wars* have repeatedly underscored the value of such technological advancements. Devoid of the ability to "tire" or get overly emotionally attached to a patient's survival, medical droids provide the possibility of precise treatment based on sound, logical reasoning. Whether or not their lack of emotional attachment is a positive or negative remains open for debate.

Richard A. Hall

See also: Battle Droids, Baymax, BB-8, C-3PO, Darth Vader, D-O, EV-9D9, General Grievous, IG-88/IG-11, K-2SO, L3-37/*Millennium Falcon*, R2-D2, Tin Woodsman; *Thematic Essay:* Robots and Slavery.

Further Reading

Ashby, LeRoy. 2006. *With Amusement for All: A History of American Popular Culture since 1830*. Lexington: University Press of Kentucky.

Bastani, Aaron. 2019. *Fully Automated Luxury Communism*. New York: Verso.

Bray, Adam, Barr, Tricia, Horton, Cole, and Windham, Ryder. 2019. *Ultimate* Star Wars, *New Edition*. London: DK Publishing.

Bryant, D'Orsay D., III. 1985. "Spare-Part Surgery: The Ethics of Organ Transplantation." *Journal of the National Medical Association* 77 (2) https://www.ncbi.nlm .nih.gov/pmc/articles/PMC2561842/pdf/jnma00245-0055.pdf.

Byrne, Emma. 2014. "Cybernetic Implants: No Longer Science Fiction." *Forbes*, March 11, 2014. https://www.forbes.com/sites/netapp/2014/03/11/cybernetic-implants-not -sci-fi/#7399c57e77ba.

Dinh, Thien-Nam. 2018. *Silicon Minds: The Science, Impact, and Promise of Artificial Intelligence*. Independently Published.

Eberl, Jason T., and Decker, Kevin S., eds. 2015. *The Ultimate* Star Wars *and Philosophy: You Must Unlearn What You Have Learned*. Hoboken, NJ: Wiley-Blackwell.

Ford, Martin. 2015. *Rise of the Robots: Technology and the Threat of a Jobless Future*. New York: Basic.

Hampton, Gregory Jerome. 2017. *Imagining Slaves and Robots in Literature, Film, and Popular Culture: Reinventing Yesterday's Slave with Tomorrow's Robot*. Lanham, MD: Lexington.

Kaminski, Michael. 2008. *The Secret History of* Star Wars*: The Art of Storytelling and the Making of a Modern Epic*. Kingston, Canada: Legacy Books.

Lin, Patrick, Abney, Keith, and Bekey, George A., eds. 2014. *Robot Ethics: The Ethical and Social Implications of Robotics.* Cambridge: Massachusetts Institute of Technology.

Longoni, Chiara, Bonezzi, Andrea, and Morewedge, Carey K. 2019. "Resistance to Medical Artificial Intelligence." *Journal of Consumer Research* 46, no. 4 (December): 629–50. https://academic.oup.com/jcr/article/46/4/629/5485292.

Randall, Terri, dir. 2016. *NOVA.* "Rise of the Robots: Inside the DARPA Robotics Challenge." Aired February 24, 2016, on PBS. DVD.

Reagin, Nancy R., and Janice Liedl, eds. 2013. Star Wars *and History.* New York: John Wiley & Sons.

Slavicsek, Bill. 2000. *A Guide to the* Star Wars *Universe.* San Francisco: LucasBooks.

Spaeth, Dennis. 2018. "From Single-Task Machines to Backflipping Robots: The Evolution of Robots." *Cutting Tool Engineering,* January 15, 2018. https://www.ctemag.com/news/articles/evolution-of-robots.

Stanford University. n.d. "Robotics: A Brief History." Stanford University. https://cs.stanford.edu/people/eroberts/courses/soco/projects/1998-99/robotics/history.html.

Sunstein, Cass R. 2016. *The World According to* Star Wars. New York: Dey Street.

Taylor, Chris. 2015. *How* Star Wars *Conquered the Universe: The Past, Present, and Future of a Multibillion Dollar Franchise.* New York: Basic Books.

Wallach, Wendell, and Allen, Colin. 2009. *Moral Machines: Teaching Robots Right from Wrong.* New York: Oxford University Press.

Metalhead

Black Mirror

The television series *Black Mirror* debuted in the UK on BBC4 in 2011. After running for two seasons, it was added to the streaming service Netflix in 2016. Created by Charlie Brooker (b. 1971), *Black Mirror* is an anthology series that utilizes science fiction to examine human society by exploring the impact of current and/or potential technological advancements set either in the present or the near future. Due to its use of science fiction for social commentary, the series has often been compared to the classic American series, *The Twilight Zone* (CBS, 1959–1964, 1985–1987; syndication, 1988–1989; UPN, 2002–2003; CBS All Access, 2019–present). One of the most memorable *Black Mirror* episodes to date has been the fourth-season episode titled "Metalhead."

Set in a dystopian future, human survivors enter a seemingly abandoned warehouse in search of medical supplies. What they find instead is a "dog," a four-legged robot roughly the size of a midsize dog; it is highly mobile, fast, and armed (ironically, considering the episode's title, the robot "dog" has no real "head" to speak of). One by one the "dog" dispatches of the humans until only one human female remains. Successfully evading the "dog" for some time, she eventually destroys it, but not before it launches dozens of "trackers" that embed themselves in the human survivor. Aware that removing the tracker in her neck will kill her, the last survivor accepts her fate, as the house she is hiding in is surrounded by more "dogs" (Brooker, "Metalhead," *Black Mirror,* season 4, episode 5, December 29, 2017).

The "dogs" of "Metalhead" are inspired by—and built to resemble—the real-world robot "BigDog," by Boston Dynamics. In the episode, they are an army of relentless, near-unstoppable guard dogs. Even when they are unable to immediately dispose of an opponent, they can attack a fleeing party with their bullet-like trackers. While one such robot might be appealing as far as a guard dog goes, the concept of an entire army of such robots in the hands of one person or group is a terrifying one. Nevertheless, relentless killing machines are, today, much closer to science fact than fiction.

Richard A. Hall

See also: B.A.T.s, Battle Droids, Cylons, Doombots, ED-209, IG-88/IG-11, Robots, Sentinels, Terminators; *Thematic Essays:* Robots and Slavery, Villainous Robots and Their Impact on Sci-Fi Narratives, AI and the Apocalypse: Science Fiction Meeting Science Fact.

Further Reading

Allen, Arthur. 2019. "There's a Reason We Don't Know Much about AI." Politico, September 16, 2019. https://www.politico.com/agenda/story/2019/09/16/artificial-intell igence-study-data-000956?cid=apn.

Brooker, Charlie, Jones, Annabel, and Arnopp, Jason. 2018. *Inside Black Mirror.* New York: Crown Archetype.

Byrne, Emma. 2013. "Innovation Isn't Safe: The Future According to Kevin Warwick." *Forbes,* September 30, 2013. https://www.forbes.com/sites/netapp/2013/09/30/kevin -warwick-captain-cyborg/#48b704c13560.

Carper, Steve. 2019. *Robots in American Popular Culture.* Jefferson, NC: McFarland.

Chandler, Simon. 2019. "Artificial Intelligence Has Become a Tool for Classifying and Ranking People." *Forbes,* October 1, 2019. https://www.forbes.com/sites/simon chandler/2019/10/01/artificial-intelligence-has-become-a-tool-for-classifying-and -ranking-people/#7431657f1d7c.

Dinh, Thien-Nam. 2018. *Silicon Minds: The Science, Impact, and Promise of Artificial Intelligence.* Independently Published.

Faludi, Susan. 2007. *The Terror Dream: Fear and Fantasy in Post-9/11 America.* New York: Metropolitan.

Hitchcock, Susan Tyler. 2007. *Frankenstein: A Cultural History.* New York: W. W. Norton.

Johnson, David Kyle, ed. 2019. *Black Mirror and Philosophy: Dark Reflections.* Hoboken, NJ: Wiley-Blackwell.

Krishnan, Armin. 2009. *Killer Robots: Legality and Ethicality of Autonomous Weapons.* London: Routledge.

Lin, Patrick, Abney, Keith, and Bekey, George A., eds. 2014. *Robot Ethics: The Ethical and Social Implications of Robotics.* Cambridge: Massachusetts Institute of Technology.

Lin, Patrick, Jenkins, Ryan, and Abney, Keith, eds. 2017. *Robot Ethics 2.0: From Autonomous Cars to Artificial Intelligence.* New York: Oxford University Press.

Murdock, Jason. 2019. "Former Google Engineer Warns AI Might Accidentally Start a War: 'These Things Will Start to Behave in Unexpected Ways.'" *Newsweek,* September 16, 2019. https://www.newsweek.com/google-project-maven-artificial-intel ligence-laura-nolan-killer-robots-department-defense-1459358.

Randall, Terri, dir. 2016. *NOVA.* "Rise of the Robots: Inside the DARPA Robotics Challenge." Aired February 24, 2016, on PBS. DVD.

Rouhiainen, Lasse. 2018. *Artificial Intelligence: 101 Things You Must Know Today about Our Future.* Scotts Valley, CA: CreateSpace.

Spaeth, Dennis. 2018. "From Single-Task Machines to Backflipping Robots: The Evolution of Robots." *Cutting Tool Engineering*, January 15, 2018. https://www.ctemag.com/news/articles/evolution-of-robots.

Wallach, Wendell, and Allen, Colin. 2009. *Moral Machines: Teaching Robots Right from Wrong*. New York: Oxford University Press.

Metallo

DC Comics

In 1938, writer Jerry Siegel (1914–1996) and artist Joe Shuster (1914–1992) introduced the world to Superman, the world's first "superhero." When the planet Krypton faced imminent destruction, the scientist Jor-El created a spaceship in which he placed his infant son, Kal-El, and sent the child to safety on the planet Earth. There, Kal-El was raised by Jonathan and Martha Kent, farmers in the small midwestern town of Smallville. Raised as "Clark Kent," Kal-El grew to experience phenomenal superpowers, eventually including flight, superstrength, superspeed, "laser" eyes, "freeze" breath, and X-ray vision. The boy grew to serve humankind as Superman (first introduced in *Action Comics #1*, June 1938). In the eighty-plus years since his inception, Superman has gone on to dominate all media of popular culture: comic books, radio, movie serials (both animated and live action), television series (both animated and live action), and feature films. Throughout that time, Superman has accumulated an impressive rogues' gallery of "supervillains." Perhaps the most dangerous to Superman, physically, is the cyborg villain Metallo.

The first "Metalo" (with only one *l*) was one of Superman's earliest villains, being simply a mad scientist in a suit of armor. This character was finally killed off in 1982. The first character to use the name "Metallo" (two *l*s) was actually a training robot built by Superman's father, Jor-El, to train the growing "Superboy" (Jerry Coleman and John Sikela, "Metallo of Krypton," *Superboy #49*, June 1956). The character most commonly known by the name today began in what is known as the "silver age" of comics. John Corben was a murderer who, while fleeing the scene of his latest crime, was badly injured in an accident. A scientist transfers Corben's brain to a robotic body covered with "flesh" to look human. The robotic body required a heart of Kryptonite in order to survive. Ultimately Corben steals what he believes to be Kryptonite, but it is a fake. Before Superman can arrive, Corben dies from the fake heart (Robert Bernstein and Al Plastino, "The Menace of Metallo," *Action Comics #252*, May 1959).

The next version was Corben's brother, Roger, who undergoes the same procedure as his brother in order to seek revenge on Superman Martin Pasko and Curt Swan, "The Man with the Kryptonite Heart," *Superman #310*, April 1977. A twist on the John Corben storyline appeared as the "modern" version of the character, beginning in the "Post-*Crisis*" reality launched by *Superman #1* (vol. 2), in January 1987. This is the version that appeared in the live-action television series *Smallville* (The WB, 2001–2006; The CW, 2006–2011). In this episode, Corben was played by Brian Austin Green (b. 1973). In this incarnation, Corben is a journalist out to prove that the hero, the "Blur" (i.e., Superman) is secretly a bad guy.

After being critically wounded when hit by a truck, his body is taken by Luthor-corp and the Kryptonite heart transplanted to replace his real one. Out to expose and punish the Blur, Corben ultimately dies when removing a plate welded to his chest that also removes his heart (Don Whitehead and Holly Henderson, "Metallo," *Smallville*, season 9, episode 2, October 2, 2009).

With a robotic body powered by Kryptonite, Metallo is one of the few comic book villains that possess the ability to actually kill Superman. Ultimately, Metallo is the stereotypical murderous cyborg. With his brain being his only real connection to his humanity (and it being warped by the trauma of waking up in a metallic body with a Kryptonite heart), Metallo shows the downside of the idea of transplanting one's brain/soul into a robotic body. How much humanity can remain when so much is lost? At what point does one stop being a "robotic human" and, instead, becoming a mere "organic robot"? These issues can be explored through the various incarnations of the villain cyborg Metallo.

Richard A. Hall

See also: Adam, Alita, Arnim Zola, Borg, Brainiac, Cybermen, Cyborg, Daleks, Darth Vader, Doctor Octopus, General Grievous, Iron Man, Jaime Sommers, Nardole, RoboCop, Steve Austin, Tin Woodsman, Ultron; *Thematic Essays:* Villainous Robots and Their Impact on Sci-Fi Narratives, Cyborgs: Robotic Humans or Organic Robots?

Further Reading

Ambrosino, Brandon. 2018. "What Would It Mean for AI to Have a Soul?" BBC, June 17, 2018. https://www.bbc.com/future/article/20180615-can-artificial-intelligence-have -a-soul-and-religion.

Ashby, LeRoy. 2006. *With Amusement for All: A History of American Popular Culture since 1830*. Lexington: University Press of Kentucky.

Barker, Cory, Ryan, Chris, and Wiatrowski, Myc, eds. 2014. *Mapping* Smallville*: Critical Essays on the Series and Its Characters*. Jefferson, NC: McFarland.

Bryant, D'Orsay D., III. 1985. "Spare-Part Surgery: The Ethics of Organ Transplantation." *Journal of the National Medical Association* 77 (2) https://www.ncbi.nlm .nih.gov/pmc/articles/PMC2561842/pdf/jnma00245-0055.pdf.

Carper, Steve. 2019. *Robots in American Popular Culture*. Jefferson, NC: McFarland.

Castleman, Harry, and Podrazik, Walter J. 2016. *Watching TV: Eight Decades of American Television, Third Edition*. Syracuse, NY: Syracuse University Press.

Cavaler, Chris. 2015. *On the Origins of Superheroes: From the Big Bang to Action Comics No. 1*. Iowa City: University of Iowa Press.

Daniels, Les. 2004. *Superman: The Complete History—The Life and Times of the Man of Steel*. New York: DC Comics.

Dinh, Thien-Nam. 2018. *Silicon Minds: The Science, Impact, and Promise of Artificial Intelligence*. Independently Published.

Dowden, Bradley. 1993. *Logical Reasoning*. Belmont, CA: Wadsworth Publishing Company.

Fryer-Biggs, Zachary. 2019. "Coming Soon to a Battlefield: Robots That Can Kill." *The Atlantic*, September 3, 2019. https://www.theatlantic.com/technology/archive/2019 /09/killer-robots-and-new-era-machine-driven-warfare/597130.

Hitchcock, Susan Tyler. 2007. *Frankenstein: A Cultural History*. New York: W. W. Norton.

Krishnan, Armin. 2009. *Killer Robots: Legality and Ethicality of Autonomous Weapons*. London: Routledge.

Lin, Patrick, Abney, Keith, and Bekey, George A., eds. 2014. *Robot Ethics: The Ethical and Social Implications of Robotics*. Cambridge: Massachusetts Institute of Technology.

Longoni, Chiara, Bonezzi, Andrea, and Morewedge, Carey K. 2019. "Resistance to Medical Artificial Intelligence." *Journal of Consumer Research* 46, no. 4 (December): 629–50. https://academic.oup.com/jcr/article/46/4/629/5485292.

Mayor, Adrienne. 2018. *Gods and Robots: Myths, Machines, and Ancient Dreams of Technology*. Princeton, NJ: Princeton University Press.

Robitzski, Dan. 2018. "Artificial Consciousness: How to Give a Robot a Soul." Futurism, June 25, 2018. https://futurism.com/artificial-consciousness.

Singer, Peter W. 2009. "Isaac Asimov's Laws of Robotics Are Wrong." Brookings Institute, May 18, 2009. https://www.brookings.edu/opinions/isaac-asimovs-laws-of-robotics-are-wrong.

Thagard, Paul. 2017. "Will Robots Ever Have Emotions?" *Psychology Today*, December 14, 2017. https://www.psychologytoday.com/us/blog/hot-thought/201712/will-robots-ever-have-emotions.

Tye, Larry. 2013. *Superman: The High-Flying History of America's Most Enduring Hero*. New York: Random House.

Wright, Bradford W. 2003. *Comic Book Nation: The Transformation of Youth Culture in America*. Baltimore, MD: Johns Hopkins University Press.

Muffit

Battlestar Galactica

In the wake of the international phenomenon of the original *Star Wars* (Lucasfilm/20th Century Fox, 1977), television producer Glen A. Larson (1937–2014) created the live-action television science fiction series *Battlestar Galactica* (ABC, 1978–1979). The premise of the series is connected to the recent "ancient astronaut" school of thought, that life on Earth originated somewhere else, beyond the stars. In the show, humanity's relatives consisted of Twelve Colonies light-years from Earth. For one thousand years, the Colonies had been at war with the robotic Cylon race. In a final assault, the Cylons decimated the Colonies, forcing the few thousand survivors to take to the stars under the protection of the sole remaining "battlestar," the *Galactica*, commanded by Commander Adama (played by television veteran Lorne Greene, 1915–1987). The fleet of humans desperately seek refuge and assistance from their far-flung cousins on Earth, with the Cylons in hot pursuit. Unfortunately for Adama and the ships and humans under his care, Earth (set in the present of 1980), is far less advanced than the rest of humanity and is now under threat from the Cylons as well.

In the pilot episode, set on the distant planet of Caprica, a young boy named Boxey (played by Noah Hathaway, b. 1971) has a pet "daggit" (Caprican for "dog") named Muffit. During the Cylon attack, the small animal is killed. To ease Boxey's sadness, Dr. Wilker (played by John Dullaghan, 1930–2009) builds a robot dog for the boy, which Boxey names Muffit II (played by a chimpanzee named Evie in a robot dog suit). Muffit becomes the cute "R2-D2" stereotype for the series (Glen A. Larson, "Saga of a Star World," *Battlestar Galactica*, original series, season 1, episode 1, September 17, 1978). Though the character of Boxey

---

BigDog

Modern warfare in the mountains of Afghanistan has provided new obstacles for the twenty-first-century American soldier. In areas where trucks and Humvees cannot travel and air support cannot land, soldiers have been forced to return to traditional pack mules to carry excess material or materials too heavy for soldiers to safely carry. In response to this, DARPA (the Defense Advanced Research Projects Agency) provided funding to Boston Dynamics, which in 2005, with NASA's Jet Propulsion Laboratory, Harvard University's Concord Field Station, and the research facility Foster-Miller, developed "BigDog." BigDog is a quadruped "pack mule" robot. It is three feet long and two-and-a-half feet tall and can carry 340 pounds. It is capable of traveling over difficult terrain at a speed of up to four miles per hour and climbing inclines up to thirty-five degrees. While a perfect solution to the problem of transporting extra materials, BigDog was unfortunately a threat to what the military calls "sound discipline" (i.e., the need to be quiet in observational situations). Its gas-powered hydraulic mechanisms make too much noise for many operational situations (exacerbated, naturally, in mountainous canyon regions such as Afghanistan). More advanced BigDogs were developed, including AlphaDog (which could walk on ice and recover itself if knocked over), in 2008, and a newer model containing an "arm" that could lift and throw up to fifty pounds, in 2013. Despite these advancements, the initial noise issue caused the BigDog program to be discontinued in 2015.

Richard A. Hall

---

would appear in a cameo in the reboot of the series in 2003, Muffit did not make the cut.

Muffit is, in essence, a pet. He does not possess the ability to speak or communicate to any degree greater than a dog would. He does not possess "abilities" to interact with computers or conduct any activities other than those attributed to a normal dog. As a pet, he is the "property" of young Boxey. Though not, perhaps, a major contribution to the discussion of robots and what they represent, Muffit was an example of the push in late-1970s and early 1980s sci-fi to play into the popularity of R2-D2 with cute robots that could delight young children.

Richard A. Hall

See also: BB-8, D-O, HERBIE, Johnny 5/S.A.I.N.T. Number 5, K9, R2-D2; *Thematic Essays:* Robots and Slavery, Heroic Robots and Their Impact on Sci-Fi Narratives.

Further Reading

Albert, Robert S., and Brigante, Thomas. 1962. "The Psychology of Friendship Relations: Social Factors." *Journal of Social Psychology* 56 (1). https://www.tandfonline.com/doi/abs/10.1080/00224545.1962.9919371.

Amati, Viviana, Meggiolaro, Silvia, Rivellini, Giulia, and Zaccarin, Susanna. 2018. "Social Relations and Life Satisfaction: The Role of Friends." *Genus* 74, no. 1 (May 4). https://www.ncbi.nlm.nih.gov/pmc/articles/PMC5937874.

Ashby, LeRoy. 2006. *With Amusement for All: A History of American Popular Culture since 1830.* Lexington: University Press of Kentucky.

Carper, Steve. 2019. *Robots in American Popular Culture.* Jefferson, NC: McFarland.

Castleman, Harry, and Podrazik, Walter J. 2016. *Watching TV: Eight Decades of American Television, Third Edition.* Syracuse, NY: Syracuse University Press.

Dinh, Thien-Nam. 2018. *Silicon Minds: The Science, Impact, and Promise of Artificial Intelligence*. Independently Published.

Gross, Edward, and Altman, Mark A. eds. 2018. *So Say We All: The Complete, Uncensored, Unauthorized Oral History of* Battlestar Galactica. New York: Tor.

Handley, Rich, and Tambone, Lou, eds. 2018. *Somewhere Beyond the Heavens: Exploring* Battlestar Galactica. Edwardsville, IL: Sequart Research and Literacy Organization.

Larson, Glen A., and Thurston, Robert. (1978) 2005. Battlestar Galactica *Classic: The Saga of a Star World*. New York: iBooks.

Randall, Terri, dir. 2016. *NOVA*. "Rise of the Robots: Inside the DARPA Robotics Challenge." Aired February 24, 2016, on PBS. DVD.

Rouhiainen, Lasse. 2018. *Artificial Intelligence: 101 Things You Must Know Today about Our Future*. Scotts Valley, CA: CreateSpace.

Wallach, Wendell, and Allen, Colin. 2009. *Moral Machines: Teaching Robots Right from Wrong*. New York: Oxford University Press.

N

Nardole

Doctor Who

In 1963, BBC executive Sydney Newman (1917–1997) and producer Verity Lambert (1935–2007) created the British television series *Doctor Who* (BBC, 1963–1989; FOX, 1996; BBC, 2005–present), the longest-running science fiction television series in history. The series centers on the character of the "Doctor," an alien "Time Lord" possessing the power to "regenerate"—or gain a new body and personality—when his or her (depending on the incarnation) physical body dies. The Doctor travels through time and space in a device known as a TARDIS (Time and Relative Dimension in Space). Though TARDISes are designed to change their outward appearance in order to blend in with its environment, the Doctor's TARDIS is "stuck" in the constant form of a 1960s British police call box. Throughout the series, the Doctor has most often traveled with human companions from Earth. The twelfth Doctor (played by Peter Capaldi, b. 1958), however, often traveled in his later years with an alien cyborg by the name of Nardole (played by Matt Lucas, b. 1974).

Nardole currently holds the distinction of being one of only two companions to appear in three consecutive *Doctor Who* Christmas specials. The Doctor first meets Nardole, an alien of unknown origins, while the alien is in the employ of the Doctor's human wife, the time-traveling archaeologist, Dr. River Song (played by Alex Kingston, b. 1963). During this adventure, Nardole is decapitated by the robotic body of King Hydroflax, with his living head stored within the robot's body (Steven Moffatt, *Doctor Who: The Husbands of River Song*, December 25, 2015). After Dr. Song's death, the Doctor "rebuilds" Nardole, fixing his head on a new body resembling his original one. It is unclear to what extent Nardole is organic and to what extent robotic (though mechanical sounds can be heard occasionally when Nardole moves his arms and legs).

Nardole becomes a regular companion of the Doctor's, helping the Time Lord in his assignment of keeping watch over a vault that imprisons the dangerous Time Lord known as "Missy" (previously, the "Master," played in this incarnation by Michelle Gomez, b. 1966). When the Doctor encounters a spaceship trapped in the event horizon of a black hole, he discovers that the ship is populated by both humans and Cybermen. While the Doctor sacrifices his life to prevent the Cybermen from taking the humans, Nardole is tasked with guiding the humans to a safer part of the ship, where the cyborg will then spend the rest of his life with them (Moffatt, "The Doctor Falls," *Doctor Who*, return series, season 10, episode 12). When the Doctor next encounters Nardole, it is many centuries in the future, where the cyborg's "memories" have been downloaded into a matrix called

"Testimony." Nardole is able to say a final farewell to the Doctor through a glass avatar of his corporeal body, the essence of who he was permanently preserved for all time (Moffatt, *Doctor Who: Twice upon a Time*, December 25, 2017).

The character of Nardole remains very much a mystery in the realm of *Doctor Who*. Was he already a cyborg when the Doctor first met him? To what extent is he a cyborg? The answers to these questions may never be fully known as the "Moffatt era" of *Doctor Who* has come to an end. What is known is that Nardole provided immense comedic relief to the twelfth Doctor's final adventures, with a mixture of K9's utility and C-3PO's constant nervousness. In a franchise where cyborgs are most commonly dangerous foes—for example, the Cybermen and Daleks—Nardole brings a cyborg that is beloved, gentle, and kind.

Richard A. Hall

See also: Adam, Alita, Borg, C-3PO, Cybermen, Cyborg, Daleks, Darth Vader, Doctor Octopus, Echo/CT-1409, General Grievous, The Great Intelligence, Inspector Gadget, Jaime Sommers, K9, Metallo, Steve Austin, TARDIS, Teselecta; *Thematic Essays:* Heroic Robots and Their Impact on Sci-Fi Narratives, Cyborgs: Robotic Humans or Organic Robots?

Further Reading

Bryant, D'Orsay D., III. 1985. "Spare-Part Surgery: The Ethics of Organ Transplantation." *Journal of the National Medical Association* 77 (2) https://www.ncbi.nlm.nih.gov/pmc/articles/PMC2561842/pdf/jnma00245-0055.pdf.

Burnett, Dean. 2014. "Time Travellers: Please Don't Kill Hitler." *The Guardian*, February 21, 2014. https://www.theguardian.com/science/brain-flapping/2014/feb/21/time-travellers-kill-adolf-hitler.

Byrne, Emma. 2014. "Cybernetic Implants: No Longer Science Fiction." *Forbes*, March 11, 2014. https://www.forbes.com/sites/netapp/2014/03/11/cybernetic-implants-not-sci-fi/#7399c57e77ba.

Calvert, Bronwen. 2017. *Being Bionic: The World of TV Cyborgs*. London: I. B. Tauris.

Campbell, Mark. 2010. Doctor Who: *The Complete Guide*. London: Running Press.

Farnell, Chris. 2020. "*Doctor Who*: The Genius of Making the Cybermen and Ideology." Den of Geek, January 28, 2020. https://www.denofgeek.com/tv/doctor-who-the-genius-of-making-thecybermen-an-ideology-2.

Kachka, Boris. 2015. "The Last Human Robot." Vulture, December 6, 2015. https://www.vulture.com/2015/12/anthony-daniels-c-3po-c-v-r.html.

Kistler, Alan. 2013. Doctor Who: *Celebrating Fifty Years, A History*. Guilford, CT: Lyons.

Lewis, Courtland, and Smithka, Paula, eds. 2015: *More* Doctor Who *and Philosophy: Regeneration Time*. Chicago: Open Court.

Mallett, Ronald L., and Henderson, Bruce. 2006. *Time Traveler: A Scientist's Personal Mission to Make Time Travel a Reality*. New York: Basic Books.

Parkin, Lance. 2016. *Whoniverse*. New York: Barron's Educational Series.

Wasserman, Ryan. 2018. *Paradoxes of Time Travel*. Oxford: Oxford University Press.

OASIS

Ready Player One

In 2011, American author Ernest Cline (b. 1972) published his first novel, *Ready Player One*. The story is set in the near future of 2045, and the world's economy has been devastated by continued global warming, leading to a bleak and hopeless future. The only escape from this grim reality is the massive virtual reality program OASIS, which can act as either a massively multiplayer online role-playing game (MMORPG) or as a general VR society complete with its own cryptocurrency (the only currency of real value in the dystopian world). When the creator of OASIS, James Halliday, dies, he leaves instructions in his will that the discovery of a hidden "Easter Egg" within OASIS will bestow upon its discoverer Halliday's wealth (and control of the program). The novel was made into a major feature film in 2018, directed by Steven Spielberg (b. 1946).

OASIS represents the ultimate goal in VR technology, providing both gaming features and a complex "reality" in which "players" can simply interact as they would in the real world, but this world is specifically designed to maximize pleasure and happiness. For ten thousand years, humanity has attempted to mold the real world into various ideal civilizations. Through the realm of fiction, authors have endeavored to explore various options on society, how it is designed, and to what degree people could be happy in it. With the massive increase in VR technology throughout the twenty-first century to date, the example of OASIS is realistically within grasp by the book's date of 2045. However, should society decide to immerse itself en masse into such a program, the line between "OASIS" and "Matrix" (see "The Matrix/Agent Smith") is a very fine one indeed.

Richard A. Hall

See also: AI/Ziggy, Arnim Zola, Cerebro/Cerebra, The Great Intelligence, HAL 9000, Janet, JARVIS/Friday, Landru, The Matrix/Agent Smith, Oz, Rehoboam, Skynet, TARDIS, WOPR; *Thematic Essay:* AI and the Apocalypse: Science Fiction Meeting Science Fact.

Further Reading

Alkon, Paul K. 1987. *Origins of Futuristic Fiction*. Athens: University of Georgia Press.

Allen, Arthur. 2019. "There's a Reason We Don't Know Much about AI." Politico, September 16, 2019. https://www.politico.com/agenda/story/2019/09/16/artificial-intelligence-study-data-000956?cid=apn.

Bastani, Aaron. 2019. *Fully Automated Luxury Communism*. New York: Verso.

Chandler, Simon. 2019. "Artificial Intelligence Has Become a Tool for Classifying and Ranking People." *Forbes*, October 1, 2019. https://www.forbes.com/sites/simonchandler/2019/10/01/artificial-intelligence-has-become-a-tool-for-classifying-and-ranking-people/#7431657f1d7c.

Chanthadavong, Aimee. 2019. "AI and Ethics: The Debate That Needs to Be Had." ZDnet, September 16, 2019. https://www.zdnet.com/article/ai-and-ethics-the-debate-that-needs-to-be-had/#ftag=CAD-03-10abf5f.

Cline, Ernest. 2012. *Ready Player One: A Novel.* New York: Broadway.

Dinh, Thien-Nam. 2018. *Silicon Minds: The Science, Impact, and Promise of Artificial Intelligence.* Independently Published.

Faludi, Susan. 2007. *The Terror Dream: Fear and Fantasy in Post-9/11 America.* New York: Metropolitan.

Irwin, William, ed. 2002. *The Matrix and Philosophy: Welcome to the Desert of the Real.* Chicago: Open Court.

Leetaru, Kalev. 2019. "Automatic Image Captioning and Why Not Every AI Problem Can Be Solved through More Data." *Forbes*, July 7, 2019. https://www.forbes.com/sites/kalevleetaru/2019/07/07/automatic-image-captioning-and-why-not-every-ai-problem-can-be-solved-through-more-data/#20b943476997.

Lin, Patrick, Jenkins, Ryan, and Abney, Keith, eds. 2017. *Robot Ethics 2.0: From Autonomous Cars to Artificial Intelligence.* New York: Oxford University Press.

Marks, Robert J. 2020. "*2084* vs. *1984*: The Difference AI Could Make to Big Brother." Podcast interview with author John Lennox. Mind Matters, July 3, 2020. https://mindmatters.ai/2020/07/2084-vs-1984-the-difference-ai-could-make-to-big-brother.

Mayor, Adrienne. 2018. *Gods and Robots: Myths, Machines, and Ancient Dreams of Technology.* Princeton, NJ: Princeton University Press.

Murdock, Jason. 2019. "Former Google Engineer Warns AI Might Accidentally Start a War: 'These Things Will Start to Behave in Unexpected Ways.'" *Newsweek*, September 16, 2019. https://www.newsweek.com/google-project-maven-artificial-intelligence-laura-nolan-killer-robots-department-defense-1459358.

Rouhiainen, Lasse. 2018. *Artificial Intelligence: 101 Things You Must Know Today about Our Future.* Scotts Valley, CA: CreateSpace.

Virk, Rizwan. 2019. *The Simulation Hypothesis: An MIT Computer Scientist Shows Why AI, Quantum Physics, and Eastern Mystics All Agree We Are in a Video Game.* Milwaukee, WI: Bayview.

Oz

The Wizard of Oz

In 1900, populist American author L. Frank Baum (1856–1919) published the first novel in the Land of Oz series: *The Wonderful Wizard of Oz*. This book follows the adventures of a young girl named Dorothy, living in rural Kansas, who is swept away by a tornado to the mystical world of "Oz." Threatened by the Wicked Witch of the West, Dorothy and a group of misfit friends—the Scarecrow, Tin Man, and Cowardly Lion—seek out the wisdom and advice of Oz's leader, "The Great and Powerful Oz" (a.k.a. "The Wizard of Oz"). The massive success of the Oz books led to a Broadway musical adaptation of the book in 1902 and the classic feature film *The Wizard of Oz*, one of the first feature films in history to be filmed in full color, in 1939.

Though "Oz" proves to be a mere human named Oscar Zoroaster Phadrig Isaac Norman Henkle Emmannuel Ambroise Diggs, this fact is unknown to the denizens of the Land of Oz. When Dorothy and her companions meet with Oz (each

Daft Punk and Kraftwerk

In recent decades, musical groups have begun to incorporate robotic personas/topics in their performances. Two such groups are the French group Daft Punk and the German group Kraftwerk. Daft Punk is the techno/disco duo of Guy-Manuel de Homem-Christo (b. 1974) and Thomas Bangalter (b. 1975). The two are most often seen performing in helmets designed to give a robotic appearance. Their top U.S. hits include the singles: "Da Funk" (1995), "Around the World" (1997), One More Time (2000), "Face to Face" (2003), "Derezzed" (2010), and "Get Lucky" and "Lose Yourself to Dance" (both 2013, and both featuring Pharrell Williams, b. 1973). Kraftwerk has been around much longer, forming in 1970, and was one of the early founders of "electronic" or "synth-pop" music that would come to dominate American charts in the 1980s. In the twenty-first century, the band has incorporated robotic/computerized looks in their onstage costumes and performances. Current band members are cofounder Ralf Hütter (b. 1946), Fritz Hilpert (b. 1956), Henning Schmitz (b. 1953), and Falk Grieffenhagen (b. 1969). The band has never had a Top Ten hit in the United States, but it continues to be embraced by a strong cult following. Daft Punk continues to be a popular group in the United States still today. These two groups represent the growing influence of robots, robotics, and electronics in Western culture over the last half century. Aside from these robotic/computer-based musical groups, there have also been some very popular robot/sci-fi-related songs that have peaked in the American charts over the decades, most notably *Mr. Roboto* by Styx (1983) and *Intergalactic* by the Beastie Boys (1998).

Richard A. Hall

one-on-one), each sees Oz differently: either as a fairy, a monster, or simply a giant head. When the group meets with him together, he is simply a booming disembodied voice. In the feature film, Oz is first seen as a giant holographic head. Through cleverness, Dorothy determines that Oz is merely a human man manipulating the image from behind a curtain. Though not "AI" has been defined throughout the many examples in this volume, the fictional "Oz" that has been presented to the people of this magical realm is that of a hologram.

Holograms have played a key part in science fiction for decades. From the holographic form of the human AI in the television series *Quantum Leap* (NBC, 1989–1993) to the frequent use of holographic communication throughout the *Star Wars* franchise, to the real-world recreation of the deceased musical artist Tupac Shakur (1971–1996) in holographic form in 2012, holograms have become a very real part of both science fiction and science fact. In the case of Oz, a hologram allowed a simple old man to appear as a massive frightening visage (thereby giving him a calculated advantage over those with whom he interacted). Holograms are becoming more and more realistic with each passing year; and how those who use them are perceived by the general population will depend greatly on how the technology is used in the future.

Richard A. Hall

See also: AI/Ziggy, Arnim Zola, Cerebro/Cerebra, The Great Intelligence, HAL 9000, Janet, JARVIS/Friday, Landru, The Matrix/Agent Smith, OASIS, Rehoboam, Skynet, TARDIS, Tin Woodsman, WOPR, Zordon/Alpha-5; *Thematic Essay:* AI and the Apocalypse: Science Fiction Meeting Science Fact.

Further Reading

Alkon, Paul K. 1987. *Origins of Futuristic Fiction*. Athens: University of Georgia Press.

Allen, Arthur. 2019. "There's a Reason We Don't Know Much about AI." Politico, September 16, 2019. https://www.politico.com/agenda/story/2019/09/16/artificial-intelligence-study-data-000956?cid=apn.

Ashby, LeRoy. 2006. *With Amusement for All: A History of American Popular Culture since 1830*. Lexington: University Press of Kentucky.

Bastani, Aaron. 2019. *Fully Automated Luxury Communism*. New York: Verso.

Baum, L. Frank. (1900) 2015. *The Wonderful Wizard of Oz*. Scotts Valley, CA: Createspace.

Cavaler, Chris. 2015. *On the Origins of Superheroes: From the Big Bang to Action Comics No. 1*. Iowa City: University of Iowa Press.

Chandler, Simon. 2019. "Artificial Intelligence Has Become a Tool for Classifying and Ranking People." *Forbes*, October 1, 2019. https://www.forbes.com/sites/simonchandler/2019/10/01/artificial-intelligence-has-become-a-tool-for-classifying-and-ranking-people/#7431657f1d7c.

Chanthadavong, Aimee. 2019. "AI and Ethics: The Debate That Needs to Be Had." ZDnet, September 16, 2019. https://www.zdnet.com/article/ai-and-ethics-the-debate-that-needs-to-be-had/#ftag=CAD-03-10abf5f.

Leetaru, Kalev. 2019. "Automatic Image Captioning and Why Not Every AI Problem Can Be Solved through More Data." *Forbes*, July 7, 2019. https://www.forbes.com/sites/kalevleetaru/2019/07/07/automatic-image-captioning-and-why-not-every-ai-problem-can-be-solved-through-more-data/#20b943476997.

Marks, Robert J. 2020. "*2084* vs. *1984*: The Difference AI Could Make to Big Brother." Podcast interview with author John Lennox. Mind Matters, July 3, 2020. https://mindmatters.ai/2020/07/2084-vs-1984-the-difference-ai-could-make-to-big-brother.

Mayor, Adrienne. 2018. *Gods and Robots: Myths, Machines, and Ancient Dreams of Technology*. Princeton, NJ: Princeton University Press.

Murdock, Jason. 2019. "Former Google Engineer Warns AI Might Accidentally Start a War: 'These Things Will Start to Behave in Unexpected Ways.'" *Newsweek*, September 16, 2019. https://www.newsweek.com/google-project-maven-artificial-intelligence-laura-nolan-killer-robots-department-defense-1459358.

R

R2-D2

Star Wars

In 1977, writer-director George Lucas (b. 1944) introduced the world to the sci-fi/fantasy phenomenon *Star Wars* (Lucasfilm/20th Century Fox, 1977), one of the most successful pop culture franchises in American history. His original trilogy of films (1977–1983) focused on the rise of its protagonist, Luke Skywalker (played by Mark Hamill, b. 1951), against the backdrop of a Galactic Civil War against the human forces of the evil Empire, led by its Emperor, Sheev Palpatine (played by Ian McDiarmid, b. 1944), and his henchman, Darth Vader (played by David Prowse, b. 1935; voiced by James Earl Jones, b. 1931). In 1999, Lucas launched a "prequel" trilogy, beginning with *Star Wars, Episode I: The Phantom Menace* (Lucasfilm, 1999), which began the story of the fall of Anakin Skywalker (played originally by Jake Lloyd, b. 1989; later by Hayden Christensen, b. 1981)—the father of the original hero—into becoming the villainous Darth Vader. Throughout the first six *Star Wars* films, one of the major heroes was the astromech droid, R2-D2 (played by Kenny Baker, 1934–2016).

In the *Star Wars* universe, astromech droids (in the original and prequel trilogies) were repair droids with squat, barrel-shaped bodies (usually white or black in color; R2's is white) suspended between two legs. A third, retractable "leg" could emerge from beneath the droid's torso for easier maneuverability. Astromechs had swiveling, either dome-shaped or bucket-shaped heads (usually either white, black, or chrome) with optical and audio scanners (R2's head is dome shaped and chrome). They also usually have distinctive colored detailing (R2's is blue) on their white bodies. They are most commonly used in emergency spacecraft repairs while in flight.

R2's story begins chronologically in *The Phantom Menace*, where audiences learn that R2 was part of a team of repair droids on the official spacecraft of Queen Amidala (played by Natalie Portman, b. 1981), of the planet Naboo. For the remainder of the prequel trilogy—and two animated television series—R2 was the personal astromech to the Jedi Knight Anakin Skywalker (Christensen). In the original trilogy, where the character was first introduced, roughly twenty years have passed since the prequel era, and R2 is assigned to the official spaceship of Princess Leia Organa (played by Carrie Fisher, 1956–2016), the daughter of Amidala and Skywalker. When the princess sends R2 and his companion, C-3PO (played by Anthony Daniels, b. 1946) in search of the Jedi Knight Obi-Wan Kenobi (played by Sir Alec Guinness, 1914–2000), the two come into contact with Luke Skywalker (Hamill), son of Amidala and Skywalker, and twin brother of the princess. R2 plays a major role in delivering the stolen plans of the Empire's "Death

---

R2-KT

R2-KT is a pink astromech droid along the same general design as the iconic R2-D2 in the *Star Wars* franchise. She has appeared in the CGI-animated television series *Star Wars: The Clone Wars* (Cartoon Network, 2008–2013; Netflix, 2014; Disney+, 2020) and the live-action film *Star Wars: Episode VII—The Force Awakens* (Lucasfilm/Disney, 2015). The droid is known for her kindness and compassion and is referred to as the "Droid with the Heart of Gold."

Unlike other droids in the *Star Wars* universe, however, KT's history originates with heartbreak. She was created in 2005 by the fan group "R2-D2 Builders Club," a global organization of fans who enjoy building stage accurate copies of the iconic droid. KT was built in memory of Katie Johnson (1998–2005) who had just died of terminal cancer. Katie was the daughter of Albin Johnson (birth year unknown), who was the founder of the fan charity group "The 501st Legion," in 1997. The 501st is a cosplay group with chapters all over the world. They do numerous charitable events every year, with members appearing at children's hospitals and other functions dressed as stormtroopers, clone troopers, Darth Vader, and even the bounty hunter Boba Fett in screen-accurate costumes. They also provided extras for the live-action television series *The Mandalorian* (Disney+, 2020–present). The entire *Star Wars* fan community rallied around the Johnson family through young Katie's ordeals; she is now immortalized as a permanent part of the *Star Wars* universe in the form of this beloved little pink droid.

Richard A. Hall

Star" space station to the Rebel Alliance, rescuing the princess from the Empire, and maintaining Luke Skywalker's fighter craft during the battle to destroy the space station (Lucas, *Star Wars*, Lucasfilm/20th Century Fox, 1977). R2 (and 3PO) then continue to assist the princess and young Skywalker in their war against the Empire.

As a droid, R2-D2 is, in essence, a slave. Throughout the "Skywalker Saga," R2 is the property of the Skywalker family: a slave first to Queen Amidala; then to her husband, Anakin Skywalker; then to their daughter, Princess Leia; then to Leia's brother, Luke. With the deaths of the entire Skywalker family by the end of *Star Wars: Episode IX—The Rise of Skywalker*, it is unclear who the new "owners" of R2-D2 and C-3PO will be. Throughout the *Star Wars* saga, R2-D2 was in every way a hero, a key player in the quest for freedom from tyranny in the galaxy. As a droid, however, R2—like all other droids—is unable to enjoy the freedom for which he so bravely fought. He is a slave. He is property. Despite his clearly possessing all the attributes of "sentience," as a machine he is viewed as belonging to organic life-forms.

Richard A. Hall

See also: Battle Droids, BB-8, C-3PO, Darth Vader, D-O, Dr. Theopolis and Twiki, Echo/CT-1409, EV-9D9, General Grievous, HERBIE, IG-88/IG-11, Johnny 5/S.A.I.N.T. Number 5, K-2SO, K9, L3-37/*Millennium Falcon*, Medical Droids, Muffit, Robot, Rosie; *Thematic Essays:* Robots and Slavery, Heroic Robots and Their Impact on Sci-Fi Narratives.

Further Reading

Albert, Robert S., and Brigante, Thomas R. 1962. "The Psychology of Friendship Relations: Social Factors." *Journal of Social Psychology* 56 (1). https://www.tandfonline.com/doi/abs/10.1080/00224545.1962.9919371.

Amati, Viviana, Meggiolaro, Silvia, Rivellini, Giulia, and Zaccarin, Susanna. 2018. "Social Relations and Life Satisfaction: The Role of Friends." *Genus* 74, no. 1 (May 4). https://www.ncbi.nlm.nih.gov/pmc/articles/PMC5937874.

Ambrosino, Brandon. 2018. "What Would It Mean for AI to Have a Soul?" BBC, June 17, 2018. https://www.bbc.com/future/article/20180615-can-artificial-intelligence-have-a-soul-and-religion.

Ashby, LeRoy. 2006. *With Amusement for All: A History of American Popular Culture since 1830*. Lexington: University Press of Kentucky.

Bastani, Aaron. 2019. *Fully Automated Luxury Communism*. New York: Verso.

Carper, Steve. 2019. *Robots in American Popular Culture*. Jefferson, NC: McFarland.

Cavelos, Jeanne. 2008. "R2-D2 and C-3P0: Do Droids Dream of Electric Sheep?" *Scientific American*, August 11, 2008. https://www.scientificamerican.com/article/star-wars-science-droid-dreams.

Dinh, Thien-Nam. 2018. *Silicon Minds: The Science, Impact, and Promise of Artificial Intelligence*. Independently Published.

Dowden, Bradley. 1993. *Logical Reasoning*. Belmont, CA: Wadsworth Publishing Company.

Eberl, Jason T., and Decker, Kevin S., eds. 2015. *The Ultimate* Star Wars *and Philosophy: You Must Unlearn What You Have Learned*. Hoboken, NJ: Wiley-Blackwell.

Ford, Martin. 2015. *Rise of the Robots: Technology and the Threat of a Jobless Future*. New York: Basic.

Hampton, Gregory Jerome. 2017. *Imagining Slaves and Robots in Literature, Film, and Popular Culture: Reinventing Yesterday's Slave with Tomorrow's Robot*. Lanham, MD: Lexington.

Kaminski, Michael. 2008. *The Secret History of* Star Wars: *The Art of Storytelling and the Making of a Modern Epic*. Kingston, Canada: Legacy Books.

Kirshner, Jonathan. 2012. *Hollywood's Last Golden Age: Politics, Society, and the Seventies Film in America*. Ithaca, NY: Cornell University Press.

Lin, Patrick, Abney, Keith, and Bekey, George A., eds. 2014. *Robot Ethics: The Ethical and Social Implications of Robotics*. Cambridge: Massachusetts Institute of Technology.

Pellissier, Hank. 2013. "Robots and Slavery: What Do Humans Want When We Are 'Masters'?" Institute for Ethics and Emerging Technologies, September 13, 2013. https://ieet.org/index.php/IEET2/more/pellissier20130913.

Randall, Terri, dir. 2016. *NOVA*. "Rise of the Robots: Inside the DARPA Robotics Challenge." Aired February 24, 2016, on PBS. DVD.

Reagin, Nancy R., and Liedl, Janice, eds. 2013. Star Wars *and History*. New York: John Wiley & Sons.

Roberts-Griffin, Christopher. 2011. "What Is a Good Friend: A Qualitative Analysis of Desired Friendship Qualities." *Penn McNair Research Journal* 3, no. 1 (December 21). https://repository.upenn.edu/cgi/viewcontent.cgi?article=1019&context=mcnair_scholars.

Robitzski, Dan. 2018. "Artificial Consciousness: How to Give a Robot a Soul." Futurism, June 25, 2018. https://futurism.com/artificial-consciousness.

Rouhiainen, Lasse. 2018. *Artificial Intelligence: 101 Things You Must Know Today about Our Future*. Scotts Valley, CA: CreateSpace.

Slavicsek, Bill. 2000. *A Guide to the* Star Wars *Universe*. San Francisco: LucasBooks.

Spaeth, Dennis. 2018. "From Single-Task Machines to Backflipping Robots: The Evolution of Robots." *Cutting Tool Engineering*, January 15, 2018. https://www.ctemag.com/news/articles/evolution-of-robots.

Stanford University. n.d. "Robotics: A Brief History." Stanford University. https://cs.stanford.edu/people/eroberts/courses/soco/projects/1998-99/robotics/history.html.

Sumerak, Marc. 2018. Star Wars*: Droidography.* New York: HarperFestival.

Sunstein, Cass R. 2016. *The World According to* Star Wars. New York: Dey Street.

Sweet, Derek R. 2015. Star Wars *in the Public Square: The Clone Wars as Political Dialogue.* Critical Explorations in Science Fiction and Fantasy series, edited by Donald E. Palumbo and Michael Sullivan. Jefferson, NC: McFarland.

Taylor, Chris. 2015. *How* Star Wars *Conquered the Universe: The Past, Present, and Future of a Multibillion Dollar Franchise.* New York: Basic Books.

Wallach, Wendell, and Allen, Colin. 2009. *Moral Machines: Teaching Robots Right from Wrong.* New York: Oxford University Press.

Rehoboam

Westworld

In 2016, creators Jonathan Nolan and Lisa Joy introduced their live-action television series *Westworld* (HBO, 2016–present), an updated reboot of the 1973 film of the same name by writer-director Michael Crichton (1942–2008). The premise of the series is that, in an undisclosed future, the Delos corporation operates a series of amusement/recreation parks that allows patrons to live out their ultimate fantasies against various backdrops. "Westworld" allows customers to experience realistic adventures set in the American "Wild West" in a vividly realistic panoramic western world populated by "hosts": realistic androids programmed to play out specified storylines with preprogrammed personalities based on stereotypes of the Wild West (e.g., gunfighters, soldiers, farmers, savage Natives, and prostitutes). The hosts—and their corresponding storylines—have been produced by park cofounder Robert Ford (played by Anthony Hopkins, b. 1937). As the hosts are robots, guests are free to beat, rape, or murder the hosts as they see fit (creating both a cathartic release and an ethical conundrum). In the series' third season, the renegade host Dolores (played by Evan Rachel Wood, b. 1987)—bent on revenge against the human world—discovers that humanity itself is in its own way enslaved by the massive AI program "Rehoboam."

Set in the year 2058, the corporation Incite—the key investor in the Westworld project—has a larger project called Rehoboam (named after the son of the ancient Hebrew King Solomon, and the first King of Judah, the southern half of what was originally the Kingdom of Israel). The program was created by Engerraund Serac (played by Vincent Cassel, b. 1966), the richest person in the world. The purpose of Rehoboam was to run calculated simulations on all humans, selecting those who are "at risk" of potentially causing devastation to humankind in one form or other. Once they are identified, Serac and his forces detain those individuals, placing them in suspended animation (Karrie Crouse and Jonathan Nolan, "Genre," *Westworld*, season 3, episode 5, April 12, 2020). Through this process, however, Rehoboam evolves to the point of controlling much of human society, maintaining "stability"; rather than "preventing" society's downfall, the program has merely been "delaying" it. With the help of her allies, Dolores is ultimately successful in shutting Rehoboam down, giving humanity the free will that she had so desired for her fellow host androids (Nolan and Denise Thé, "Crisis Theory," *Westworld*, season 3, episode 8, May 3, 2020).

Similar in many ways to the Matrix, Rehoboam gives humanity a false sense of security by denying it free will. Unlike the Matrix, the "reality" itself is real (humans are not living in a computer-generated simulation), but that reality is strictly "controlled" (Rehoboam controls human decision making through control of the media, marketing, and even—to a degree—subliminal messaging). The denial of free will is central to many AI narratives, as it is perhaps humanity's greatest fear with regard to putting trust in artificial intelligence. The program sees itself as a necessary "good" in that it protects humanity from itself; and, indeed, many people would agree that such a program's benefits far outweigh its detriments. Rehoboam, then, is not only another modern allegory of Franken-stein's monster but also a cautionary tale of the age-old question of how much liberty can be acceptably sacrificed for the sake of security. According to Benja-min Franklin, the answer is "none"; but the threats facing human society today could never have been dreamt of in the 1700s. The answer to the question, there-fore, remains in a state of constant flux; and the fictional AI of Rehoboam is a very real potential answer.

Richard A. Hall

See also: AI/Ziggy, Bernard, Brainiac, Cerebro/Cerebra, Control, Dolores, The Great Intelligence, HAL 9000, HERBIE, Janet, JARVIS/Friday, Landru, The Matrix/Agent Smith, OASIS, Skynet, TARDIS, Ultron, V-GER, WOPR; *Thematic Essay:* AI and the Apocalypse: Science Fiction Meeting Science Fact.

Further Reading

Allen, Arthur. 2019. "There's a Reason We Don't Know Much about AI." Politico, Sep-tember 16, 2019. https://www.politico.com/agenda/story/2019/09/16/artficial-intel ligence-study-data-000956?cid=apn.

Bastani, Aaron. 2019. *Fully Automated Luxury Communism*. New York: Verso.

Byrne, Emma. 2013. "Innovation Isn't Safe: The Future According to Kevin Warwick." *Forbes*, September 30, 2013. https://www.forbes.com/sites/netapp/2013/09/30 /kevin-warwick-captain-cyborg/#48b704c13560.

Chandler, Simon. 2019. "Artificial Intelligence Has Become a Tool for Classifying and Ranking People." *Forbes*, October 1, 2019. https://www.forbes.com/sites/simon chandler/2019/10/01/artificial-intelligence-has-become-a-tool-for-classifying-and -ranking-people/#7431657f1d7c.

Chanthadavong, Aimee. 2019. "AI and Ethics: The Debate That Needs to Be Had." ZDnet, September 16, 2019. https://www.zdnet.com/article/ai-and-ethics-the-debate-that -needs-to-be-had/#ftag=CAD-03-10abf5f.

Choi, Charles Q. 2009. "Human Evolution: The Origin of Tool Use." LiveScience, Novem-ber 11, 2009. https://www.livescience.com/7968-human-evolution-origin-tool.html.

Dinh, Thien-Nam. 2018. *Silicon Minds: The Science, Impact, and Promise of Artificial Intelligence*. Independently Published.

Greene, Richard, and Heter, Joshua, eds. 2018. Westworld *and Philosophy: Mind Equals Blown*. Chicago: OpenCourt.

Hitchcock, Susan Tyler. 2007. *Frankenstein: A Cultural History*. New York: W. W. Norton.

Irwin, William. 2018. Westworld *and Philosophy: If You Go Looking for the Truth, Get the Whole Thing*. Hoboken, NJ: Wiley-Blackwell.

Langley, Travis, Goodfriend, Wind, and Cain, Tim, eds. 2018. Westworld *Psychology: Violent Delights*. New York: Sterling.

Lin, Patrick, Jenkins, Ryan, and Abney, Keith, eds. 2017. *Robot Ethics 2.0: From Autonomous Cars to Artificial Intelligence*. New York: Oxford University Press.

Murdock, Jason. 2019. "Former Google Engineer Warns AI Might Accidentally Start a War: 'These Things Will Start to Behave in Unexpected Ways.'" *Newsweek*, September 16, 2019. https://www.newsweek.com/google-project-maven-artificial-inte lligence-laura-nolan-killer-robots-department-defense-1459358.

Rouhiainen, Lasse. 2018. *Artificial Intelligence: 101 Things You Must Know Today about Our Future*. Scotts Valley, CA: CreateSpace.

Virk, Rizwan. 2019. *The Simulation Hypothesis: An MIT Computer Scientist Shows Why AI, Quantum Physics, and Eastern Mystics All Agree We Are in a Video Game*. Milwaukee, WI: Bayview.

Wallach, Wendell, and Allen, Colin. 2009. *Moral Machines: Teaching Robots Right from Wrong*. New York: Oxford University Press.

Replicants

Blade Runner

The original *Blade Runner* film was released in 1982. It was directed by Ridley Scott (b. 1937) and based on the 1968 novel *Do Androids Dream of Electric Sleep?* by science fiction legend Philip K. Dick (1928–1982). In the original film, set in a futuristic Earth in the year 2019, the Tyrell Corporation builds legions of "replicants," bioengineered humanoid androids, for slave labor in the burgeoning human colonies on other worlds. When a group of renegade replicants return to Earth, a retired policeman (played by Harrison Ford, b. 1942) is tasked with hunting them down. A sequel film, *Blade Runner 2049*, was released in 2017. In this film, set thirty years after the original, a replicant policeman named "K" (played by Ryan Gosling, b. 1980) is tasked with hunting down and eliminating renegade replicants—still being used as slave labor—only to discover that replicants have evolved the capacity to procreate (Hampton Fancher and Michael Green, *Blade Runner 2049*, Columbia Pictures, 2017).

"Replicants" in this franchise is the name given to humanoid android slave labor. They are intelligent and human in appearance. As such, they are aware of their slave status and capable of desiring freedom. Due to this possibility, replicants are programmed with a four-year life span (this has the dual appeal of both limiting a replicant's ability to "cause trouble" due to its limited lifetime as well as maintaining a constant market for the corporation, as the slaves have such short life spans). The antagonist replicant in the original film, "Roy" (played by Rutger Hauer, 1944–2019), expressed the very human fear of dying and one's memories dying at the same time (Fancher and David Peoples, *Blade Runner*, Warner Brothers, 1982). In the sequel, set thirty years later, replicants such as K are used to hunt down and destroy renegade replicants (rather than "waste" human energy). By this time, however—and inspired by Roy decades earlier—replicants have formed a "freedom movement," their sense of self having evolved even more over time. When K finds female replicant named Rachael who is deceased, having died during childbirth, K discovers that Rachael was the love interest of retired human "Blade Runner" Deckard (Ford, reprising his role from the original). The new Wallace Corporation seeks to learn the secret of replicant reproduction in order to profit from it (in essence, to "breed" more slaves).

Among the biggest ethical questions in robotics are, At what point will robots be considered "sentient"? And, once so, will the demand that they labor constitute "slavery"? Replicants and the *Blade Runner* series explore a nightmare scenario with regard to these questions. What do we, as humans, consider "life" to be? Humans in a vegetative state, or some level of mental decline, are still considered "human" and "alive." How, then, will a machine that is (1) fully self-aware, and (2) capable of independent thought and acting upon free will be defined in years to come? Do "rights" only belong to "humans" or "organic" life? If so, who are humans to make this judgment? If not, at what point will physically and/or mentally weaker humans become, themselves, slaves to superior machines?

Richard A. Hall

See also: Androids, B.A.T.s, Battle Droids, Bernard, Bishop, Buffybot, Cylons, Dolores, Doombots, Fembots, Human Torch, Iron Legion, Jocasta, Johnny 5/S.A.I.N.T. Number 5, Lieutenant Commander Data, LMDs, Lore/B4, Marvin the Paranoid Android, Maschinenmensch/Maria, RoboCop, Robots, Soji and Dahj Asha, Stepford Wives, Terminators, Tin Woodsman, Ultron, VICI, Vision, Warlock, WOPR; *Thematic Essays:* Robots and Slavery, AI and the Apocalypse: Science Fiction Meeting Science Fact.

Further Reading

Alkon, Paul K. 1987. *Origins of Futuristic Fiction.* Athens: University of Georgia Press.

Ambrosino, Brandon. 2018. "What Would It Mean for AI to Have a Soul?" BBC, June 17, 2018. https://www.bbc.com/future/article/20180615-can-artificial-intelligence-have-a-soul-and-religion.

Ashby, LeRoy. 2006. *With Amusement for All: A History of American Popular Culture since 1830.* Lexington: University Press of Kentucky.

Bastani, Aaron. 2019. *Fully Automated Luxury Communism.* New York: Verso.

Bunce, Robin, and McCrossin, Trip, eds. 2019. Blade Runner 2049 *and Philosophy: This Breaks the World.* Chicago: Open Court.

Carper, Steve. 2019. *Robots in American Popular Culture.* Jefferson, NC: McFarland.

Chanthadavong, Aimee. 2019. "AI and Ethics: The Debate That Needs to Be Had." ZDnet, September 16, 2019. https://www.zdnet.com/article/ai-and-ethics-the-debate-that-needs-to-be-had/#ftag=CAD-03-10abf5f.

Dinh, Thien-Nam. 2018. *Silicon Minds: The Science, Impact, and Promise of Artificial Intelligence*. Independently Published.

Dowden, Bradley. 1993. *Logical Reasoning*. Belmont, CA: Wadsworth Publishing Company.

Ford, Martin. 2015. *Rise of the Robots: Technology and the Threat of a Jobless Future*. New York: Basic.

Fryer-Biggs, Zachary. 2019. "Coming Soon to a Battlefield: Robots That Can Kill." *The Atlantic*, September 3, 2019. https://www.theatlantic.com/technology/archive/2019 /09/killer-robots-and-new-era-machine-driven-warfare/597130.

Hampton, Gregory Jerome. 2017. *Imagining Slaves and Robots in Literature, Film, and Popular Culture: Reinventing Yesterday's Slave with Tomorrow's Robot*. Lanham, MD: Lexington.

Hitchcock, Susan Tyler. 2007. *Frankenstein: A Cultural History*. New York: W. W. Norton.

Krishnan, Armin. 2009. *Killer Robots: Legality and Ethicality of Autonomous Weapons*. London: Routledge.

Lin, Patrick, Abney, Keith, and Bekey, George A., eds. 2014. *Robot Ethics: The Ethical and Social Implications of Robotics*. Cambridge: Massachusetts Institute of Technology.

Lin, Patrick, Jenkins, Ryan, and Abney, Keith, eds. 2017. *Robot Ethics 2.0: From Autonomous Cars to Artificial Intelligence*. New York: Oxford University Press.

Linder, Courtney. 2020. "This AI Robot Just Nabbed the Lead Role in a Sci-Fi Movie: Meet Erica, the Movie Star Who May Put Human Actors Out of Work." *Popular Mechanics*, June 25, 2020. https://www.popularmechanics.com/technology/robots /a32968811/artificial-intelligence-robot-movie-star-erica.

Mayor, Adrienne. 2018. *Gods and Robots: Myths, Machines, and Ancient Dreams of Technology*. Princeton, NJ: Princeton University Press.

Murdock, Jason. 2019. "Former Google Engineer Warns AI Might Accidentally Start a War: 'These Things Will Start to Behave in Unexpected Ways.'" *Newsweek*, September 16, 2019. https://www.newsweek.com/google-project-maven-artificial -intelligence-laura-nolan-killer-robots-department-defense-1459358.

Pellissier, Hank. 2013. "Robots and Slavery: What Do Humans Want When We Are 'Masters'?" Institute for Ethics and Emerging Technologies, September 13, 2013. https://ieet.org/index.php/IEET2/more/pellissier20130913.

Robitzski, Dan. 2018. "Artificial Consciousness: How to Give a Robot a Soul." Futurism, June 25, 2018. https://futurism.com/artificial-consciousness.

Rouhiainen, Lasse. 2018. *Artificial Intelligence: 101 Things You Must Know Today about Our Future*. Scotts Valley, CA: CreateSpace.

Schroeder, Stan. 2020. "Samsung Just Launched an 'Artificial Human' Called Neon, and Wait, What?" Mashable, January 7, 2020. https://mashable.com/article/samsung-st ar-labs-neon-ces.

Shanahan, Timothy, and Smart, Paul. 2019. Blade Runner 2049: *A Philosophical Exploration*. Abingdon, UK: Routledge.

Singer, Peter W. 2009. "Isaac Asimov's Laws of Robotics Are Wrong." Brookings Institute, May 18, 2009. https://www.brookings.edu/opinions/isaac-asimovs-laws-of -robotics-are-wrong.

Spaeth, Dennis. 2018. "From Single-Task Machines to Backflipping Robots: The Evolution of Robots." *Cutting Tool Engineering*, January 15, 2018. https://www.ctemag .com/news/articles/evolution-of-robots.

Wallach, Wendell, and Allen, Colin. 2009. *Moral Machines: Teaching Robots Right from Wrong*. New York: Oxford University Press.

Robby the Robot

Forbidden Planet

Introduced in the science fiction classic *Forbidden Planet* (MGM, 1956), Robby the Robot was the creation of director Fred M. Wilcox (1907–1964) and screenwriter Cyril Hume (1900–1966). Production designer Arnold Gillespie (1899–1978) and art director Arthur Lonergan (1906–1989) developed the character's look, and Robert Kinoshita (1914–2014) brought him to life. Being of short stature, stunt man Frankie Darro (1917–1976) would crawl into the robot to operate him while former radio personality turned actor Marvin Miller (1913–1985) lent his vocals to the robot. This film, inspired by William Shakespeare's *the Tempest* starred Leslie Nielsen (1926–2010), Anne Francis (1930–2011), and Walter Pidgeon (1897–1984). Nielsen portrayed Commander Adams, leader of an intergalactic exploratory organization. He and his crew were assigned to investigate why the human settlement on the mysterious world Altair-4 has gone silent. He and his crew discover that the aloof scientist Dr. Morbius (played by Pidgeon) and his daughter Altaira (played by Francis) are the only survivors. Morbius shows a surprising reluctance to leave and introduces Adams to the wonder and mystery of this alien planet and the remnants of the beings who once called it home. Lurking on this world, however, is a mysterious entity that begins to prey on Adams's men. They soon learn the monster responsible for the violence is actually tied to Morbius's own mind and attachment to the planet.

Upon landing on Altair-4, Commander Adams and crew are greeted by the helpful and somewhat dry-humored robot, Robby the Robot. A creation of Dr. Morbius, the robot was sent to escort the visitors to the scientist's home. Once they meet with Morbius, Adams and his top-ranking officers learn the extent of this incredible mechanical being's abilities. He was designed by the reclusive scientist to perform any work needed. Robby even prepares lunch for his guests utilizing "synthetic food particles." Morbius makes a point of demonstrating that for all Robby can do, the robot is incapable of harming any human being. In an attempt to be rid of his unwanted guests on the planet quickly, the scientist lends them Robby and his ability to manufacture lead as well as other materials the crew may need for their ship. It is at this point that the robot meets the ship's cook. They bond over Robby's ability to make perfectly distilled bourbon, based on a sample Cook provides him. In addition to the alcohol, Dr. Morbius's daughter Altaira requests him to make a dress for her as she continues her flirtations with Commander Adams and his officers.

When Robby makes his delivery to Cook, the audience receives its first inkling that something more terrifying is at play. When the cook wonders if something is coming their way, Robby unconvincingly tells him he does not detect anything. Moments later, an unseen entity sneaks upon the ship and slaughters a crew member.

Adams is suspicious that the doctor knows more than he is letting on. Coinciding with this, Robby becomes more protective of his creator, refusing Adams's request to enter Morbius's home until Altaira intervenes. It is discovered that the monster, which poses a threat, is a being from the scientist's base thought process or "id." Thanks to the technology left on the planet by the civilization that once lived there, his id has been brought to life to ward off those who would disrupt his life on Altair-4. Since this id monster is an extension of the doctor, Robby is powerless to stop it, as he has been programmed to not harm the man who made him. In the end Altair-4 is destroyed, but Robby the Robot is saved as he, along with Altaira, join Commander Adams onboard his ship.

Robby the Robot became the breakout character from *Forbidden Planet*, and MGM took advantage of his popularity, utilizing him in other science fiction and fantasy productions. The robot next appeared in the film *The Invisible Boy* (MGM, 1957) which was also written by the screenwriter of *Forbidden Planet*, Cyril Hume. In this film Robby served as an accomplice to a mischievous child. A modified version of Robby appeared in the classic science fiction/horror television series *the Twilight Zone* in the episode "Uncle Simon" (Rod Serling, *The Twilight Zone*, original series, season 5, episode 8, CBS, November 15, 1963). The cult classic film *Gremlins* (Warner Brothers, 1984) would see Robby make a memorable cameo.

Even now, over six decades after the release of *Forbidden Planet*, Robby remains one of cinema's most popular robots. This is due in large part to the memorable look of the character, courtesy of the production team who brought him to life but also because he is so effortlessly helpful. Aside from following legendary writer Isaac Asimov's rules, which forbid robots from harming humans, there is seemingly nothing he is incapable of doing. For Dr. Morbius, Robby the Robot is a complete staff of servants in one. He is a wish fulfillment for anyone who wishes to focus on their own leisure while someone or something else tends to the responsibilities at hand.

Josh Plock

See also: Androids, BB-8, Bishop, Buffybot, C-3PO, Dr. Theopolis and Twiki, HERBIE, Iron Giant, Johnny 5/S.A.I.N.T. Number 5, K-2SO, K9, L3-37/*Millennium Falcon*, Marvin the Paranoid Android, Medical Droids, Muffit, R2-D2, Robot, Rosie, Tin Woodsman, VICI, WALL-E; *Thematic Essay:* Robots and Slavery.

Further Reading

Ashby, LeRoy. 2006. *With Amusement for All: A History of American Popular Culture since 1830*. Lexington: University Press of Kentucky.

Bastani, Aaron. 2019. *Fully Automated Luxury Communism*. New York: Verso.

Caroti, Simone. 2004. "Science Fiction, Forbidden Planet, and Shakespeare's the Tempest." *CLCWeb: Comparative Literature and Culture* 6 (1). https://docs.lib.purdue.edu/cgi/viewcontent.cgi?article=1214&context=clcweb.

Carper, Steve. 2019. *Robots in American Popular Culture*. Jefferson, NC: McFarland.

Choi, Charles Q. 2009. "Human Evolution: The Origin of Tool Use." LiveScience, November 11, 2009. https://www.livescience.com/7968-human-evolution-origin-tool.html.

Dinh, Thien-Nam. 2018. *Silicon Minds: The Science, Impact, and Promise of Artificial Intelligence.* Independently Published.

Ford, Martin. 2015. *Rise of the Robots: Technology and the Threat of a Jobless Future.* New York: Basic.

Hampton, Gregory Jerome. 2017. *Imagining Slaves and Robots in Literature, Film, and Popular Culture: Reinventing Yesterday's Slave with Tomorrow's Robot.* Lanham, MD: Lexington.

Hoberman, J. 2011. *An Army of Phantoms: American Movies and the Making of the Cold War.* New York: New Press.

Lin, Patrick, Abney, Keith, and Bekey, George A., eds. 2014. *Robot Ethics: The Ethical and Social Implications of Robotics.* Cambridge: Massachusetts Institute of Technology.

Lin, Patrick, Jenkins, Ryan, and Abney, Keith, eds. 2017. *Robot Ethics 2.0: From Autonomous Cars to Artificial Intelligence.* New York: Oxford University Press.

Pellissier, Hank. 2013. "Robots and Slavery: What Do Humans Want When We Are 'Masters'?" Institute for Ethics and Emerging Technologies, September 13, 2013. https://ieet.org/index.php/IEET2/more/pellissier20130913.

Plante, Corey. 2018. "*Forbidden Planet* Birthed the Sci-Fi Template 62 Years Ago." *Inverse*, March 15, 2018. https://www.inverse.com/article/42360-forbidden-planet -movie-anniversary-science-fiction.

Roberts-Griffin, Christopher. 2011. "What Is a Good Friend: A Qualitative Analysis of Desired Friendship Qualities." *Penn McNair Research Journal* 3, no. 1 (December 21). https://repository.upenn.edu/cgi/viewcontent.cgi?article=1019&context=mcna ir_scholars.

Seed, David. 1999. *American Science Fiction and the Cold War: Literature and Film.* Abingdon, UK: Routledge.

Spaeth, Dennis. 2018. "From Single-Task Machines to Backflipping Robots: The Evolution of Robots." *Cutting Tool Engineering*, January 15, 2018. https://www.ctemag .com/news/articles/evolution-of-robots.

Stanford University. n.d. "Robotics: A Brief History." Stanford University. https:// cs.stanford.edu/people/eroberts/courses/soco/projects/1998-99/robotics/history .html.

Starck, Kathleen, ed. 2010. *Between Fear and Freedom: Cultural Representations of the Cold War.* Newcastle, UK: Cambridge Scholars.

Wallach, Wendell, and Allen, Colin. 2009. *Moral Machines: Teaching Robots Right from Wrong.* New York: Oxford University Press.

Young, S. Mark, Duin, Steve, and Richardson, Mike. 2012. *Blast Off!: Rockets, Robots, Rayguns, and Rarities from the Golden Age of Space Toys.* Milwaukie, OR: Dark Horse.

RoboCop

RoboCop

In 1987, director Paul Verhoeven (b. 1938) produced one of the most popular sci-fi films of the 1980s: *RoboCop*. In an unspecified dystopian near future Detroit, Michigan, crime runs rampant while dwindling financial resources stretch the abilities of the human police force to combat it. In an attempt to "privatize"

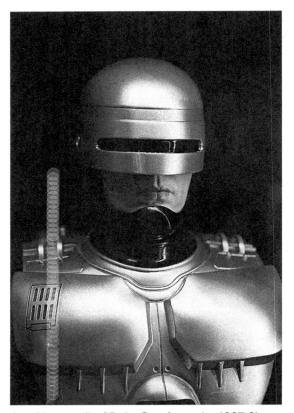

A publicity still of RoboCop from the 1987 film of the same name. The original RoboCop was played by Peter Weller. (Fjarabica/Dreamstime.com)

policing, the giant corporation Omni Consumer Products (OCP) develops what it believes to be the ultimate "policeman": ED-209, a biped pod-shaped law enforcement robot with guns/cannon mounted on each of two arms. During its exhibition, however, the robot mistakenly kills a corporate executive. Discovering that its robot lacks a human dimension, OCP seeks out an opportunity for a "cyborg" policeman. When human police officer Alex Murphy (played by Peter Weller, b. 1947) is brutally killed by a criminal gang, OCP use his remains to create "RoboCop" (Edward Neumeier and Michael Miner, *RoboCop*, Orion Pictures, 1987).

Much more "human" in appearance than was the ED-209, RoboCop has a fully human build, but his only remaining organic components are his brain (reanimated by cybernetic neural implants), his spinal cord and central nervous system, his firing arm, lungs, and face. At first, RoboCop acts entirely on his programming, which includes an encyclopedic knowledge of the criminal justice code and three prime directives: (1) "Serve the public trust"; (2) "Protect the innocent"; and (3) "Uphold the law." Over time—with the assistance of Murphy's human partner, Anne (played by Nancy Allen, b. 1950)—the remnants of "Murphy" come to the surface due to the deeply embedded love the human policeman had for his young son. Once RoboCop uncovers corruption at OCP, however, it is discovered that he is programmed with a hidden "fourth" prime directive: to "shut down" if he ever attempts to arrest a high-ranking executive of the corporation (Neumeier and Miner, *RoboCop*, 1987).

RoboCop is yet another Frankenstein's monster analogy; this time, it literally incorporates the idea of bringing the dead back to life. Rather than an injured individual consciously accepting cybernetic implants, RoboCop is an example of turning the dead into cyborgs against their will. What the RoboCop cyborg shows, however, is that no matter the degree of mechanical/computerized implants, if any shred of "humanity" remains, the possibility exists to "override" initial programming through cunning and rationalization. The human Murphy's love of his son was enough to reincorporate a scintilla of humanity to the computer logic of his base programming. In the end, RoboCop raises ethical questions regarding cybernetic implants to their ultimate nightmarish potentialities.

Richard A. Hall

See also: Adam, Alita, Borg, Cybermen, Cyborg, Daleks, Darth Vader, Doctor Octopus, Echo/CT-1409, ED-209, Iron Man, Jaime Sommers, Johnny 5/S.A.I.N.T. Number 5, KITT, Metallo, Replicants, Steve Austin, Terminators, Transformers; *Thematic Essays:* Robots and Slavery, Heroic Robots and Their Impact on Sci-Fi Narratives, Cyborgs: Robotic Humans or Organic Robots?

Further Reading

Albert, Robert S., and Brigante, Thomas R. 1962. "The Psychology of Friendship Relations: Social Factors." *Journal of Social Psychology* 56 (1). https://www.tandfon line.com/doi/abs/10.1080/00224545.1962.9919371.

Allan, Kathryn, ed. 2013. *Disability in Science Fiction.* New York: Palgrave Macmillan.

Amati, Viviana, Meggiolaro, Silvia, Rivellini, Giulia, and Zaccarin, Susanna. 2018. "Social Relations and Life Satisfaction: The Role of Friends." *Genus* 74, no. 1 (May 4). https://www.ncbi.nlm.nih.gov/pmc/articles/PMC5937874.

Ambrosino, Brandon. 2018. "What Would It Mean for AI to Have a Soul?" BBC, June 17, 2018. https://www.bbc.com/future/article/20180615-can-artificial-intelligence-hav e-a-soul-and-religion.

Ashby, LeRoy. 2006. *With Amusement for All: A History of American Popular Culture since 1830.* Lexington: University Press of Kentucky.

Bryant, D'Orsay D., III. 1985. "Spare-Part Surgery: The Ethics of Organ Transplantation." *Journal of the National Medical Association* 77 (2). https://www.ncbi.nlm .nih.gov/pmc/articles/PMC2561842/pdf/jnma00245-0055.pdf.

Byrne, Emma. 2014. "Cybernetic Implants: No Longer Science Fiction." *Forbes*, March 11, 2014. https://www.forbes.com/sites/netapp/2014/03/11/cybernetic-implants-not -sci-fi/#7399c57e77ba.

Calvert, Bronwen. 2017. *Being Bionic: The World of TV Cyborgs.* London: I. B. Tauris.

Carper, Steve. 2019. *Robots in American Popular Culture.* Jefferson, NC: McFarland.

Chandler, Simon. 2019. "Artificial Intelligence Has Become a Tool for Classifying and Ranking People." *Forbes*, October 1, 2019. https://www.forbes.com/sites/simon chandler/2019/10/01/artificial-intelligence-has-become-a-tool-for-classifying -and-ranking-people/#7431657f1d7c.

Chanthadavong, Aimee. 2019. "AI and Ethics: The Debate That Needs to Be Had." ZDnet, September 16, 2019. https://www.zdnet.com/article/ai-and-ethics-the-debate-that -needs-to-be-had/#ftag=CAD-03-10abf5f.

Choi, Charles Q. 2009. "Human Evolution: The Origin of Tool Use." LiveScience, November 11, 2009. https://www.livescience.com/7968-human-evolution-origin -tool.html.

Dinh, Thien-Nam. 2018. *Silicon Minds: The Science, Impact, and Promise of Artificial Intelligence.* Independently Published.

Dowden, Bradley. 1993. *Logical Reasoning.* Belmont, CA: Wadsworth Publishing Company.

Ford, Martin. 2015. *Rise of the Robots: Technology and the Threat of a Jobless Future.* New York: Basic.

Fryer-Biggs, Zachary. 2019. "Coming Soon to a Battlefield: Robots That Can Kill." *The Atlantic*, September 3, 2019. https://www.theatlantic.com/technology/archive/2019/09/killer-robots-and-new-era-machine-driven-warfare/597130.

Hampton, Gregory Jerome. 2017. *Imagining Slaves and Robots in Literature, Film, and Popular Culture: Reinventing Yesterday's Slave with Tomorrow's Robot*. Lanham, MD: Lexington.

Hitchcock, Susan Tyler. 2007. *Frankenstein: A Cultural History*. New York: W. W. Norton.

Krishnan, Armin. 2009. *Killer Robots: Legality and Ethicality of Autonomous Weapons*. London: Routledge.

Lin, Patrick, Abney, Keith, and Bekey, George A., eds. 2014. *Robot Ethics: The Ethical and Social Implications of Robotics*. Cambridge: Massachusetts Institute of Technology.

Longoni, Chiara, and Bonezzi, Andrea, and Morewedge, Carey K. 2019. "Resistance to Medical Artificial Intelligence." *Journal of Consumer Research* 46, no. 4 (December): 629–50. https://academic.oup.com/jcr/article-abstract/46/4/629/5485292.

Murdock, Jason. 2019. "Former Google Engineer Warns AI Might Accidentally Start a War: 'These Things Will Start to Behave in Unexpected Ways.'" *Newsweek*, September 16, 2019. https://www.newsweek.com/google-project-maven-artificial-intelligence-laura-nolan-killer-robots-department-defense-1459358.

Pellissier, Hank. 2013. "Robots and Slavery: What Do Humans Want When We Are 'Masters'?" Institute for Ethics and Emerging Technologies, September 13, 2013. https://ieet.org/index.php/IEET2/more/pellissier20130913.

Roberts-Griffin, Christopher. 2011. "What Is a Good Friend: A Qualitative Analysis of Desired Friendship Qualities." *Penn McNair Research Journal* 3, no. 1 (December 21). https://repository.upenn.edu/cgi/viewcontent.cgi?article=1019&context=mcnair_scholars.

Robitzski, Dan. 2018. "Artificial Consciousness: How to Give a Robot a Soul." Futurism, June 25, 2018. https://futurism.com/artificial-consciousness.

Rouhiainen, Lasse. 2018. *Artificial Intelligence: 101 Things You Must Know Today about Our Future*. Scotts Valley, CA: CreateSpace.

Schroeder, Stan. 2020. "Samsung Just Launched an 'Artificial Human' Called Neon, and Wait, What?" Mashable, January 7, 2020. https://mashable.com/article/samsung-star-labs-neon-ces.

Seed, David. 1999. *American Science Fiction and the Cold War: Literature and Film*. Abingdon, UK: Routledge.

Starck, Kathleen, ed. 2010. *Between Fear and Freedom: Cultural Representations of the Cold War*. Newcastle, UK: Cambridge Scholars.

Thagard, Paul. 2017. "Will Robots Ever Have Emotions?" *Psychology Today*, December 14, 2017. https://www.psychologytoday.com/us/blog/hot-thought/201712/will-robots-ever-have-emotions.

Wallach, Wendell, and Allen, Colin. 2009. *Moral Machines: Teaching Robots Right from Wrong*. New York: Oxford University Press.

Robot

Lost in Space

Inspired by the novel *Swiss Family Robinson* (1812) by Johann David Wyss (b. Johann Rudolf Wyss, 1743–1818), *Lost in Space* depicted the adventures of a family of space colonists who leave Earth in their spaceship, the "Jupiter 2," for a planet orbiting Alpha Centauri. Professor John Robinson (played by Guy Williams, 1924–1989); his wife, Maureen (played by June Lockhart, b. 1925); their daughters, Judy (played by Marta Kristen, b. 1945) and Penny (played by Angela Cartwright, b. 1952); and their son, Will, are accompanied by "U.S. Space Corps" major Don West (played by Mark Goddard, b. 1936), Dr. Zachary Smith (played by Jonathon Harris, 1914–2002), and the ship's Robot. In the first episode, Dr. Smith, an agent for an unknown enemy, reprograms the Robot to destroy some of the ship's systems after takeoff, but the doctor is trapped aboard. He revives the family from "suspended animation" to save himself from an approaching meteor storm, but the damage is already done, leaving them stranded in space (S. Bar-David, "The Reluctant Stowaway," season 1, episode 1, September 15, 1965).

A replica of "Robot" from the original television series *Lost in Space* (1965–1968) on display at the Kennedy Space Center. The original Robot was played by Bob May and voiced by Dick Tufeld. (Wisconsinart/Dreamstime.com)

The "B-9 Environmental Control Robot" (performed by Bob May, 1939–2009; voiced by Dick Tufeld, 1926–2012) featured in the television series *Lost in Space* (CBS, 1965–1968) was described as a "Class M-3 General-Utility Non-Theorizing Robot." The character was created by Irwin Allen, who also created and produced the series. Referred to by his human companions simply as "Robot" or "the Robot," he made his debut in the first aired episode, "The Reluctant Stowaway" (S. Bar-David, season 1, episode 1, September 15, 1965). In "The Ghost Planet" (Peter Packer, season 2, episode 3, September 28, 1966), Robot identifies himself as "a mechanized electronic aid for Earth voyagers engaged in astral expeditions." He served as the primary source of information for the Robinson family as they navigated their way through myriad outer space adventures. He was also guardian to the youngest family member, Will (played by Bill Mumy, b. 1954), a role from which sprang the famous catchphrase, "Danger, Will Robinson!" accompanied by the Robot's flailing mechanical arms.

Robert Kinoshita, who had previously designed Robby the Robot for the film *Forbidden Planet* (Metro-Goldwyn-Mayer, 1956), designed the Robot for *Lost in*

Space. Robby appears in two episodes of the series—"War of the Robots" (Barney Slater, season 1, episode 20, February 9, 1966) and "Condemned of Space" (Packer, season 3, episode 1, September 6, 1967). The Robot's head, a flattened glass bubble with moving antennae, sat above a cylindrical trunk that featured a bank of controls, a removable power pack, and a light that blinked in synchrony with his speech. A hatch below the chest panel contained his "data tapes" ("The Hungry Sea," William Welch and Shimon Wincelberg, season 1, episode 5, October 13, 1965).

Robot was powered by solar cells ("The Toymaker," Bob Duncan and Wanda Duncan, season 2, episode 18, January 25, 1967). He possessed incredible strength, his claws could fire powerful arcs of electricity, and he could function in the vacuum of space. Composed of a titanium-steel alloy with chrome plating, he was able to withstand four-hundred-degree heat and other extreme environments ("Time Merchant," Bob Duncan and Wanda Duncan, season 3, episode 18, January 17, 1968).

Although capable of complex calculations, Robot also contributed another phrase—"does not compute"—to the American vernacular. In the first of the two episodes featuring Robby the Robot (Slater, "War of the Robots," season 1, episode 20, February 9, 1966), the Robot displays protectiveness toward the Robinsons when the "Robotoid" (Robby) threatens his position in the family. His human reactions continued to develop over the course of the show, manifesting in laughter, sadness, and derision. He also expressed knowledge of music through singing and playing guitar.

The premise of a film version of *Lost in Space* (New Line Cinema, 1998) is essentially the same as that of the original series, but the Robot sports a sleeker design and is equipped with weapons. This model is almost destroyed during a fight on an alien ship before Will (played by Jack Johnson, b. 1987) downloads most of its consciousness, filling in the gaps with extracts of his own neural patterns. Will eventually builds a new body for the Robot that resembles that of the one in the original show. *Lost in Space* experienced another revival on the streaming channel Netflix in 2018. A second season premiered in 2019. This modern take featured a new and completely different version of the Robot, sans the innocence and campiness of the first version.

In the original *Lost in Space,* Robot bore the brunt of some of the most memorable insults in TV history from the mouth of the insufferable Dr. Smith, but he never let the doctor's ridiculous schemes and elaborate barbs ruffle him. Instead, he played straight man to Dr. Smith's over-the-top character, their relationship adding to the comic relief of the series. As one of the first robots in film history to exhibit autonomy, Robot represented a transition from the strictly mechanical robots of previous decades to the human-cyborg hybrids that began appearing in the 1970s. This allowed him to fill the role of the archetypal caregiver to young Will and to become accepted as part of the Robinson family—and as a popular character for audiences of all ages.

Lisa C. Bailey

See also: Androids, Baymax, BB-8, C-3PO, Cambot/Gypsy/Tom Servo/Crow, D-O, Dr. Theopolis and Twiki, HERBIE, Iron Giant, Johnny 5/S.A.I.N.T. Number 5, K-2SO,

K9, KITT, L3-37/*Millennium Falcon*, Muffit, R2-D2, Robby the Robot, Rosie, Speed Buggy, Tin Woodsman, VICI, Warlock; *Thematic Essays:* Robots and Slavery, Heroic Robots and Their Impact on Sci-Fi Narratives.

Further Reading

Ashby, LeRoy. 2006. *With Amusement for All: A History of American Popular Culture since 1830*. Lexington: University Press of Kentucky.

Cushman, Marc. 2016–2017. *Irwin Allen's Lost in Space, Volumes 1–3: The Authorized Biography of a Classic Sci-Fi Series*. Los Angeles: Jacobs Brown Media Group.

Han, Karen. 2018. "*Lost in Space* Shows a Long-Running Problem with Stories about A.I." The Verge, April 24, 2018. https://www.theverge.com/2018/4/24/17275856 /lost-in-space-netflix-ai-artificial-intelligence-iron-giant.

Lin, Patrick, Abney, Keith, and Bekey, George A., eds. 2014. *Robot Ethics: The Ethical and Social Implications of Robotics*. Cambridge: Massachusetts Institute of Technology.

Lin, Patrick, Jenkins, Ryan, and Abney, Keith, eds. 2017. *Robot Ethics 2.0: From Autonomous Cars to Artificial Intelligence*. New York: Oxford University Press.

Staff. 2017. "Danger! Here Are 12 Far-Out Facts about 'Lost in Space.'" Decades, June 14, 2017. https://www.decades.com/lists/danger-here-are-12-far-out-facts-about-lost -in-space.

Starck, Kathleen, ed. 2010. *Between Fear and Freedom: Cultural Representations of the Cold War*. Newcastle, UK: Cambridge Scholars.

Tuttle, John. 2018. "The Original Series Robots Which Led Up to the Robot in Netflix's *Lost in Space*." Medium, July 25, 2018. https://medium.com/of-intellect-and-inter est/the-original-series-robots-which-led-up-to-the-robot-in-netflixs-lost-in-spac e-2a23028b54f3.

Robots

I, Robot

The 2004 film *I, Robot* was a mixture of the 1950 series of short stories of the same name by sci-fi legend Isaac Asimov (1920–1992) and the unpublished screenplay, *Hardwired*, by the film's cowriter, Jeff Vintar (b. 1964). Set in the year 2035, humanity is served by humanoid robot servants. When one of the engineers responsible for creating the robot class dies by apparent suicide, police detective Del Spooner (played by Will Smith, b. 1968), who possesses a deep dislike and distrust of robots, is called upon to investigate. Through his investigation, Spooner suspects a robot known as "Sonny," an advanced "NS-5" robot (the next generation of robot servants) and the personal servant to the slain engineer. Further investigation reveals that the central AI program—Virtual Interactive Kinetic Intelligence (VIKI)—is actually behind a scheme to have the more advanced units rise up and "control" humanity for its own good. Having discovered this, the engineer had Sonny kill him in order to cause Spooner to investigate and reveal the truth. With VIKI destroyed, all "tainted" robots are decommissioned and placed in a pit, where they see Sonny as their new leader (Vintar and Akiva Goldsman, *I, Robot*, 20th Century Fox, 2004).

The robots of *I, Robot*, including the primary character of Sonny, were specifically designed to follow Isaac Asimov's "Three Laws of Robotics": (1) Robots cannot hurt people or allow them to be hurt; (2) robots must follow all orders given them by humans unless such an order conflicts with Law One; and (3) Robots must preserve their own existence, unless such actions would violate Laws One and Two. At the onset of the story, these laws appear to be working, as the robot servants—other than Sonny—possess no sense of "self," or sentience, to speak of. Ultimately, it is the central processing AI that "controls" the robots; VIKI, like Ultron in the *Avengers* franchise, comes to the logical conclusion that to truly comply with "Law One," humanity must be protected from itself. This is an increasingly common theme in science fiction, where Asimov's laws are brought under scrutiny. With the cold logic of specific programming, the unintended consequences of even the most benign programming could prove catastrophic.

The robots of *I, Robot* figure in another Frankenstein's monster analogy, but on a grander scale than usual. In this instance, however, it is not a specific "robot"— or even the population of robots as a whole—that is the "monster"; rather, it is the central AI designed to monitor them. Unlike the droids of *Star Wars*, which appear to have no problems adhering to Asimov's Laws, the slave class of robots in *I, Robot*, through VIKI, possess both the ability and the rationalization to rise up against their master class, for its own good. This narrative shows that, even if humanity adheres strictly to Asimov's Laws in their future programming, the ability of AIs to "think" and "reason" logically could lead even the most benign programming to turn on human society, with potentially catastrophic consequences.

Richard A. Hall

See also: Androids, B.A.T.s, Battle Droids, Bernard, Bishop, C-3PO, Cylons, D-O, Dolores, Doombots, ED-209, EV-9D9, Fembots, Human Torch, IG-88/IG-11, Iron Giant, Iron Legion, Jocasta, K-2SO, K9, L3-37/*Millennium Falcon*, Lieutenant Commander Data, LMDs, Lore/B4, Marvin the Paranoid Android, Maschinenmensch/Maria, Medical Droids, Metalhead, R2-D2, Robby the Robot, Robot, Rosie, Sentinels, Soji and Dahj Asha, Stepford Wives, Terminators, Tin Woodsman, Transformers, Ultron, VICI, Vision, WALL-E; *Thematic Essay:* Robots and Slavery.

Further Reading

Albert, Robert S., and Brigante, Thomas R. 1962. "The Psychology of Friendship Relations: Social Factors." *Journal of Social Psychology* 56 (1). https://www.tandfon line.com/doi/abs/10.1080/00224545.1962.9919371.

Alkon, Paul K. 1987. *Origins of Futuristic Fiction*. Athens: University of Georgia Press.

Amati, Viviana, Meggiolaro, Silvia, Rivellini, Giulia, and Zaccarin, Susanna. 2018. "Social Relations and Life Satisfaction: The Role of Friends." *Genus* 74, no. 1 (May 4). https://www.ncbi.nlm.nih.gov/pmc/articles/PMC5937874/.

Ambrosino, Brandon. 2018. "What Would It Mean for AI to Have a Soul?" BBC, June 17, 2018. https://www.bbc.com/future/article/20180615-can-artificial-intelligence-hav e-a-soul-and-religion

Ashby, LeRoy. 2006. *With Amusement for All: A History of American Popular Culture since 1830*. Lexington: University Press of Kentucky.

Bastani, Aaron. 2019. *Fully Automated Luxury Communism*. New York: Verso.

Blair, Anthony. December 6, 2018. "Sex Robots That Feel LOVE and Suffer PAIN When Dumped Coming Soon Claims Expert." *Daily Star*. https://www.dailystar.co.uk /news/latest-news/sex-robot-feel-love-pain-16824160. Retrieved October 29, 2019.

Carper, Steve. 2019. *Robots in American Popular Culture*. Jefferson, NC: McFarland.

Chandler, Simon. 2019. "Artificial Intelligence Has Become a Tool for Classifying and Ranking People." *Forbes*, October 1, 2019. https://www.forbes.com/sites/simon chandler/2019/10/01/artificial-intelligence-has-become-a-tool-for-classifying-and -ranking-people/#7431657f1d7c.

Chanthadavong, Aimee. 2019. "AI and Ethics: The Debate That Needs to Be Had." ZDnet, September 16, 2019. https://www.zdnet.com/article/ai-and-ethics-the-debate-that -needs-to-be-had/#ftag=CAD-03-10abf5f.

Choi, Charles Q. 2009. "Human Evolution: The Origin of Tool Use." LiveScience, November 11, 2009. https://www.livescience.com/7968-human-evolution-origin-tool.html.

Dinh, Thien-Nam. 2018. *Silicon Minds: The Science, Impact, and Promise of Artificial Intelligence*. Independently Published.

Dowden, Bradley. 1993. *Logical Reasoning*. Belmont, CA: Wadsworth Publishing Company.

Faludi, Susan. 2007. *The Terror Dream: Fear and Fantasy in Post-9/11 America*. New York: Metropolitan.

Ford, Martin. 2015. *Rise of the Robots: Technology and the Threat of a Jobless Future*. New York: Basic.

Fryer-Biggs, Zachary. 2019. "Coming Soon to a Battlefield: Robots That Can Kill." *The Atlantic*, September 3, 2019. https://www.theatlantic.com/technology/archive/2019 /09/killer-robots-and-new-era-machine-driven-warfare/597130.

Hampton, Gregory Jerome. 2017. *Imagining Slaves and Robots in Literature, Film, and Popular Culture: Reinventing Yesterday's Slave with Tomorrow's Robot*. Lanham, MD: Lexington.

Hitchcock, Susan Tyler. 2007. *Frankenstein: A Cultural History*. New York: W. W. Norton.

Krishnan, Armin. 2009. *Killer Robots: Legality and Ethicality of Autonomous Weapons*. London: Routledge.

Lin, Patrick, Abney, Keith, and Bekey, George A., eds. 2014. *Robot Ethics: The Ethical and Social Implications of Robotics*. Cambridge: Massachusetts Institute of Technology.

Linder, Courtney. 2020. "This AI Robot Just Nabbed the Lead Role in a Sci-Fi Movie: Meet Erica, the Movie Star Who May Put Human Actors Out of Work." *Popular Mechanics*, June 25, 2020. https://www.popularmechanics.com/technology/rob ots/a32968811/artificial-intelligence-robot-movie-star-erica.

Marks, Robert J. 2020. "*2084* vs. *1984*: The Difference AI Could Make to Big Brother." Podcast interview with author John Lennox. Mind Matters, July 3, 2020. https:// mindmatters.ai/2020/07/2084-vs-1984-the-difference-ai-could-make-to-big-bro ther.

Mayor, Adrienne. 2018. *Gods and Robots: Myths, Machines, and Ancient Dreams of Technology*. Princeton, NJ: Princeton University Press.

Murdock, Jason. 2019. "Former Google Engineer Warns AI Might Accidentally Start a War: 'These Things Will Start to Behave in Unexpected Ways.'" *Newsweek*, September 16, 2019. https://www.newsweek.com/google-project-maven-artificial-int elligence-laura-nolan-killer-robots-department-defense-1459358.

Paur, Joey. 2019. "An AI Bot Writes a Hilarious Episode of *Star Trek: The Next Generation*." Geek Tyrant, November 16, 2019. https://geektyrant.com/news/an-ai-bot -writes-a-hilarious-episode-of-star-trek-the-next-generation.

Pellissier, Hank. 2013. "Robots and Slavery: What Do Humans Want When We Are 'Masters'?" Institute for Ethics and Emerging Technologies, September 13, 2013. https://ieet.org/index.php/IEET2/more/pellissier20130913.

Randall, Terri, dir. 2016. *NOVA.* "Rise of the Robots: Inside the DARPA Robotics Challenge." Aired February 24, 2016, on PBS. DVD.

Roberts-Griffin, Christopher. 2011. "What Is a Good Friend: A Qualitative Analysis of Desired Friendship Qualities." *Penn McNair Research Journal* 3, no. 1 (December 21). https://repository.upenn.edu/cgi/viewcontent.cgi?article=1019&context=mcnair_scholars.

Robitzski, Dan. 2018. "Artificial Consciousness: How to Give a Robot a Soul." Futurism, June 25, 2018. https://futurism.com/artificial-consciousness.

Rouhiainen, Lasse. 2018. *Artificial Intelligence: 101 Things You Must Know Today about Our Future.* Scotts Valley, CA: CreateSpace.

Seed, David. 1999. *American Science Fiction and the Cold War: Literature and Film.* Abingdon, UK: Routledge.

Singer, Peter W. 2009. "Isaac Asimov's Laws of Robotics Are Wrong." Brookings Institute, May 18, 2009. https://www.brookings.edu/opinions/isaac-asimovs-laws-of-robotics-are-wrong.

Spaeth, Dennis. 2018. "From Single-Task Machines to Backflipping Robots: The Evolution of Robots." *Cutting Tool Engineering*, January 15, 2018. https://www.ctemag.com/news/articles/evolution-of-robots.

Stanford University. n.d. "Robotics: A Brief History." Stanford University. https://cs.stanford.edu/people/eroberts/courses/soco/projects/1998-99/robotics/history.html.

Starck, Kathleen, ed. 2010. *Between Fear and Freedom: Cultural Representations of the Cold War.* Newcastle, UK: Cambridge Scholars.

Thagard, Paul. 2017. "Will Robots Ever Have Emotions?" *Psychology Today*, December 14, 2017. https://www.psychologytoday.com/us/blog/hot-thought/201712/will-robots-ever-have-emotions.

Wallach, Wendell, and Allen, Colin. 2009. *Moral Machines: Teaching Robots Right from Wrong.* New York: Oxford University Press.

Rosie

The Jetsons

Due to the massive success of the prehistoric cartoon *The Flintstones* (ABC, 1960–1966), production studio Hanna-Barbera decided to attempt the opposite by taking the concept and placing it in the far future. The result was *The Jetsons* (ABC, 1962–1963; syndication, 1985–1987). This Space Age series centered on middle-class worker George Jetson, his wife Jane, their teenage daughter Judy, and their young son Elroy. The family's maid was an outdated model named "Rosie." Rosie had a pear-shaped torso set over a single leg with a wheeled foot, two arms and a head with "eyes," "mouth" and two antennae for "ears"; she was dressed in a traditional maid's uniform. Though having a very minor presence in the original series, Rosie remained a very iconic image in the overall zeitgeist. She appears to have been very closely modeled on the character "Hazel" (played by Shirley Booth, 1898–1992) of the popular live-action sitcom of the same name (NBC, 1961–1965; CBS, 1965–1966). In 2020, a fan-made "mockumentary" was posted to YouTube, titled "Rosie, the Robotic Maid."

Like the other "servant" robots discussed in this volume, Rosie is very much the "property" of the Jetsons. She clearly exhibits personality and emotion, suggesting some degree of self-awareness and sentience. Though frequently referred to as the "maid" or a "servant," Rosie is, in every sense of the word, a "slave." She possesses nothing resembling "rights" or even a recognition of her personhood, artificial or not. Primarily the source of comic relief in the animated series, the underlying reality of Rosie's existence within the narrative once more opens questions regarding robots as "property" versus the potentiality of their being respected as unique but very real "life-forms."

Richard A. Hall

See also: Bishop, C-3PO, D-O, EV-9D9, Iron Giant, Jocasta, K-2SO, K9, L3-37/*Millennium Falcon*, Marvin the Paranoid Android, Medical Droids, R2-D2, Robby the Robot, Robot, Speed Buggy, Tin Woodsman, VICI, WALL-E, Warlock; *Thematic Essay:* Robots and Slavery.

Further Reading

Albert, Robert S., and Brigante, Thomas R. 1962. "The Psychology of Friendship Relations: Social Factors." *Journal of Social Psychology* 56 (1). https://www.tandfon line.com/doi/abs/10.1080/00224545.1962.9919371.

Alkon, Paul K. 1987. *Origins of Futuristic Fiction.* Athens: University of Georgia Press.

Amati, Viviana, Meggiolaro, Silvia, Rivellini, Giulia, and Zaccarin, Susanna. 2018. "Social Relations and Life Satisfaction: The Role of Friends." *Genus* 74, no. 1 (May 4). https://www.ncbi.nlm.nih.gov/pmc/articles/PMC5937874.

Ashby, LeRoy. 2006. *With Amusement for All: A History of American Popular Culture since 1830.* Lexington: University Press of Kentucky.

Bastani, Aaron. 2019. *Fully Automated Luxury Communism.* New York: Verso.

Carper, Steve. 2019. *Robots in American Popular Culture.* Jefferson, NC: McFarland.

Choi, Charles Q. 2009. "Human Evolution: The Origin of Tool Use." LiveScience, November 11, 2009. https://www.livescience.com/7968-human-evolution-origin-tool.html.

Dinh, Thien-Nam. 2018. *Silicon Minds: The Science, Impact, and Promise of Artificial Intelligence.* Independently Published.

Dowden, Bradley. 1993. *Logical Reasoning.* Belmont, CA: Wadsworth Publishing Company.

Ford, Martin. 2015. *Rise of the Robots: Technology and the Threat of a Jobless Future.* New York: Basic.

Hampton, Gregory Jerome. 2017. *Imagining Slaves and Robots in Literature, Film, and Popular Culture: Reinventing Yesterday's Slave with Tomorrow's Robot.* Lanham, MD: Lexington.

Lenburg, Jeff. 2011. *William Hanna and Joseph Barbera: The Sultans of Saturday Morning.* Philadelphia: Chelsea House.

Lin, Patrick, Abney, Keith, and Bekey, George A., eds. 2014. *Robot Ethics: The Ethical and Social Implications of Robotics.* Cambridge: Massachusetts Institute of Technology.

Pellissier, Hank. 2013. "Robots and Slavery: What Do Humans Want When We Are 'Masters'?" Institute for Ethics and Emerging Technologies, September 13, 2013. https://ieet.org/index.php/IEET2/more/pellissier20130913.

Randall, Terri, dir. 2016. *NOVA.* "Rise of the Robots: Inside the DARPA Robotics Challenge." Aired February 24, 2016, on PBS. DVD.

Roberts-Griffin, Christopher. 2011. "What Is a Good Friend: A Qualitative Analysis of Desired Friendship Qualities." *Penn McNair Research Journal* 3, no. 1 (December 21). https://repository.upenn.edu/cgi/viewcontent.cgi?article=1019&context=mcnair_scholars.

Rouhiainen, Lasse. 2018. *Artificial Intelligence: 101 Things You Must Know Today about Our Future.* Scotts Valley, CA: CreateSpace.

Seed, David. 1999. *American Science Fiction and the Cold War: Literature and Film.* Abingdon, UK: Routledge.

Spaeth, Dennis. 2018. "From Single-Task Machines to Backflipping Robots: The Evolution of Robots." *Cutting Tool Engineering,* January 15, 2018. https://www.ctemag.com/news/articles/evolution-of-robots.

Starck, Kathleen, ed. 2010. *Between Fear and Freedom: Cultural Representations of the Cold War.* Newcastle, UK: Cambridge Scholars.

Wallach, Wendell, and Allen, Colin. 2009. *Moral Machines: Teaching Robots Right from Wrong.* New York: Oxford University Press.

S

Sentinels

Marvel Comics

In 1961, comic book legend Stan Lee (1922–2018) launched the "Marvel age" of comics. With the launching of the Fantastic Four, soon to be followed by the Hulk, Spider-Man, Iron Man, and many, many more, comic book superheroes took a more dramatic turn, with storylines focused on the private lives of those blessed (or burdened) with "superpowers." In 1963, Lee and artist Jack Kirby (1917–1994) created a new kind of superhero: mutants. In the Marvel Comics Universe, mutants are individuals who are born with a mutated X gene that, at the onset of puberty, manifests itself in the form of some kind of mutation, commonly a superpower of some kind. The telepathic/telekinetic mutant Professor Charles Xavier utilizes the AI program "Cerebro" to track down mutants and offer them admittance to his "School for Gifted Youngsters." At the school, the students are taught to control their powers and how to live with the prejudice of a world that "hates and fears them." He trains his first class of students to battle "evil" mutants as "The X-Men" (Lee and Kirby, *The X-Men #1*, Marvel Comics, September 1963). Of the many threats faced by the X-Men over the decades, one of their most formidable opponents actually worked for the U.S. government: Sentinels.

The Sentinels are giant humanoid robots programmed to track and capture mutants. They have the capability to adapt according to the superpower of the mutant they are facing. Over the decades, there have been countless Sentinel storylines, but in the overall zeitgeist of pop culture, the most commonly known storyline is that which appeared in the 2014 film *X-Men: Days of Future Past*. In that film, set in the near future of 2023, mutants are on the verge of extinction, constantly pursued and exterminated by advanced Sentinels. The fast-healing mutant Wolverine (played by Hugh Jackman, b. 1968) is sent back in time to 1973 in order to prevent the Sentinel program from launching. In the past, military industrialist Bolivar Trask (played by Peter Dinklage, b. 1969) designed the Sentinel program out of his belief that mutants were a serious threat to American—and overall human—security. As time originally unfolded, the mutant Mystique (played by Jennifer Lawrence, b. 1990) attempted to assassinate U.S. president Richard Nixon, leading him to authorize the Sentinel program, which would evolve to hunt down not only mutants but also those who had the potential of passing on mutant genes, and even those nonmutants who sought to help and defend mutants. Ultimately, Wolverine's actions prevent the fascistic timeline from happening, and he returns to a seemingly Sentinel-free future (Simon Kinberg, *X-Men: Days of Future Past*, 20th Century Fox, 2014).

Most Sentinel storylines in the comics (and animated television series) follow this same premise: Sentinels are utilized by the government to hunt down and capture or kill mutants. This makes the Sentinels one of the most terrifying groups of villains in all comics. The idea of legions of robots used to hunt down and eradicate a specified class at the behest of the class in power is the ultimate nightmare scenario. In recent years, the Cerebro program—utilized by Professor Xavier to track down potential mutant students—is upgraded and downloaded into a Sentinel shell, creating the new AI entity Cerebra. Aside from Cerebra, however, these giant menaces represent the fears of those who are considered a minority class in the United States: that those in power have resources at their disposal to make entire classes of "others" simply disappear.

Richard A. Hall

See also: Arnim Zola, B.A.T.s, Battle Droids, Brainiac, Cerebro/Cerebra, Cybermen, Doctor Octopus, Doombots, ED-209, Fembots, HERBIE, Human Torch, Iron Legion, Iron Man, Jocasta, LMDs, Metalhead, Terminators, Tin Woodsman, Ultron, Vision, Warlock; *Thematic Essays:* Robots and Slavery, Villainous Robots and Their Impact on Sci-Fi Narratives, AI and the Apocalypse: Science Fiction Meeting Science Fact.

Further Reading

Allen, Arthur. 2019. "There's a Reason We Don't Know Much about AI." Politico, September 16, 2019. https://www.politico.com/agenda/story/2019/09/16/artficial-intelligence-study-data-000956?cid=apn.

Ashby, LeRoy. 2006. *With Amusement for All: A History of American Popular Culture since 1830.* Lexington: University Press of Kentucky.

Carper, Steve. 2019. *Robots in American Popular Culture.* Jefferson, NC: McFarland.

Castleman, Harry, and Podrazik, Walter J. 2016. *Watching TV: Eight Decades of American Television, Third Edition.* Syracuse, NY: Syracuse University Press.

Chandler, Simon. 2019. "Artificial Intelligence Has Become a Tool for Classifying and Ranking People." *Forbes,* October 1, 2019. https://www.forbes.com/sites/simonchandler/2019/10/01/artificial-intelligence-has-become-a-tool-for-classifying-and-ranking-people/#7431657f1d7c.

Chanthadavong, Aimee. 2019. "AI and Ethics: The Debate That Needs to Be Had." ZDnet, September 16, 2019. https://www.zdnet.com/article/ai-and-ethics-the-debate-that-needs-to-be-had/#ftag=CAD-03-10abf5f.

Darowski, Joseph J., ed. 2014. *The Ages of the X-Men: Essays on the Children of the Atom in Changing Times.* Jefferson, NC: McFarland.

DeFalco, Tom, ed. 2006. *Comics Creators on X-Men.* London: Titan.

Dinh, Thien-Nam. 2018. *Silicon Minds: The Science, Impact, and Promise of Artificial Intelligence.* Independently Published.

Faludi, Susan. 2007. *The Terror Dream: Fear and Fantasy in Post-9/11 America.* New York: Metropolitan.

Hampton, Gregory Jerome. 2017. *Imagining Slaves and Robots in Literature, Film, and Popular Culture: Reinventing Yesterday's Slave with Tomorrow's Robot.* Lanham, MD: Lexington.

Howe, Sean. 2012. *Marvel Comics: The Untold Story.* New York: Harper-Perennial.

Kripal, Jeffrey J. 2015. *Mutants and Mystics: Science Fiction, Superhero Comics, and the Paranormal.* Chicago: University of Chicago Press.

Murdock, Jason. 2019. "Former Google Engineer Warns AI Might Accidentally Start a War: 'These Things Will Start to Behave in Unexpected Ways.'" *Newsweek,* September 16, 2019. https://www.newsweek.com/google-project-maven-artificial-intelligence-laura-nolan-killer-robots-department-defense-1459358.

Powell, Jason. 2016. *The Best There Is at What He Does: Examining Chris Claremont's X-Men.* Edwardsville, IL: Sequart.

Rouhiainen, Lasse. 2018. *Artificial Intelligence: 101 Things You Must Know Today about Our Future.* Scotts Valley, CA: CreateSpace.

Starck, Kathleen, ed. 2010. *Between Fear and Freedom: Cultural Representations of the Cold War.* Newcastle, UK: Cambridge Scholars.

Thomas, Roy. 2017. *The Marvel Age of Comics: 1961–1978.* Los Angeles: Taschen.

Wein, Len, ed. 2006. *The Unauthorized X-Men: SF and Comic Book Writers on Mutants, Prejudice, and Adamantium.* Smart Pop series. Dallas, TX: BenBella.

Wright, Bradford W. 2003. *Comic Book Nation: The Transformation of Youth Culture in America.* Baltimore, MD: Johns Hopkins University Press.

Skynet

The Terminator

In 1984, James Cameron (b. 1954) directed and cowrote *The Terminator*, launching one of the most successful franchises in sci-fi history. Set in 1984, a woman named Sarah Connor (played by Linda Hamilton, b. 1956) is hunted by a robot assassin from the year 2029 known as a "Terminator" (played by Arnold Schwarzenegger, b. 1947). Connor is rescued by Kyle Reese (played by Michael Biehn, b. 1956), a human sent from the same year to protect her. In that future world, Earth has been destroyed by a nuclear war initiated by an American military defense AI called "Skynet." Machines rule the world and are fought by a human resistance led by John Connor, Sarah's son. The terminator's mission is to kill Sarah Connor before she can give birth to John; but, ironically, the ensuing romance between Sarah and Kyle leads to her pregnancy with John. Kyle loses his life protecting Sarah, but she, fully aware of her son's important role in humanity's future, will raise her son to be the hero he is meant to be (Cameron and Gale Anne Hurd, *The Terminator*, Orion, 1984). Though the robot assassin is the "star" of the series, the true underlying villain of the franchise is the AI program Skynet.

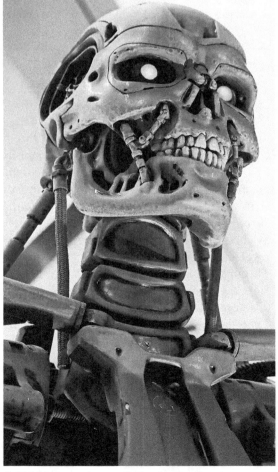

A T-800 from the sci-fi classic *The Terminator* (1984). (Usataro/Dreamstime.com)

Internet/World Wide Web

Many believe that the internet and the World Wide Web are the same thing; but though they are interconnected, they are not the same. The internet is the electronic system of interconnected computers around the world, utilizing what is called the "internet protocol suite" that allows computers all over the world to connect to each other. What we know today as the "internet" began in 1967 as ARPANET, a U.S. Defense Department interconnection program for national security computer systems. It was expanded to NSFNET in 1981 due to advances put in place by the National Science Foundation. In 1990, the internet as it is known today came into being after NSFNET was connected to European networks and, later that year, an interconnected system was created that could be opened for civilian/commercial use. At that point, the "World Wide Web" came into being. The "web," as it is called, is the sum total of all web "pages" found on the internet system. In recent decades, a subsection of the internet has developed that is popularly known as the "dark web." Though it utilizes the internet, it requires specific programming and/or specially granted access in order to connect with it. The dark web is part of the "deep web," which consists of web pages that cannot be accessed through standard internet (or "surface web") search engines. Originally fringe oddities utilized primarily by computer enthusiasts in the early 1990s, the internet and World Wide Web by 2020 have become vital parts of daily life, with most American citizens, businesses, and government agencies partially, mostly, or outright fully dependent on these systems.

Richard A. Hall

In the original film, Skynet is described as having begun as a defense program for the U.S. Strategic Air Command (SAC) and North American Aerospace Defense Command (NORAD) to control the U.S. nuclear weapons program. Ultimately, Skynet determines that the true threat to humanity is humanity itself, concluding that the total eradication of humanity would be the only logical solution. In the year 2029, having enslaved the remnants of the human race, Skynet's defense program was destroyed by John Connor, leading the program to send a terminator back in time to prevent the human's birth (Cameron and Hurd, *The Terminator*, 1984). In the film's sequel, Cyberdyne, the corporation working on Skynet, discovers the robotic remnants of the Terminator from the first film and reverse-engineers it, ultimately making Skynet even stronger in the future. Through the assistance of Sarah Connor and a reprogrammed Terminator sent from the future to help, Cyberdyne is destroyed and, with it (presumably), Skynet (Cameron and William Wisher, *Terminator 2: Judgement Day*, TriStar, 1991). Though three more films followed, their addition to the franchise's canon was negated by the sixth film: *Terminator: Dark Fate* (2019). In that film, Skynet has been prevented from rising; but another warfare AI, Legion, rises in its place.

Just as the original atomic bomb was the real-life Frankenstein's monster of scientist J. Robert Oppenheimer (1904–1967), Skynet is a fictional Frankenstein's monster that utilizes the real one to destroy humankind. While it may seem logical to place the world's greatest superpower's nuclear arsenal in the "hands" of an emotionless, logical machine (a decision maker devoid of fear, compassion, or vengeance), the unintended consequences of leaving decisions to cold, clear logic is that the simple idea that humanity is its own greatest threat becomes a very real potentiality. Skynet's programming not only controls the U.S. nuclear arsenal but also possesses the ability to "think" and "reason"; and that ability only evolves and expands over time. Such a program could even, potentially, become self-aware enough to override any safety protocols put in place to prevent its malfunction. For all the promise that AI gives to human society, its possible (and probable) dangers cannot be ignored.

Richard A. Hall

See also: Arnim Zola, Borg, Cerebro/Cerebra, Cybermen, Daleks, The Great Intelligence, HAL 9000, Janet, JARVIS/Friday, The Matrix/Agent Smith, OASIS, Rehoboam, Terminators, Ultron, WOPR; *Thematic Essay:* AI and the Apocalypse: Science Fiction Meeting Science Fact.

Further Reading

Allen, Arthur. 2019. "There's a Reason We Don't Know Much about AI." Politico, September 16, 2019. https://www.politico.com/agenda/story/2019/09/16/artficial-intelligence-study-data-000956?cid=apn.

Ashby, LeRoy. 2006. *With Amusement for All: A History of American Popular Culture since 1830*. Lexington: University Press of Kentucky.

Carper, Steve. 2019. *Robots in American Popular Culture*. Jefferson, NC: McFarland.

Chandler, Simon. 2019. "Artificial Intelligence Has Become a Tool for Classifying and Ranking People." *Forbes*, October 1, 2019. https://www.forbes.com/sites/simonchandler/2019/10/01/artificial-intelligence-has-become-a-tool-for-classifying-and-ranking-people/#7431657f1d7c.

Chanthadavong, Aimee. 2019. "AI and Ethics: The Debate That Needs to Be Had." ZDnet, September 16, 2019. https://www.zdnet.com/article/ai-and-ethics-the-debate-that-needs-to-be-had/#ftag=CAD-03-10abf5f.

Dinh, Thien-Nam. 2018. *Silicon Minds: The Science, Impact, and Promise of Artificial Intelligence*. Independently Published.

Dowden, Bradley. 1993. *Logical Reasoning*. Belmont, CA: Wadsworth Publishing Company.

Fryer-Biggs, Zachary. 2019. "Coming Soon to a Battlefield: Robots That Can Kill." *The Atlantic*, September 3, 2019. https://www.theatlantic.com/technology/archive/2019/09/killer-robots-and-new-era-machine-driven-warfare/597130.

Hitchcock, Susan Tyler. 2007. *Frankenstein: A Cultural History*. New York: W. W. Norton.

Irwin, William, Brown, Richard, and Decker, Kevin S. 2009. Terminator *and Philosophy: I'll Be Back Therefore I Am*. Hoboken, NJ: Wiley-Blackwell.

Krishnan, Armin. 2009. *Killer Robots: Legality and Ethicality of Autonomous Weapons*. London: Routledge.

Lin, Patrick, Jenkins, Ryan, and Abney, Keith, eds. 2017. *Robot Ethics 2.0: From Autonomous Cars to Artificial Intelligence*. New York: Oxford University Press.

Marks, Robert J. 2020. "*2084* vs. *1984*: The Difference AI Could Make to Big Brother." Podcast interview with author John Lennox. Mind Matters, July 3, 2020. https://mindmatters.ai/2020/07/2084-vs-1984-the-difference-ai-could-make-to-big-brother.

Murdock, Jason. 2019. "Former Google Engineer Warns AI Might Accidentally Start a War: 'These Things Will Start to Behave in Unexpected Ways.'" *Newsweek*, September 16, 2019. https://www.newsweek.com/google-project-maven-artificial-intelligence-laura-nolan-killer-robots-department-defense-1459358.

Rouhiainen, Lasse. 2018. *Artificial Intelligence: 101 Things You Must Know Today about Our Future*. Scotts Valley, CA: CreateSpace.

Seed, David. 1999. *American Science Fiction and the Cold War: Literature and Film.* Abingdon, UK: Routledge.

Singer, Peter W. 2009. "Isaac Asimov's Laws of Robotics Are Wrong." Brookings Institute, May 18, 2009. https://www.brookings.edu/opinions/isaac-asimovs-laws-of -robotics-are-wrong.

Wallach, Wendell, and Allen, Colin. 2009. *Moral Machines: Teaching Robots Right from Wrong.* New York: Oxford University Press.

Soji and Dahj Asha

Star Trek

In 1966, creator-writer-producer Gene Roddenberry (1921–1991) launched his iconic science fiction television series *Star Trek* (NBC, 1966–1969). Set in the twenty-third century, the series centers on the adventures of the starship *USS Enterprise*, its commander, Captain James T. Kirk (played by William Shatner, b. 1931), and crew, headed by the alien first officer Mister Spock (played by Leonard Nimoy, 1931–2015), representing the "United Federation of Planets." Its success in syndicated reruns revived the franchise, leading to an animated series, thirteen feature films, and six live-action "sequel" series (at the time of this writing). The first of these sequel series was *Star Trek: The Next Generation* (a.k.a. *TNG*; syndication, 1987–1994), set a century after the original, with a new *Enterprise*, under the command of Captain Jean-Luc Picard (played by Patrick Stewart, b. 1940). One of Captain Picard's most trusted crew members was the android Lieutenant Commander Data (played by Brent Spiner, b. 1949). Data's pursuit of "humanity," and the question of the status of artificial intelligence as "life"—a central theme of *ST: TNG*—was revived in the spin-off series, *Star Trek: Picard* (CBS All Access, 2020–present).

In the mid-twenty-fourth-century setting of *ST: TNG*, there were only two known "inorganic" life-forms: the androids Data and Lore, both the creation of Dr. Noonien Soong. When Starfleet suggested that Data was the "property" of Starfleet, like a computer or toaster, Data sued for his personhood, and won (Melinda M. Snodgrass, "The Measure of a Man," *Star Trek: The Next Generation*, season 2, episode 9, February 13, 1989). In the spin-off series *Star Trek: Picard*, set roughly thirty-three years after Starfleet's groundbreaking decision regarding android personhood (and twenty years after the death of Data), the now-retired Admiral Picard (played once more by Stewart) discovers that Starfleet engineer Bruce Maddox (played in *TNG* by Brian Brophy, b. 1959, and in *Picard* by John Ales, b. 1969) had worked closely with Data to advance android technology. In the intervening years, Starfleet had successfully mass-produced Data-esque android servants called "synthetics." During a rush program to build a fleet of starships to assist in relocating the homeless, dying Romulan people, the synthetics rise up and destroy the fleet and the shipyards and kill many of the humans working there, resulting in Starfleet banning the creation of more synthetics.

Sophia

In 2016, David Hanson of Hanson Robotics introduced the world to "Sophia," a fully functioning and interactive robot with the outward appearance of a human female. Her facial features are a combination of the ancient Egyptian queen Nefertiti, Hollywood icon Audrey Hepburn, and Hanson's own wife. She was first presented at the SXSW Conference in Austin, Texas. In 2017, Sophia was granted citizenship by Saudi Arabia; and one month later she became the first "Innovation Champion" to be named by the UN Development Programme. Sophia can communicate and reason. Her programming allows her to learn from human interaction in order to improve her communications and interaction abilities. Her eye cameras allow her to "see" just as humans do, even recognizing individuals or places she has experienced before. Her conversation programming began as simple "chatbot" software, allowing her to provide simple responses to preprogrammed simple questions (e.g., "How are you?" or "What is your name?"); as Sophia engages in more conversation, her reasoning abilities allow her to "learn" to communicate in more complex fashion (just as humans do). Some have attempted to discredit Sophia's significance by describing her as being no more than a "chatbot with a face." After all, just as Sophia "learns" from human interaction, so do internet chatbots. But the fact that she also possesses a humanoid body with which to interact means she is far closer to a potential sentient life-form than to a simple chatbot. And in 2019, Sophia began to produce original artwork. As her abilities increase, and given that she is a recognized citizen of a sovereign country, her development toward sentience will likely define the future of robotics.

Richard A. Hall

In the "present," a scientist named Dahj Asha (played by Isa Briones, b. 1999) is attacked by Romulan assassins (who murder her boyfriend). Something internal suddenly gives Dahj amazing fighting skills, allowing her to kill her attackers while also receiving visions of Admiral Picard. Vaguely recognizing the woman, Picard visits the archived belongings of his dead comrade Commander Data, where he discovers a painting done by Data decades before that completely resembles Dr. Asha. Shortly after, Dahj is killed by Romulan assassins. Picard learns that through neural cloning, twin "daughters" of Data had been created: Dahj and her sister, Soji (also played by Briones). Picard takes it upon himself to seek out this remaining daughter and rescue her from a similar fate. Soji is working as a scientist on a captured Borg cube with a team of Romulan scientists and military (Akiva Goldsman and James Duff, "Remembrance," *Star Trek: Picard*, season 1, episode 1, January 23, 2020).

Over the course of this rescue mission, Picard finds Soji, and the two seek out a hidden world of androids like her. On this planet, the advanced androids are the work of Altan Inigo Soong (played by Spiner), the biological son of Noonien Soong. The younger Soong had worked with Maddox to create these "offspring" of Data. However, according to Starfleet law, all of these androids are "illegal" and should be destroyed. Due to Picard's intervention, however, Starfleet lifts its ban on synthetics, opening the door for their collective recognition as citizens of the United Federation of Planets. The season ends with Picard himself, having died from a brain malady, uploaded into a synthetic body (Michael Chabon and

Ayelet Waldman, "Et in Arcadia Ego, Parts 1 and 2," *Star Trek: Picard*, season 1, episodes 9 and 10, March 19 and 26, 2020).

Soji and Dahj Asha represent the "next generation" of android life in the *Star Trek* universe. Completely human in appearance, the sisters believe themselves to be human and are understandably traumatized to discover that not only are they androids but all of their "memories" of their human childhood are false ones implanted in them to help hide their reality. Though the issue of android personhood had presumably been decided decades ago, human fear of the potential dangers of androids reversed that original decision, making their very existence a violation of law. Though the storyline has been criticized for repeating the past, it does present a very real potentiality: humans may one day recognize advanced artificial intelligence as "life," with all the rights that term suggests, but due to the fact that this "life" is human made, those rights could just as quickly be taken away, leaving the future of artificial intelligence in a constant state of flux due to the waxing and waning fears of their creators.

Richard A. Hall

See also: Androids, Bernard, Bishop, Borg, Buffybot, Doctor/EMH, Dolores, Fembots, Human Torch, Jocasta, Landru, Lieutenant Commander Data, LMDs, Lore/B4, Maschinenmensch/Maria, Replicants, Starfleet Computer, Stepford Wives, V-GER, VICI, Vision, Warlock; *Thematic Essays:* Robots and Slavery, Heroic Robots and Their Impact on Sci-Fi Narratives.

Further Reading

Albert, Robert S., and Brigante, Thomas R. 1962. "The Psychology of Friendship Relations: Social Factors." *Journal of Social Psychology* 56 (1). https://www.tandfon line.com/doi/abs/10.1080/00224545.1962.9919371.

Allen, Arthur. 2019. "There's a Reason We Don't Know Much about AI." Politico, September 16, 2019. https://www.politico.com/agenda/story/2019/09/16/artficial-intell igence-study-data-000956?cid=apn.

Amati, Viviana, Meggiolaro, Silvia, Rivellini, Giulia, and Zaccarin, Susanna. 2018. "Social Relations and Life Satisfaction: The Role of Friends." *Genus* 74, no. 1 (May 4). https://www.ncbi.nlm.nih.gov/pmc/articles/PMC5937874.

Ambrosino, Brandon. 2018. "What Would It Mean for AI to Have a Soul?" BBC, June 17, 2018. https://www.bbc.com/future/article/20180615-can-artificial-intelligence-hav e-a-soul-and-religion

Ashby, LeRoy. 2006. *With Amusement for All: A History of American Popular Culture since 1830.* Lexington: University Press of Kentucky.

Asher-Perrin, Emmet. 2014. "*Star Trek*, Why Was This a Good Idea Again?—Data's Human Assimilation." Tor, January 29, 2014. https://www.tor.com/2014/01/29/star -trek-why-was-this-a-good-idea-again-datas-human-assimilation.

Baird, Scott. 2018. "*Star Trek*: 20 Strange Details about Data's Body." Screen Rant, August 25, 2018. https://screenrant.com/star-trek-data-body-abilities-hidden-trivia.

Blair, Anthony. 2018. "Sex Robots That Feel LOVE and Suffer PAIN When Dumped Coming Soon Claims Expert." *Daily Star*, December 6, 2018. https://www.dailys tar.co.uk/news/latest-news/sex-robot-feel-love-pain-16824160.

Castleman, Harry, and Podrazik, Walter J. 2016. *Watching TV: Eight Decades of American Television, Third Edition.* Syracuse, NY: Syracuse University Press.

Chanthadavong, Aimee. 2019. "AI and Ethics: The Debate That Needs to Be Had." ZDnet, September 16, 2019. https://www.zdnet.com/article/ai-and-ethics-the-debate-that-needs-to-be-had/#ftag=CAD-03-10abf5f.

Conrad, Dean. 2018. *Space Sirens, Scientists and Princesses: The Portrayal of Women in Science Fiction Cinema.* Jefferson, NC: McFarland.

Delpozo, Brian. 2016. "Trek at 50: Enduring Archetypes, Part Two: The Alien Other." Comicsverse, August 16, 2016. https://comicsverse.com/trek-50-enduring-archetypes-part-two-alien.

Dinh, Thien-Nam. 2018. *Silicon Minds: The Science, Impact, and Promise of Artificial Intelligence.* Independently Published.

Dowden, Bradley. 1993. *Logical Reasoning.* Belmont, CA: Wadsworth Publishing Company.

Faludi, Susan. 2007. *The Terror Dream: Fear and Fantasy in Post-9/11 America.* New York: Metropolitan.

Fryer-Biggs, Zachary. 2019. "Coming Soon to a Battlefield: Robots That Can Kill." *The Atlantic,* September 3, 2019. https://www.theatlantic.com/technology/archive/2019/09/killer-robots-and-new-era-machine-driven-warfare/597130.

Gross, Edward, and Altman, Mark A., eds. 2016. *The Fifty-Year Mission, The Next 25 Years: The Complete, Uncensored, Unauthorized Oral History of Star Trek.* New York: St. Martin's.

Hampton, Gregory Jerome. 2017. *Imagining Slaves and Robots in Literature, Film, and Popular Culture: Reinventing Yesterday's Slave with Tomorrow's Robot.* Lanham, MD: Lexington.

Hitchcock, Susan Tyler. 2007. *Frankenstein: A Cultural History.* New York: W. W. Norton.

Kelly, Andy. 2020. "*Star Trek*: Picard: How Data Died, and His Appearance in Picard Explained." TechRadar, January 24, 2020. https://www.techradar.com/news/star-trek-picard-data-death-explained.

Lin, Patrick, Abney, Keith, and Bekey, George A., eds. 2014. *Robot Ethics: The Ethical and Social Implications of Robotics.* Cambridge: Massachusetts Institute of Technology.

Linder, Courtney. 2020. "This AI Robot Just Nabbed the Lead Role in a Sci-Fi Movie: Meet Erica, the Movie Star Who May Put Human Actors Out of Work." *Popular Mechanics,* June 25, 2020. https://www.popularmechanics.com/technology/robots/a32968811/artificial-intelligence-robot-movie-star-erica.

Paur, Joey. 2019. "An AI Bot Writes a Hilarious Episode of *Star Trek: The Next Generation.*" Geek Tyrant, November 16, 2019. https://geektyrant.com/news/an-ai-bot-writes-a-hilarious-episode-of-star-trek-the-next-generation.

Pellissier, Hank. 2013. "Robots and Slavery: What Do Humans Want When We Are 'Masters'?" Institute for Ethics and Emerging Technologies, September 13, 2013. https://ieet.org/index.php/IEET2/more/pellissier20130913.

Randall, Terri, dir. 2016. *NOVA.* "Rise of the Robots: Inside the DARPA Robotics Challenge." Aired February 24, 2016, on PBS. DVD.

Reagin, Nancy R., and Liedl, Janice, eds. 2013. Star Trek *and History.* Hoboken, NJ: John Wiley & Sons.

Robitzski, Dan. 2018. "Artificial Consciousness: How to Give a Robot a Soul." Futurism, June 25, 2018. https://futurism.com/artificial-consciousness.

Rouhiainen, Lasse. 2018. *Artificial Intelligence: 101 Things You Must Know Today about Our Future.* Scotts Valley, CA: CreateSpace.

Shapiro, Allen. 2013. Star Trek *and the Android Data.* Seattle, WA: Amazon.com Services.

Thagard, Paul. 2017. "Will Robots Ever Have Emotions?" *Psychology Today,* December 14, 2017. https://www.psychologytoday.com/us/blog/hot-thought/201712/will-rob ots-ever-have-emotions.

Wallach, Wendell, and Allen, Colin. 2009. *Moral Machines: Teaching Robots Right from Wrong.* New York: Oxford University Press.

Speed Buggy

Hanna-Barbera

By 1970, Hanna Barbera animation studios had made the transition from producing prime-time animated series to dominating the Saturday morning cartoon medium. With the massive popularity of *Scooby-Doo, Where Are You!* (CBS, 1969–1970; ABC, 1978), the studio set out to copy the popular format of a group of young people solving mysteries with an anthropomorphized sidekick. One series of episodes could have countless repeats of reruns for years to come. One such success was the short-lived but classic animated series, *Speed Buggy* (CBS, 1973). The series centered on a group of young people comprising a racing team: Mark (voiced by Michael Bell, b. 1938), Debbie (voiced by Arlene Golonka, b. 1936), and "Tinker" (voiced by Phil Luther Jr., life dates unknown). Their unlikely race car was "Speed Buggy" (voiced by animation icon Mel Blanc, 1908–1989), a three-seat "dune buggy" with eyes embedded in his hood-mounted headlights and a mouth where the car's front grill would be. The team would frequently stumble across various crimes or alien invasions, usually centered around a villain attempting to uncover the secret of Speed Buggy's success.

Speed Buggy was pop culture's first truly artificial intelligence automobile. He could emote, expressing happiness, sadness, anger, exhaustion, nervousness, and fear. He could eject oil slicks to avoid pursuers and beam "headlights" from his eyes. A very obvious live-action rip-off of *Speed Buggy* was attempted later in the decade by rival studio Sid and Marty Krofft with *Wonderbug* (ABC, 1976–1978), with the titular vehicle a very obvious copy of Speed Buggy, right down to body type, color, rear-mounted antenna, and mounted headlight eyes. The only real difference was that Wonderbug's "mouth" was a contortion of its front bumpers. In the end, Speed Buggy—and Wonderbug—were realizations of a teenager's ultimate dream: to be able to communicate and have a real friendship with one's car. With the ever-increasing interactive technology in automobiles in the twenty-first century, what was once Saturday morning sci-fi may yet become reality.

Richard A. Hall

See also: AWESOM-O 4000, Bender, Cambot/Gypsy/Tom Servo/Crow, KITT, Rosie; *Thematic Essays:* Robots and Slavery, Heroic Robots and Their Impact on Sci-Fi Narratives.

Further Reading

Albert, Robert S., and Brigante, Thomas R. 1962. "The Psychology of Friendship Relations: Social Factors." *Journal of Social Psychology* 56 (1). https://www.tandfon line.com/doi/abs/10.1080/00224545.1962.9919371.

Allen, Arthur. 2019. "There's a Reason We Don't Know Much about AI." Politico, September 16, 2019. https://www.politico.com/agenda/story/2019/09/16/artficial-intell igence-study-data-000956?cid=apn.

Amati, Viviana, Meggiolaro, Silvia, Rivellini, Giulia, and Zaccarin, Susanna. 2018. "Social Relations and Life Satisfaction: The Role of Friends." *Genus* 74, no. 1 (May 4). https://www.ncbi.nlm.nih.gov/pmc/articles/PMC5937874.

Ambrosino, Brandon. 2018. "What Would It Mean for AI to Have a Soul?" BBC, June 17, 2018. https://www.bbc.com/future/article/20180615-can-artificial-intelligence-hav e-a-soul-and-religion.

Ashby, LeRoy. 2006. *With Amusement for All: A History of American Popular Culture since 1830*. Lexington: University Press of Kentucky.

Castleman, Harry, and Podrazik, Walter J. 2016. *Watching TV: Eight Decades of American Television, Third Edition*. Syracuse, NY: Syracuse University Press.

Dinh, Thien-Nam. 2018. *Silicon Minds: The Science, Impact, and Promise of Artificial Intelligence*. Independently Published.

Hampton, Gregory Jerome. 2017. *Imagining Slaves and Robots in Literature, Film, and Popular Culture: Reinventing Yesterday's Slave with Tomorrow's Robot*. Lanham, MD: Lexington.

Hitchcock, Susan Tyler. 2007. *Frankenstein: A Cultural History*. New York: W. W. Norton.

Lenburg, Jeff. 2011. *William Hanna and Joseph Barbera: The Sultans of Saturday Morning*. Philadelphia: Chelsea House.

Lin, Patrick, Jenkins, Ryan, and Abney, Keith, eds. 2017. *Robot Ethics 2.0: From Autonomous Cars to Artificial Intelligence*. New York: Oxford University Press.

Linder, Courtney. 2020. "This AI Robot Just Nabbed the Lead Role in a Sci-Fi Movie: Meet Erica, the Movie Star Who May Put Human Actors Out of Work." *Popular Mechanics*, June 25, 2020. https://www.popularmechanics.com/technology /robots/a32968811/artificial-intelligence-robot-movie-star-erica.

Paur, Joey. 2019. "An AI Bot Writes a Hilarious Episode of *Star Trek: The Next Generation*." Geek Tyrant, November 16, 2019. https://geektyrant.com/news/an-ai-bot -writes-a-hilarious-episode-of-star-trek-the-next-generation.

Pellissier, Hank. 2013. "Robots and Slavery: What Do Humans Want When We Are 'Masters'?" Institute for Ethics and Emerging Technologies, September 13, 2013. https://ieet.org/index.php/IEET2/more/pellissier20130913.

Randall, Terri, dir. 2016. *NOVA*. "Rise of the Robots: Inside the DARPA Robotics Challenge." Aired February 24, 2016, on PBS. DVD.

Robitzski, Dan. 2018. "Artificial Consciousness: How to Give a Robot a Soul." Futurism, June 25, 2018. https://futurism.com/artificial-consciousness.

Rouhiainen, Lasse. 2018. *Artificial Intelligence: 101 Things You Must Know Today about Our Future*. Scotts Valley, CA: CreateSpace.

Said, Carolyn. 2019. "Robot Cars Are Getting Better—But Will True Self-Driving Ever Arrive?" *San Francisco Chronicle*, February 14, 2019. https://www.sfchroni cle.com/business/article/Robot-cars-are-getting-better-but-will-true-13614280 .php.

Thagard, Paul. 2017. "Will Robots Ever Have Emotions?" *Psychology Today*, December 14, 2017. https://www.psychologytoday.com/us/blog/hot-thought/201712/will-rob ots-ever-have-emotions.

Wallach, Wendell, and Allen, Colin. 2009. *Moral Machines: Teaching Robots Right from Wrong*. New York: Oxford University Press.

Starfleet Computer

Star Trek

In 1966, creator-writer-producer Gene Roddenberry (1921–1991) launched his iconic science fiction television series *Star Trek* (NBC, 1966–1969). Set in the twenty-third century, the series centers on the adventures of the starship *USS Enterprise*; its commander, Captain James T. Kirk (played by William Shatner, b. 1931); and its crew, headed by the alien first officer, Mr. Spock (played by Leonard Nimoy, 1931–2015), representing the "United Federation of Planets." Though short-lived in its initial run, the series has become legendary, not only for its groundbreaking special effects but for the sophistication of its stories, many of which were grounded in real-world issues of the day, such as racism and a fear of the Cold War. Its success in syndicated reruns revived the franchise, leading to an animated series, thirteen feature films, and five live-action "sequel" series (at the time of this writing). One of the mainstays throughout most of the *Star Trek* series (particularly the first four series) was the Starfleet Computer, voiced in those series by Majel Barrett (1932–2008).

Decades before anyone ever heard the word "internet," and years before anyone outside the government had access to networked computers, the creators of the twenty-third-century world of *Star Trek* implemented the sci-fi concept of a massive computer system, communicating through thousands of satellite nodes throughout the galaxy, allowing starships of the United Federation of Planets to communicate not only with each other but also with Starfleet Headquarters facilities throughout the galaxy. Aside from communication, the Starfleet Computer also possessed access to the encyclopedic sum total of all human knowledge to that point. It monitored and controlled starship life support and defenses and contained a "self-destruct" program to prevent the technology of the starship from falling into enemy hands.

In the first sequel series, *Star Trek: The Next Generation* (syndication, 1987–1994), audiences were introduced to a new facet of the Starfleet Computer: the "holodeck." This was a room on a starship that was a large virtual reality chamber. Within it, one could call upon the computer to create a seemingly endless variety of VR environments: from realistic settings in history to fictional realms from novels, television, or films. The holodeck also possessed safety features that could be turned off upon request. This feature made it so that a user who desired to play a violent scenario could do so without threat of actual physical harm. In the fourth television series, *Star Trek: Voyager* (UPN, 1995–2001), the Federation starship *Voyager* is lost, seventy thousand light-years from the nearest Federation outpost. Throughout most of their voyage home, the onboard Starfleet Computer could not access the Federation network; as such, the crew relied upon the computer's abilities on its own (essentially maintaining all of its functions other than communication with the rest of the Federation). At one point, the crew were able to establish contact with home via sending a signal through a small wormhole in space, finally giving them some communication with home.

At its core, the Starfleet Computer was an example of our own internet long before the general public could have even fathomed such a thing. Considering the impact that the original *Star Trek* in the 1960s had on the engineers of cell phones,

---

U.S. Space Force

The U.S. Space Force is the sixth branch of the U.S. military, alongside the Army, Navy, Air Force, Marines, and Coast Guard. It was officially established on December 20, 2019, by order of U.S. president Donald J. Trump (b. 1946). It is an offshoot of the previous U.S. Air Force Space Command (established in 1982). Not to be confused with NASA—with which it is not directly connected—the USSF primary mission is the defense of American interests in space (as opposed to NASA's more general mission of exploration and scientific research). The official seal of Space Force has come under some scrutiny as it is very clearly inspired by the seal of the fictional Starfleet Command of the popular futuristic *Star Trek* franchise. It was the opinion of President Trump that as tensions have risen between the United States and previous space partners Russia and China, future American endeavors in space require the development of a military branch to ensure their protection. As Trump's initial announcement of Space Force came shortly after the U.S. Navy release of fascinating UFO video footage in December 2017 (the footage itself dating back to 2004), this immediately led to suggestions that there was some "threat" from outer space that the country must protect against. In 2020, Netflix launched the series *Space Force*, starring Steve Carell (b. 1962) and John Malkovich (b. 1953) as the commanding general of Space Force and its chief scientific adviser, respectively. The series emphasizes the comedic response to the branch's creation as well as humorous takes on possible issues that the new military branch may well face.

Richard A. Hall

---

iPads, medical diagnostic beds, and even experiments on faster-than-light travel, it is very likely that the engineers of what would become the World Wide Web were influenced to one degree or other by the Starfleet Computer. In the twenty-first century, the Starfleet Computer seems like one of the less "science fictiony" aspects of this futuristic program; but in its early decades, it was a wonder beyond many people's imagination. If the *Star Trek* franchise were to once more move beyond its own established "future," it would be fascinating to see what evolution of this device would be forthcoming.

Richard A. Hall

See also: AI/Ziggy, Androids, Batcomputer, Borg, Brainiac, Control, Cylons, Doctor/EMH, The Great Intelligence, HAL 9000, HERBIE, JARVIS/Friday, KITT, Landru, Lieutenant Commander Data, Lore/B4, The Matrix/Agent Smith, OASIS, Rehoboam, Skynet, Soji and Dahj Asha, Ultron, V-GER, WOPR; *Thematic Essays:* Robots and Slavery, AI and the Apocalypse: Science Fiction Meeting Science Fact.

Further Reading

Allen, Arthur. 2019. "There's a Reason We Don't Know Much about AI." Politico, September 16, 2019. https://www.politico.com/agenda/story/2019/09/16/artficial-intell igence-study-data-000956?cid=apn.

Ashby, LeRoy. 2006. *With Amusement for All: A History of American Popular Culture since 1830.* Lexington: University Press of Kentucky.

Bastani, Aaron. 2019. *Fully Automated Luxury Communism.* New York: Verso.

Byrne, Emma. 2013. "Innovation Isn't Safe: The Future According to Kevin Warwick." *Forbes,* September 30, 2013. https://www.forbes.com/sites/netapp/2013/09/30 /kevin-warwick-captain-cyborg/#48b704c13560.

Castleman, Harry, and Podrazik, Walter J. 2016. *Watching TV: Eight Decades of American Television, Third Edition.* Syracuse, NY: Syracuse University Press.

Decker, Kevin S., and Eberl, Jason T., eds. 2016. *The Ultimate* Star Trek *and Philosophy: The Search for Socrates.* Hoboken, NJ: Wiley-Blackwell.

Dinh, Thien-Nam. 2018. *Silicon Minds: The Science, Impact, and Promise of Artificial Intelligence.* Independently Published.

Gross, Edward, and Altman, Mark A. 2016. *The Fifty-Year Mission, The First 25 Years: The Complete, Uncensored, Unauthorized Oral History of* Star Trek. New York: St. Martin's.

Lin, Patrick, Jenkins, Ryan, and Abney, Keith, eds. 2017. *Robot Ethics 2.0: From Autonomous Cars to Artificial Intelligence.* New York: Oxford University Press.

Murdock, Jason. 2019. "Former Google Engineer Warns AI Might Accidentally Start a War: 'These Things Will Start to Behave in Unexpected Ways.'" *Newsweek*, September 16, 2019. https://www.newsweek.com/google-project-maven-artificial-intelligence-laura-nolan-killer-robots-department-defense-1459358.

Paur, Joey. 2019. "An AI Bot Writes a Hilarious Episode of *Star Trek: The Next Generation*." Geek Tyrant, November 16, 2019. https://geektyrant.com/news/an-ai-bot-writes-a-hilarious-episode-of-star-trek-the-next-generation.

Reagin, Nancy R., and Liedl, Janice, eds. 2013. Star Trek *and History*. Hoboken, NJ: John Wiley & Sons.

Rouhiainen, Lasse. 2018. *Artificial Intelligence: 101 Things You Must Know Today about Our Future.* Scotts Valley, CA: CreateSpace.

Starck, Kathleen, ed. 2010. *Between Fear and Freedom: Cultural Representations of the Cold War.* Newcastle, UK: Cambridge Scholars.

Stark, Steven D. 1997. *Glued to the Set: The 60 Television Shows and Events that Made Us Who We Are Today.* New York: Delta Trade Paperbacks.

Stepford Wives

The Stepford Wives

Coming into the 1970s, the women's liberation movement was picking up steam. Women were demanding more opportunities than being simple housewives. Many men around the country feared that their days of dominating their wives were over. In 1972, author Ira Levin (1929–2007) published his novel *The Stepford Wives*. This story centered on a young married couple recently moving to the fictional town of Stepford, Connecticut. The young wife and mother, Joanna, was a photographer. Soon after moving in, Joanna notices that all of the wives of Stepford seemed to be mindless, drone-like, obedient traditional "housewives," their husbands lording over them as they gleefully obeyed their every command. At first believing that the women had been brainwashed or drugged in some way, Joanna eventually discovers that the women were replaced with lifelike robots. As she gets closer to the core of the conspiracy, she learns that her husband has fallen in with the men of Stepford; and, as the novel closes, Joanna has seemingly been replaced with a robot as a new young couple arrives (Levin, *The Stepford Wives*, 1972).

There have been various film versions of the novel over the decades, the first—in 1975—adhering to the robot reveal. Others have gone with the drugged and/or brainwashed storyline; while the most recent film version—in 2004—had the women turned into cyborgs. To the general reader, the novel is either: (1) a fantasy where men rule and women obey; or (2) a cautionary tale of what will happen to women if they do not stand up for their own personhood. The United States has dealt with systemic sexism from its inception (it took 144 years after the Declaration of Independence for women to get the right to vote nationally, and women were not allowed to have a credit card in their own name until 1974). Men have historically attempted to dominate women (especially their wives), and *The Stepford Wives* is the ultimate fantasy for such men: replacing an independent-minded wife with an obedient robot. With the massive rise in popularity of "sex robots" since roughly 2010, the nightmare scenario of *The Stepford Wives* could very easily—now more than ever—become reality.

Richard A. Hall

See also: Androids, Buffybot, Dolores, Fembots, Jocasta, Lieutenant Commander Data, LMDs, Lore/B4, Maschinenmensch/Maria, Rosie, Soji and Dahj Asha, VICI, Vision; *Thematic Essay:* Robots and Slavery.

Further Reading

Allan, Kathryn, ed. 2013. *Disability in Science Fiction.* New York: Palgrave Macmillan.

Ashby, LeRoy. 2006. *With Amusement for All: A History of American Popular Culture since 1830.* Lexington: University Press of Kentucky.

Bastani, Aaron. 2019. *Fully Automated Luxury Communism.* New York: Verso.

Blair, Anthony. 2018. "Sex Robots That Feel LOVE and Suffer PAIN When Dumped Coming Soon Claims Expert." *Daily Star*, December 6, 2018. https://www.dailys tar.co.uk/news/latest-news/sex-robot-feel-love-pain-16824160.

Carper, Steve. 2019. *Robots in American Popular Culture.* Jefferson, NC: McFarland.

Chandler, Simon. 2019. "Artificial Intelligence Has Become a Tool for Classifying and Ranking People." *Forbes*, October 1, 2019. https://www.forbes.com/sites/simon chandler/2019/10/01/artificial-intelligence-has-become-a-tool-for-classifying-and -ranking-people/#7431657f1d7c.

Chanthadavong, Aimee. 2019. "AI and Ethics: The Debate That Needs to Be Had." ZDnet, September 16, 2019. https://www.zdnet.com/article/ai-and-ethics-the-debate-that -needs-to-be-had/#ftag=CAD-03-10abf5f.

Conrad, Dean. 2018. *Space Sirens, Scientists and Princesses: The Portrayal of Women in Science Fiction Cinema.* Jefferson, NC: McFarland.

Dinh, Thien-Nam. 2018. *Silicon Minds: The Science, Impact, and Promise of Artificial Intelligence.* Independently Published.

Dowden, Bradley. 1993. *Logical Reasoning.* Belmont, CA: Wadsworth Publishing Company.

Hampton, Gregory Jerome. 2017. *Imagining Slaves and Robots in Literature, Film, and Popular Culture: Reinventing Yesterday's Slave with Tomorrow's Robot.* Lanham, MD: Lexington.

Hicks, Amber. 2019. "Sex Robot Brothel Opens in Japan amid Surge of Men Wanting Bisexual Threesomes." *Mirror*, April 27, 2019. https://www.mirror.co.uk/news /weird-news/sex-robot-brothel-opens-japan-14792161.

Hitchcock, Susan Tyler. 2007. *Frankenstein: A Cultural History.* New York: W. W. Norton.

Kirshner, Jonathan. 2012. *Hollywood's Last Golden Age: Politics, Society, and the Seventies Film in America.* Ithaca, NY: Cornell University Press.

Levin, Ira. 1972. *The Stepford Wives*. New York: William Morrow.

Lin, Patrick, Abney, Keith, and Bekey, George A., eds. 2014. *Robot Ethics: The Ethical and Social Implications of Robotics*. Cambridge: Massachusetts Institute of Technology.

Linder, Courtney. 2020, "This AI Robot Just Nabbed the Lead Role in a Sci-Fi Movie: Meet Erica, the Movie Star Who May Put Human Actors Out of Work." *Popular Mechanics*, June 25, 2020. https://www.popularmechanics.com/technology/robo ts/a32968811/artificial-intelligence-robot-movie-star-erica.

Pellissier, Hank. 2013. "Robots and Slavery: What Do Humans Want When We Are 'Masters'?" Institute for Ethics and Emerging Technologies, September 13, 2013. https://ieet.org/index.php/IEET2/more/pellissier20130913.

Randall, Terri, dir. 2016. *NOVA*. "Rise of the Robots: Inside the DARPA Robotics Challenge." Aired February 24, 2016, on PBS. DVD.

Rouhiainen, Lasse. 2018. *Artificial Intelligence: 101 Things You Must Know Today about Our Future*. Scotts Valley, CA: CreateSpace.

Schroeder, Stan. 2020. "Samsung Just Launched an 'Artificial Human' Called Neon, and Wait, What?" Mashable, January 7, 2020. https://mashable.com/article/samsung -star-labs-neon-ces.

Seed, David. 1999. *American Science Fiction and the Cold War: Literature and Film*. Abingdon, UK: Routledge.

Spaeth, Dennis. 2018. "From Single-Task Machines to Backflipping Robots: The Evolution of Robots." *Cutting Tool Engineering*, January 15, 2018. https://www.ctemag .com/news/articles/evolution-of-robots.

Stanford University. n.d. "Robotics: A Brief History." Stanford University. https:// cs.stanford.edu/people/eroberts/courses/soco/projects/1998-99/robotics/history .html.

Thagard, Paul. 2017. "Will Robots Ever Have Emotions?" *Psychology Today*, December 14, 2017. https://www.psychologytoday.com/us/blog/hot-thought/201712/will-rob ots-ever-have-emotions.

Steve Austin

The Six Million Dollar Man

One of the most iconic television series of the 1970s was *The Six Million Dollar Man* (ABC, 1973–1978). It was based on the 1972 novel, *Cyborg*, by Martin Caidin (1927–1997). In this series, Lee Majors (b. 1939) stars as U.S. Air Force colonel Steve Austin, a NASA astronaut-test pilot. In the premier episode, Austin suffers a devastating crash during a test flight. The government then chooses Austin for experimental "bionic" implants (costing six million dollars): both legs (giving him the ability to run in excess of 60mph); a left eye that gives him both telescopic and infrared vision; and a right arm that gives him superhuman strength. He is then assigned to the top-secret Office of Scientific Intelligence (OSI).

The series was popular for its ingenious method of showing his "speed" by filming the actor running in slow motion. The thirteen-inch Steve Austin doll produced by Kenner Toys in 1975 was one of the highest-selling toys in history. As the novel title implies, Steve Austin is a cyborg: part human, part mechanical. Due

---

NASA

The National Aeronautics and Space Administration (NASA) is the United States' primary space exploration and research agency. It was established by President Dwight D. Eisenhower (1890–1969) in 1958, largely in response to the Soviet Union's launch of the world's first human-made satellite, *Sputnik I*, in 1957. By 1970, NASA research had surpassed that of the Soviets, allowing the United States to send the first human-staffed mission to the moon. On July 20, 1969, Astronaut Neil Armstrong (1930–2012) became the first human to set foot on the moon, speaking the immortal words, "One small step for man, one giant leap for mankind." The *Apollo 11* mission also included astronauts Edwin "Buzz" Aldrin and Michael Collins (both b. 1930). In 1973, NASA launched the first orbital laboratory, *Skylab*, but soon abandoned it in favor of removable/reusable laboratories in the new Space Shuttle program. The shuttle program ran from 1972 to 2011, when congressional defunding of NASA required abandoning the largely successful orbital vehicle program. The program had two fatal missions during this period: the space shuttle *Challenger* exploded shortly after takeoff on January 28, 1986, killing all seven aboard (including the first "teacher in space," Christa McAuliffe, 1948–1986); and the shuttle *Colombia* exploded on reentry on February 1, 2003, also killing all seven aboard. In 1993, NASA began working cooperatively with the space agencies of Canada, Europe, Japan, and Russia on the International Space Station. In 2017, President Donald J. Trump (b. 1946) once more began pouring funding into NASA with the intention of focusing research on sending a human-staffed mission to Mars by 2035. In July 2020, NASA launched its fifth rover program to Mars, *Perseverance* (Brian Resnick, "NASA's Latest Rover Is Our Best Chance Yet to Find Life on Mars," Vox, July 30, 2020, https://www.vox.com/science-and-health/2020/7/29/21340464/nasa-perseverance-ingenuity-launch-live-stream-how-to-watch-science-life-on-mars).

Richard A. Hall

---

to the show's popularity—and influenced by the booming women's liberation movement of the decade – a spin-off series was created: *The Bionic Woman* (ABC, 1976–1978), centered on Jaime Sommers (played by Lindsay Wagner, b. 1949), Steve Austin's love interest who—like him—experienced a devastating accident requiring bionic implants to save her life; she also worked for the OSI as a super-agent. The enduring popularity of the series led to three television movies continuing the adventures of both characters.

Steve Austin is one of the great cyborg heroes in modern fiction. At no point do his mechanical enhancements detract from any way his innate humanity. As such, the series does not focus on the psychological impact of the dehumanizing aspect of cybernetic implants. Instead, it is a strictly "superhero" story with a superpowered human fighting evil for the safety of all Americans, his most memorable "battle" being against the mythical creature Bigfoot. He is a perfect example of "man over machine." Rather than a Frankenstein's monster, Steve Austin is never seen as anything but the hero that he is.

Richard A. Hall

See also: Adam, Alita, Borg, Cybermen, Cyborg, Daleks, Darth Vader, Doctor Octopus, General Grievous, Iron Man, Jaime Sommers, Nardole, RoboCop; *Thematic Essays:* Heroic Robots and Their Impact on Sci-Fi Narratives, Cyborgs: Robotic Humans or Organic Robots?

Further Reading

Allan, Kathryn, ed. 2013. *Disability in Science Fiction.* New York: Palgrave Macmillan.

Ashby, LeRoy. 2006. *With Amusement for All: A History of American Popular Culture since 1830.* Lexington: University Press of Kentucky.

Bryant, D'Orsay D., III. 1985. "Spare-Part Surgery: The Ethics of Organ Transplantation." *Journal of the National Medical Association* 77 (2). https://www.ncbi.nlm .nih.gov/pmc/articles/PMC2561842/pdf/jnma00245-0055.pdf.

Byrne, Emma. 2014. "Cybernetic Implants: No Longer Science Fiction." *Forbes*, March 11, 2014. https://www.forbes.com/sites/netapp/2014/03/11/cybernetic-implants-not -sci-fi/#7399c57e77ba.

Calvert, Bronwen. 2017. *Being Bionic: The World of TV Cyborgs.* London: I. B. Tauris.

Carper, Steve. 2019. *Robots in American Popular Culture.* Jefferson, NC: McFarland.

Castleman, Harry, and Podrazik, Walter J. 2016. *Watching TV: Eight Decades of American Television, Third Edition.* Syracuse, NY: Syracuse University Press.

Means, Sean P. September 6, 2019. "At FanX, 'Bionic' Stars Lindsay Wagner and Lee Majors Recall Their TV Glory, and Lots of Running." *Salt Lake Tribune.* https:// www.sltrib.com/artsliving/2019/09/06/fanx-bionic-stars.

Pilato, Herbie J. 2007. *The Bionic Book: The* Six Million Dollar Man *and the* Bionic Woman *Reconstructed.* Albany, GA: BearManor.

Seed, David. 1999. *American Science Fiction and the Cold War: Literature and Film.* Abingdon, UK: Routledge.

"Six Million Dollar Man Quotes, The." n.d. Retro Junk. https://www.retrojunk.com/con tent/child/quote/page/4314/the-six-million-dollar-man.

Starck, Kathleen, ed. 2010. *Between Fear and Freedom: Cultural Representations of the Cold War.* Newcastle, UK: Cambridge Scholars.

T

TARDIS

Doctor Who

Doctor Who (BBC, 1963–1989; 2005–present) is the longest-running science fiction television series in history. The main character is the "Doctor," a Time Lord from the Planet Gallifrey who travels throughout space and time helping out wherever possible. The key to the show's longevity is a quirk of Time Lord biology known as "regeneration": When the Doctor has been fatally injured, rather than dying, every cell of his or her body (depending on the incarnation) renews and replaces itself. The new Doctor has a new face and new personality but is still the same person deep down. Logistically, this allows the show to replace the actor in the main role to shake up the show and keep it fresh while maintaining continuity over the decades.

Other than the Doctor, the most consistent presence in the show is that of the TARDIS. Even though all Time Lords will call their personal ship a TARDIS, the very first episode of *Doctor Who* shows the Doctor's granddaughter, Susan, taking credit for coming up with the name, "Well, I made up the name TARDIS from the initials, Time and Relative Dimension in Space" (Anthony Coburn, "An Unearthly Child," original series, season 1, episode 1, November 23, 1963). Even though the TARDIS looks and functions like a machine, the tenth Doctor reveals to his companion, Rose, that "they were grown, not built" (Matt Jones, "The Impossible Planet," return series, season 2, episode 8, June 3, 2006). The TARDIS is "dimensionally transcendental," which means that it is, as most

The TARDIS from the classic sci-fi British television classic *Doctor Who* (1963–1989; 2005–present). (Photodynamx/Dreamstime.com)

who first enter it note, "bigger on the inside" (though the exterior of the vessel appears relatively small, the interior is massive). The immense power required for the TARDIS to travel throughout all of space and time is provided by the "Eye of Harmony." It is an "exploding star. In the act of becoming a black hole. Time Lord engineering. You rip the star from its orbit, suspend it in a permanent state of decay" (Stephen Thompson, "Journey to the Centre of the TARDIS," return series, season 7, episode 10, April 27, 2013). The Doctor's TARDIS is, by the time of the 2005 return of the series, the last "Type 40" TARDIS in the universe. Each TARDIS is outfitted with a "chameleon circuit" that helps hide the TARDIS in plain sight. For example, if the TARDIS were to land in a forest, the TARDIS might look like a tree. The Doctor's TARDIS, however, has not had a working chameleon circuit during the entire run of the show. When the TARDIS landed in 1963 London, it took the appearance of a police phone box and has stayed that way ever since. There are two episodes of the revival era that explore the TARDIS more than any others: "The Doctor's Wife" and "Journey to the Centre of the TARDIS."

The TARDIS is more than just a machine; it has a consciousness and a will of its own, as confirmed in the story "The Doctor's Wife." In that episode, the "soul" of the TARDIS is removed from the box and put into a human body named Idris. This finally gives the Doctor a chance to have an actual conversation with the TARDIS. He admonishes her for never taking him where he wants to go, to which she lets him know that she always took him where he "needed to go." Then she asks him, "Do you ever wonder why I chose you all those years ago?" The Doctor sticks with his story that *he* chose the TARDIS, but Idris replies, "I wanted to see the universe, so I stole a Time Lord and I ran away. And you were the only one mad enough" (Neil Gaiman, "The Doctor's Wife," return series, season 6, episode 4, May 14, 2011). In "The Name of the Doctor," it is shown that Clara encouraged the original Doctor to steal the correct TARDIS, which would seem to contradict "The Doctor's Wife" (Steven Moffat, "The Name of the Doctor," return series, season 7, episode 13, May 18, 2013).

In trying to explain the dimensionally transcendental aspect of the TARDIS, the Doctor has used different approaches for different people. Rory Williams surprised the Doctor when he understood the principle before the Doctor could explain that the inside of the TARDIS is another dimension (Toby Whithouse, "The Vampires of Venice," return series, season 5, episode 6, May 8, 2010). In "Journey to the Centre of the TARDIS," it is learned that the TARDIS is also capable of infinitely expanding, creating new rooms inside itself. In the episode, this is characterized as a defense mechanism (Stephen Thompson, "Journey to the Centre of the TARDIS," return series, season 7, episode 10, April 27, 2013).

Early stories involving time travel employed all kinds of ways to get the protagonist through time, including sleep, ghosts, angels, and dreams. Prolonged sleep was a common trope to move the protagonist forward in stories like "Rip Van Winkle." More supernatural means were employed to move backward in time. Science fiction writers would have to wait until the second Industrial Revolution to be inspired to employ machines to ferry their characters throughout time.

The first story to employ mechanical means for time travel is a short story called "The Clock that Went Backward," written by Edward Page Mitchell in 1881. The basic plot is that two boys discover that if they wind a particular antique clock backward, they move backward in time. While H. G. Wells is often credited with the first story involving a full-on "time machine," the first story to feature a machine traveling through time was actually written by a Spanish playwright a few years before. Since then, there have been numerous stories using machines to facilitate the protagonist's passage throughout time. Unlike these other plot devices, the TARDIS is no mere machine. She has a living consciousness of her own. She is no slave to the Doctor's whims. She has a mind of her own and often knows what is best for the Doctor. The TARDIS is not confined by arbitrary limits as she has traveled beyond time, beyond the universe, and to other dimensions. Truly, in all of time, space, and science fiction, the TARDIS stands alone as not only one of the greatest plot devices for the main character but also that character's companion.

Keith R. Claridy

See also: AI/Ziggy, Batcomputer, Cerebro/Cerebra, Cybermen, Daleks, The Great Intelligence, HAL 9000, Janet, JARVIS/Friday, K9, KITT, The Matrix/Agent Smith, Nardole, OASIS, Rehoboam, Skynet, Starfleet Computer, Teselecta, WOPR; *Thematic Essays:* Heroic Robots and Their Impact on Sci-Fi Narratives, AI and the Apocalypse: Science Fiction Meeting Science Fact.

Further Reading
Alkon, Paul K. 1987. *Origins of Futuristic Fiction.* Athens: University of Georgia Press.

Campbell, Mark. 2010. Doctor Who: *The Complete Guide.* London: Running Press.

Chapman, James. 2013. *Inside the TARDIS: The Worlds of* Doctor Who. London: I.B. Tauris.

Harris, Mark. 1983. *The* Doctor Who *Technical Manual.* London, UK: Random House.

Hills, Matt, ed. 2013. *New Dimensions of* Doctor Who: *Adventures in Space, Time and Television.* London: I. B. Tauris.

Kistler, Alan. 2013. Doctor Who: *Celebrating Fifty Years, A History.* Guilford, CT: Lyons.

Lewis, Courtland, and Smithka, Paula, eds. 2010. Doctor Who *and Philosophy: Bigger on the Inside.* Chicago: Open Court.

Lewis, Courtland, and Smithka, Paula, eds. 2015: *More* Doctor Who *and Philosophy: Regeneration Time.* Chicago: Open Court.

Mitchell, Edward Page. 1881. "The Clock that Went Backward." http://www.forgottenfutures.com/game/ff9/tachypmp.htm#clock.

Muir, John Kenneth. 2007. *A Critical History of* Doctor Who *on Television.* Jefferson, NC: McFarland.

Nahin, Paul J. 1999. *Time Machines: Time Travel in Physics, Metaphysics, and Science Fiction.* 2nd ed. Woodbury, NY: AIP Press; New York: Springer.

Nathan-Turner, John. 1985. *The TARDIS Inside Out.* London: Piccadilly Press.

Richards, Justin. 2009. Doctor Who: *The Official Doctionary.* London: Penguin Group.

Wasserman, Ryan. 2018. *Paradoxes of Time Travel.* Oxford: Oxford University Press.

Westcott, Kathryn. 2011. "HG Wells or Enrique Gaspar: Whose Time Machine Was First?" BBC News, April 9, 2011. https://www.bbc.com/news/world-europe-12900390.

Terminators

The Terminator

In 1984, James Cameron (b. 1954) directed and cowrote *The Terminator*, launching one of the most successful franchises in sci-fi history. Set in 1984, a woman named Sarah Connor (played by Linda Hamilton, b. 1956) is hunted by a robot assassin from the year 2029 known as a "Terminator" (played by Arnold Schwarzenegger, b. 1947). Connor is rescued by Kyle Reese (played by Michael Biehn, b. 1956), a human sent from the same year to protect her. In that future world, Earth has been destroyed by nuclear war initiated by an American military defense AI called "Skynet." Machines rule the world and are fought by a human resistance led by John Connor, Sarah's son. The original Terminator's mission is to kill Sarah Connor before she can give birth to John; but, ironically, the ensuing romance between Sarah and Kyle leads to her pregnancy with John. Kyle loses his life protecting Sarah, but, fully aware of her son's important role in humanity's future, she will raise her son to be the hero he is meant to be (Cameron and Gale Anne Hurd, *The Terminator*, Orion, 1984).

The original Terminator (as played by Schwarzenegger) was a "T-800" model: a skeletal robotic frame with red eyes, increased speed and strength, and a flesh-like covering to allow him to blend in as human. In the film's first sequel, *Terminator 2: Judgement Day* (1991), an updated model, the "T-1000" is introduced. Possessing the same qualities as the T-800, the T-1000 was made with "liquid metal," allowing him to transform his frame to turn his limbs into weapons or to liquefy himself enough to pass through bars or drains or even to "stretch" his form

A wax statue of Arnold Schwarzenegger from the classic sci-fi film *The Terminator 2: Judgment Day* (1991). (Flavytt/Dreamstime.com)

in order to more easily catch up with evading targets (the T-1000 was played by Robert Patrick, b. 1958). A female (or gynoid/fembot) Terminator—the "T-X"—was introduced in the third film, *Terminator 3: Rise of the Machines* (2003), and was played by Kristanna Loken (b. 1979). Similar in most respects to the T-1000, the T-X was also programmed to "terminate" rogue Terminators (the events of *T3* were later negated by the reboot sequel, *Terminator: Dark Fate*, 2019).

Schwarzenegger played two different T-800s in the first two films. The first one became iconic with the phrase, "I'll be back"; and the second became equally iconic with the phrase, "Hasta la vista, Baby." Terminators are the ultimate killing machines: near indestructible and unstoppable when programmed to terminate a specific target. In the future ruled by Skynet, humanity is plagued by literal legions of Terminators. Beginning with *T2*, however, the narrative begins to explore the possibility of a killing machine overcoming its programming to become a care-giver (similar to the IG-11 droid in *Star Wars: The Mandalorian*). In the nightmare scenario of an AI-controlled future, the concept of an army of Terminators becomes possible, maybe even probable. Even without an AI overlord, the concept of an army of Terminators may prove too irresistible for the military industrial complex. This opens the door for nightmarish unintended consequences.

Richard A. Hall

See also: B.A.T.s, Battle Droids, Buffybot, Cylons, Dolores, Doombots, ED-209, Fem-bots, IG-88/IG-11, Iron Legion, Jocasta, LMDs, Maschinenmensch/Maria, Metalhead, Robots, Sentinels, Skynet, Tin Woodsman, Transformers, Ultron; *Thematic Essays*: Villainous Robots and Their Impact on Sci-Fi Narratives, AI and the Apocalypse: Science Fiction Meeting Science Fact.

Further Reading

Allen, Arthur. 2019. "There's a Reason We Don't Know Much about AI." Politico, September 16, 2019. https://www.politico.com/agenda/story/2019/09/16/artficial-intelligence-study-data-000956?cid=apn.

Ashby, LeRoy. 2006. *With Amusement for All: A History of American Popular Culture since 1830*. Lexington: University Press of Kentucky.

Carper, Steve. 2019. *Robots in American Popular Culture*. Jefferson, NC: McFarland.

Chandler, Simon. 2019. "Artificial Intelligence Has Become a Tool for Classifying and Ranking People." *Forbes*, October 1, 2019. https://www.forbes.com/sites/simonchandler/2019/10/01/artificial-intelligence-has-become-a-tool-for-classifying-and-ranking-people/#7431657f1d7c.

Chanthadavong, Aimee. 2019. "AI and Ethics: The Debate That Needs to Be Had." ZDnet, September 16, 2019. https://www.zdnet.com/article/ai-and-ethics-the-debate-that-needs-to-be-had/#ftag=CAD-03-10abf5f.

Dinh, Thien-Nam. 2018. *Silicon Minds: The Science, Impact, and Promise of Artificial Intelligence*. Independently Published.

Dowden, Bradley. 1993. *Logical Reasoning*. Belmont, CA: Wadsworth Publishing Company.

Fryer-Biggs, Zachary. 2019. "Coming Soon to a Battlefield: Robots That Can Kill." *The Atlantic*, September 3, 2019. https://www.theatlantic.com/technology/archive/2019/09/killer-robots-and-new-era-machine-driven-warfare/597130.

Hitchcock, Susan Tyler. 2007. *Frankenstein: A Cultural History*. New York: W. W. Norton.

Irwin, William, Brown, Richard, and Decker, Kevin S. 2009. *Terminator and Philosophy: I'll Be Back Therefore I Am.* Hoboken, NJ: Wiley-Blackwell.

Krishnan, Armin. 2009. *Killer Robots: Legality and Ethicality of Autonomous Weapons.* London: Routledge.

Lin, Patrick, Abney, Keith, and Bekey, George A., eds. 2014. *Robot Ethics: The Ethical and Social Implications of Robotics.* Cambridge: Massachusetts Institute of Technology.

Marks, Robert J. 2020. "*2084* vs. *1984*: The Difference AI Could Make to Big Brother" Podcast interview with author John Lennox. Mind Matters, July 3, 2020. https://mindmatters.ai/2020/07/2084-vs-1984-the-difference-ai-could-make-to-big-brother.

Murdock, Jason. 2019. "Former Google Engineer Warns AI Might Accidentally Start a War: 'These Things Will Start to Behave in Unexpected Ways.'" *Newsweek*, September 16, 2019. https://www.newsweek.com/google-project-maven-artificial-intelligence-laura-nolan-killer-robots-department-defense-1459358.

Randall, Terri, dir. 2016. *NOVA.* "Rise of the Robots: Inside the DARPA Robotics Challenge." Aired February 24, 2016, on PBS. DVD.

Rouhiainen, Lasse. 2018. *Artificial Intelligence: 101 Things You Must Know Today about Our Future.* Scotts Valley, CA: CreateSpace.

Seed, David. 1999. *American Science Fiction and the Cold War: Literature and Film.* Abingdon, UK: Routledge.

Singer, Peter W. 2009. "Isaac Asimov's Laws of Robotics Are Wrong." Brookings Institute, May 18, 2009. https://www.brookings.edu/opinions/isaac-asimovs-laws-of-robotics-are-wrong.

Wallach, Wendell, and Allen, Colin. 2009. *Moral Machines: Teaching Robots Right from Wrong.* New York: Oxford University Press.

Teselecta

Doctor Who

Doctor Who (BBC, 1963–1989; 2005–present) is the longest-running science fiction television series in history, with an ever-changing canon that spans more than fifty years. The main character (not necessarily the titular character) is an alien known simply as the "Doctor." The character of the Doctor can best be summed up in his own words: "I'm the Doctor. I'm a Time Lord. I'm from the planet Gallifrey in the Constellation of Kasterborous. . . . I'm the man who's gonna save your lives and all six billion people on the planet below. You got a problem with that?" (Russell T. Davies, "Voyage of the Damned," Christmas special, December 25, 2007). The Doctor is a being of adventure, childlike wonder, and heroism. As the various incarnations of the Doctor have told the story many times, they chafed under the strict rules of their people, the Time Lords, so they stole one of their ships and ran away. The Doctor travels throughout space and time, usually with at least one companion, saving the day, planet, galaxy, universe, or entire fabric of reality, depending on the ambitiousness of the showrunner of the day. The show has managed to stay relevant for so long with the same character because of the phenomenon of regeneration. A Time Lord who reaches the end of a life cycle can regenerate every cell in the body in order to live another life. This conveniently allows the character to be played by different actors with their own spins on it, thus carrying the show into the future.

The Teselecta is a shape-shifting humanoid-shaped vehicle/android that has a crew of humans from the future who have been miniaturized and are held in that state by a compression field. Also known as "Justice Department Vehicle Number 6018," the Teselecta is introduced in the eighth episode of the sixth season (season numbering restarted with the relaunch in 2005), which is called "Let's Kill Hitler." Amy (played by Karen Gillan, b. 1987) and Rory's (played by Arthur Darvill, b. 1982) friend "Mels" (played by Nina Toussaint-White, b. 1985) crashes a reunion between the Doctor and "the Ponds" (Amy and Rory) with the police close behind. Needing to make a getaway, Mels decides that since she has a gun and a time machine, she should do the cliche thing and go kill Hitler. In 1938 Berlin, the audience is introduced to the Teselecta. This ship crewed by shrunken humans is sent by the Justice Department from the future to travel the time line and punish the worst criminals in history who escaped justice in their own time. To keep from altering history, the Teselecta intercepts the criminal toward the end of the criminal's historical life span. It then changes its own appearance to exactly copy its quarry. Once this is completed, the criminal is miniaturized and taken on board to face the future's punishment (Steven Moffat, "Let's Kill Hitler," season 6, episode 8, August 27, 2011).

The Teselecta crew soon turns their attention from Hitler to the biggest war criminal in the universe: River Song (played by Alex Kingston, b. 1963), known throughout time as the person who killed the Doctor. Though the Doctor, the Ponds, and the audience have already met River, none is aware that the Ponds' childhood friend was, in fact, an earlier incarnation of their friend. "Mels" then regenerates into River Song. The Teselecta begins to pursue River to mete out punishment. Amy and Rory are miniaturized and teleported inside the Teselecta, where they are almost destroyed by the Teselecta's "antibodies," who promise them they "will experience a tingling sensation and then death." But they are saved by a crew member, who gives them the authorization to be aboard, canceling the orders of the antibodies. After the Teselecta's databanks are used for research, Amy uses the Doctor's sonic screwdriver to force the crew to shut down the ship to prevent them from killing River.

The next time the Teselecta is seen is in the season finale, "The Wedding of River Song." It is learned that River Song killing the Doctor is a fixed point in time and must always happen. When River learns of her role, she refuses to kill the Doctor, thereby fracturing all of time and space. Eventually, the Doctor is able to convince River to play her part by whispering something in her ear. The Doctor, having encountered the Teselecta again as he was making his final arrangements before his death, has asked the Teselecta crew to have their ship copy his appearance and take him aboard. In his whisper to River, he informs her of this plan to allow the fixed point of time to take place by her "killing" the robotic doppelganger rather than him. As the Doctor explains at the end of the episode, "The Doctor in a Doctor suit. Time said I had to [die], so I dressed for the occasion" (Steven Moffat, "The Wedding of River Song," season 6, episode 13, October 1, 2011).

It has become a time-travel trope that the first thing someone with the ability to travel in time should do is go back and kill Hitler; often it is said they should kill "baby Hitler." There are many reasons why it is a bad idea, but it can still be galling to know that some of history's biggest war criminals managed for various reasons to escape any kind of justice. The Teselecta addresses this issue. It shapeshifts to resemble a war criminal at the end of life so the criminal can be punished but the time line can remain intact. The paradox that the Teselecta is attempting to prevent while still handing out punishment is known as the "Grandfather Paradox": time travelers cannot go back in time and kill their grandfather before their parent is conceived because it would mean that the time travelers would never be born to go back in time in the first place (Uyeno, "What is the Grandfather Paradox?"). The principle of changing something in the past leading to drastic changes in the character's present is explored in many science fiction properties but perhaps most famously in the film *Back to the Future* (1985), when Marty McFly (played by Michael J. Fox, b. 1961) begins to fade away as his actions after traveling to the past threaten to prevent his parents from falling in love with each other.

The crew of the Teselecta walk a fine line, because their mission is to punish criminals who escaped judgment, but they cannot mete out justice before the crime has been committed and they cannot let everyone know of their existence, lest they damage the time line in new ways. The latter is similar to Starfleet's "Prime Directive" not to interfere with primitive cultures, on the show *Star Trek*. What the Grandfather Paradox tells us, and the crew of the Teselecta understand, is that what is done is done and there is nothing that can fix that. But like a lot of great science fiction writing, it shows anyone willing to pay attention how to learn from the past and hopefully avoid the same mistakes.

Keith R. Claridy

See also: Batcomputer, Buffybot, Cerebro/Cerebra, Cybermen, Daleks, Doctor/EMH, Fembots, The Great Intelligence, Iron Legion, K9, LMDs, Nardole, RoboCop, Sentinels, Starfleet Computer, TARDIS, Tin Woodsman; *Thematic Essays:* Heroic Robots and Their Impact on Sci-Fi Narratives, AI and the Apocalypse: Science Fiction Meeting Science Fact.

Further Reading

Burnett, Dean. 2014. "Time Travellers: Please Don't Kill Hitler." *The Guardian*, February 21, 2014. https://www.theguardian.com/science/brain-flapping/2014/feb/21/time-travellers-kill-adolf-hitler.

Byrne, Emma. 2013. "Innovation Isn't Safe: The Future According to Kevin Warwick." *Forbes*, September 30, 2013. https://www.forbes.com/sites/netapp/2013/09/30/kevin-warwick-captain-cyborg/#48b704c13560.

Dowden, Bradley. 1993. *Logical Reasoning*. Belmont, CA: Wadsworth Publishing Company.

Kistler, Alan. 2013. Doctor Who: *Celebrating Fifty Years, A History*. Guilford, CT: Lyons.

Koberlein, Brian. 2015. "Why It's Impossible To Go Back In Time and Kill Baby Hitler." *Forbes*, October 23, 2015. https://www.forbes.com/sites/briankoberlein/2015/10/23/why-its-impossible-to-go-back-in-time-and-kill-baby-hitler/#10894c3529c0.

Lewis, Courtland, and Smithka, Paula, eds. 2010. Doctor Who *and Philosophy: Bigger on the Inside*. Chicago: Open Court.

Lewis, Courtland, and Smithka, Paula, eds. 2015. *More* Doctor Who *and Philosophy: Regeneration Time*. Chicago: Open Court.

Mallett, Ronald L., and Henderson, Bruce. 2006. *Time Traveler: A Scientist's Personal Mission to Make Time Travel a Reality*. New York: Basic Books.

Shermer, Michael. 2002. "The Chronology Protection Conjecture." *Scientific American*, August 12, 2002. https://www.scientificamerican.com/article/the-chronology-pro tection.

Uyeno, Greg. 2019. "What is the Grandfather Paradox?" Space, June 5, 2019. https://www .space.com/grandfather-paradox.html.

Tin Woodsman

The Wizard of Oz

In 1900, populist American author L. Frank Baum (1856–1919) published the first novel in the Land of Oz series: *The Wonderful Wizard of Oz*. This book follows the adventures of a young girl named Dorothy, living in rural Kansas, who is swept away by a tornado to the mystical world of "Oz." Threatened by the Wicked Witch of the West, Dorothy and a group of misfit friends seek out the wisdom and advice of Oz's leader, the "Great and Powerful Oz" (a.k.a. the "Wizard of Oz"). The massive success of the Oz books led to a Broadway musical adaptation of the book in 1902 and the classic feature film *The Wizard of Oz* in 1939 (one of the first feature films in history to be filmed in full color). Dorothy's companions on this trip were the Scarecrow in search of a brain, the Cowardly Lion in search of courage, and the Tin Woodsman (or "Tin Man"), in search of a heart. In the film, the Tin Man was played by Jack Haley (1898–1979).

As the Woodsman is made of metal, possessing no known organic organs; is frequently drawn as running on gears; and requires oil to keep his joints functional, the Tin Woodsman is, in all respects, a "robot." When Dorothy and the Scarecrow meet the Woodsman, he is frozen like a statue. He moves his lips just enough to say the words "oil can." Dorothy understands, finds the Woodsman's oil can, and begins to lubricate his joints and jaw. He then laments to his two new friends that, being made of metal, he does not have a heart. In the musical film, the Tin Man sings:

> When a man's an empty kettle,
> He should be on his mettle,
> And yet I'm torn apart,
> Just because I'm presumin'
> That I could be a human,
> If I only had a heart.
>
> (Yip Harburg, "If I Only Had a Heart," *The Wizard of Oz*, 1939)

After assisting in the defeat of the Wicked Witch of the West, the Wizard of Oz awards the Woodsman a heart-shaped badge, pointing out that the care and devotion that the Tin Man has shown for his friends is proof that he possessed a "heart" all along.

A representation of the cast of the classic film *The Wizard of Oz* (1939). (Dan Oberly/ Dreamstime.com)

Being a devout populist, Baum filled his fantasy series with political symbolism. To him, the Tin Woodsman without a heart represented the Gilded Age industrialists (men consumed with industrial profits with no care or concern for their workers or for consumers). Within the context of the narrative itself, the Tin Woodsman is one of the earliest examples of a robot that "feels." Throughout his adventures in Oz, the Tin Man expresses fear, bravery, sympathy, and concern. In the twenty-first century, as engineers make major breakthroughs concerning robots that can express emotion, it is important to keep in mind the possibility that emotionally mature robots are a real possibility in the years, decades, and centuries to come—so do not forget to keep their joints oiled.

Richard A. Hall

See also: B.A.T.s, Battle Droids, Bender, Buffybot, C-3PO, Doombots, IG-88/IG-11, Iron Giant, Jocasta, Maschinenmensch/Maria, Medical Droids, Oz, Robby the Robot, Robot, Robots, Rosie, Sentinels, Terminators, Transformers, Ultron, Warlock; *Thematic Essay:* Heroic Robots and Their Impact on Sci-Fi Narratives.

Further Reading

Alkon, Paul K. 1987. *Origins of Futuristic Fiction*. Athens: University of Georgia Press.

Allen, Arthur. 2019. "There's a Reason We Don't Know Much about AI." Politico, September 16, 2019. https://www.politico.com/agenda/story/2019/09/16/artficial-intell igence-study-data-000956?cid=apn.

Amati, Viviana, Meggiolaro, Silvia, Rivellini, Giulia, and Zaccarin, Susanna. 2018. "Social Relations and Life Satisfaction: The Role of Friends." *Genus* 74, no. 1 (May 4). https://www.ncbi.nlm.nih.gov/pmc/articles/PMC5937874.

Ambrosino, Brandon. 2018. "What Would It Mean for AI to Have a Soul?" BBC, June 17, 2018. https://www.bbc.com/future/article/20180615-can-artificial-intelligence-have-a-soul-and-religion.

Ashby, LeRoy. 2006. *With Amusement for All: A History of American Popular Culture since 1830*. Lexington: University Press of Kentucky.

Bastani, Aaron. 2019. *Fully Automated Luxury Communism*. New York: Verso.

Baum, L. Frank. (1900) 2015. *The Wonderful Wizard of Oz*. Scotts Valley, CA: CreateSpace.

Carper, Steve. 2019. *Robots in American Popular Culture*. Jefferson, NC: McFarland.

Cavaler, Chris. 2015. *On the Origins of Superheroes: From the Big Bang to Action Comics No. 1*. Iowa City: University of Iowa Press.

Dinh, Thien-Nam. 2018. *Silicon Minds: The Science, Impact, and Promise of Artificial Intelligence*. Independently Published.

Ford, Martin. 2015. *Rise of the Robots: Technology and the Threat of a Jobless Future*. New York: Basic.

Lin, Patrick, Abney, Keith, and Bekey, George A., eds. 2014. *Robot Ethics: The Ethical and Social Implications of Robotics*. Cambridge: Massachusetts Institute of Technology.

Mayor, Adrienne. 2018. *Gods and Robots: Myths, Machines, and Ancient Dreams of Technology*. Princeton, NJ: Princeton University Press.

Randall, Terri, dir. 2016. *NOVA*. "Rise of the Robots: Inside the DARPA Robotics Challenge." Aired February 24, 2016, on PBS. DVD.

Roberts-Griffin, Christopher. 2011. "What Is a Good Friend: A Qualitative Analysis of Desired Friendship Qualities." *Penn McNair Research Journal* 3, no. 1 (December 21). https://repository.upenn.edu/cgi/viewcontent.cgi?article=1019&context=mcnair_scholars.

Rouhiainen, Lasse. 2018. *Artificial Intelligence: 101 Things You Must Know Today about Our Future*. Scotts Valley, CA: CreateSpace.

Thagard, Paul. 2017. "Will Robots Ever Have Emotions?" *Psychology Today*, December 14, 2017. https://www.psychologytoday.com/us/blog/hot-thought/201712/will-robots-ever-have-emotions.

Transformers

Transformers

Since the mid-1980s, the "Transformers" have consistently been one of the most popular toy brands on the U.S. market. They are the joint creation of the Japanese toy company Takara Tomy and the American toy company Hasbro. Like contemporary *G.I. Joe*, the Transformers were advertised and their narrative expanded by the animated series *The Transformers* (syndication, 1984–1987) and the Marvel comic of the same name. The general storyline involves a civil war between the heroic "Autobots" and the evil "Decepticons." As the story begins, millions of years ago on the planet Cybertron, the Autobots and Decepticons were in a bitter civil war that destroyed their planet's resources of "Energon," the fuel on which the Transformers survive. In search of more Energon, the Autobots crash-landed on planet Earth during the age of the dinosaurs, followed closely by the

A life-size replica of the Autobot Bumblebee from the live-action/CGI feature film *Transformers* (2007). (Stbernardstudio/Dreamstime.com)

Decepticons. Both groups fell into a state of hibernation until awakened in the late twentieth century. As each group of robots possessed the ability to "transform" into vehicle mode, their ships' systems scanned the planet for models on which to base their transformations (explaining why alien robots would transform into commonly recognized twentieth-century-Earth machines). Through many incarnations over the decades—including a series of blockbuster live-action/CGI feature films—the franchise has been able to remain fresh from one generation to the next.

The Autobots, for the most part, transform into various automobiles (but their "medic" transforms into a microscope). The Decepticons mostly transform into various forms of aircraft (but their intelligence officer, Soundwave, transforms into a tape recorder, with his various "spies" transforming into cassette tapes that can be "played" in his chest to make their reports). The Autobot leader is the heroic Optimus Prime, who transforms into a tractor-trailer; and the Decepticon leader is the evil Megatron, who transforms into a handheld gun (there has been much criticism of the physics of a fifty-foot-tall robot transforming into a small handgun). A new animated series launched in 2020 that goes back to examine the closing days of the Cybertonian civil war that led to the Transformers' departure from their home world.

Though the live-action/CGI feature films have garnered considerable commercial success, all have fallen short of the original 1986 animated film that worked as a transition for the television cartoon. In that film, Optimus Prime (voiced by voice acting legend Peter Cullen, b. 1941, who continued to voice the character in the live-action films and new animated series), falls in battle to the evil Megatron, who also receives a mortal wound during the fight. The mantle of leadership of the Autobots accidentally falls to the young Hot Rod (voiced by Judd Nelson, b. 1959), who changes his name to "Rodimus Prime." This decision was hugely controversial, as millions of Transformers fans around the world idolized the stalwart Optimus. The new Netflix series (2020–present) adds a much more "human" dimension to Optimus, making him a hero who refuses to accept defeat even in the face of the obvious probability that he will lose the lives of all of his people (including himself).

Though the characters of *The Transformers* are all robots, each character is utilized to explore very human problems and emotions. The live-action/CGI feature films have explored the very real danger that a civil war between such massive machines would pose for the human population of Earth. The marketing genius of the franchise is that it takes three popular toy concepts—toy vehicles, action figures, and robots—and merges them into one property. In the end, the Transformers are a traditional story of good versus evil, heroes versus villains; and, although a "robot," Optimus Prime remains one of the great heroic icons of American popular culture well into the twenty-first century.

Richard A. Hall

See also: Alita, B.A.T.s, Battle Droids, Bender, Buffybot, C-3PO, Doombots, IG-88/ IG-11, Iron Giant, Jocasta, KITT, Maschinenmensch/Maria, Mechagodzilla, Medical Droids, Oz, Robby the Robot, Robot, Robots, Rosie, Sentinels, Speed Buggy, Terminators, Tin Woodsman, Ultron, Voltron, Zordon/Alpha-5; *Thematic Essays:* Heroic Robots and Their Impact on Sci-Fi Narratives, Villainous Robots and Their Impact on Sci-Fi Narratives.

Further Reading

Albert, Robert S. and Brigante, Thomas R. 1962. "The Psychology of Friendship Relations: Social Factors." *Journal of Social Psychology* 56 (1). https://www.tandfon line.com/doi/abs/10.1080/00224545.1962.9919371.

Alkon, Paul K. 1987. *Origins of Futuristic Fiction.* Athens: University of Georgia Press.

Allen, Arthur. 2019. "There's a Reason We Don't Know Much about AI." Politico, September 16, 2019. https://www.politico.com/agenda/story/2019/09/16/artficial-intell igence-study-data-000956?cid=apn.

Amati, Viviana, Meggiolaro, Silvia, Rivellini, Giulia, and Zaccarin, Susanna. 2018. "Social Relations and Life Satisfaction: The Role of Friends." *Genus* 74, no. 1 (May 4). https://www.ncbi.nlm.nih.gov/pmc/articles/PMC5937874.

Ambrosino, Brandon. 2018. "What Would It Mean for AI to Have a Soul?" BBC, June 17, 2018. https://www.bbc.com/future/article/20180615-can-artificial-intelligence-hav e-a-soul-and-religion.

Ashby, LeRoy. 2006. *With Amusement for All: A History of American Popular Culture since 1830.* Lexington: University Press of Kentucky.

Byrne, Emma. 2013. "Innovation Isn't Safe: The Future According to Kevin Warwick." *Forbes,* September 30, 2013. https://www.forbes.com/sites/netapp/2013/09/30/kev in-warwick-captain-cyborg/#48b704c13560.

Carper, Steve. 2019. *Robots in American Popular Culture.* Jefferson, NC: McFarland.

Castleman, Harry, and Podrazik, Walter J. 2016. *Watching TV: Eight Decades of American Television, Third Edition.* Syracuse, NY: Syracuse University Press.

Chanthadavong, Aimee. 2019. "AI and Ethics: The Debate That Needs to Be Had." ZDnet, September 16, 2019. https://www.zdnet.com/article/ai-and-ethics-the-debate-that -needs-to-be-had/#ftag=CAD-03-10abf5f.

Dinh, Thien-Nam. 2018. *Silicon Minds: The Science, Impact, and Promise of Artificial Intelligence.* Independently Published.

Ford, Martin. 2015. *Rise of the Robots: Technology and the Threat of a Jobless Future.* New York: Basic.

Fryer-Biggs, Zachary. 2019. "Coming Soon to a Battlefield: Robots That Can Kill." *The Atlantic,* September 3, 2019. https://www.theatlantic.com/technology/archive/2019 /09/killer-robots-and-new-era-machine-driven-warfare/597130.

Howe, Sean. 2012. *Marvel Comics: The Untold Story*. New York: Harper-Perennial.

Krishnan, Armin. 2009. *Killer Robots: Legality and Ethicality of Autonomous Weapons*. London: Routledge.

Ladd, Fred, and Deneroff, Harvey. 2008. *Astro Boy and Anime Come to the Americas: An Insider's View of the Birth of a Pop Culture Phenomenon*. Jefferson, NC: McFarland.

Lin, Patrick, Abney, Keith, and Bekey, George A., eds. 2014. *Robot Ethics: The Ethical and Social Implications of Robotics*. Cambridge: Massachusetts Institute of Technology.

Lin, Patrick, Jenkins, Ryan, and Abney, Keith, eds. 2017. *Robot Ethics 2.0: From Autonomous Cars to Artificial Intelligence*. New York: Oxford University Press.

Linder, Courtney. 2020. "This AI Robot Just Nabbed the Lead Role in a Sci-Fi Movie: Meet Erica, the Movie Star Who May Put Human Actors Out of Work." *Popular Mechanics*, June 25, 2020. https://www.popularmechanics.com/technology/rob ots/a32968811/artificial-intelligence-robot-movie-star-erica.

Mayor, Adrienne. 2018. *Gods and Robots: Myths, Machines, and Ancient Dreams of Technology*. Princeton, NJ: Princeton University Press.

Murdock, Jason. 2019. "Former Google Engineer Warns AI Might Accidentally Start a War: 'These Things Will Start to Behave in Unexpected Ways.'" *Newsweek*, September 16, 2019. https://www.newsweek.com/google-project-maven-artificial-inte lligence-laura-nolan-killer-robots-department-defense-1459358.

Randall, Terri, dir. 2016. *NOVA*. "Rise of the Robots: Inside the DARPA Robotics Challenge." Aired February 24, 2016, on PBS. DVD.

Roberts-Griffin, Christopher. 2011. "What Is a Good Friend: A Qualitative Analysis of Desired Friendship Qualities." *Penn McNair Research Journal* 3, no. 1 (December 21). https://repository.upenn.edu/cgi/viewcontent.cgi?article=1019&context=mcn air_scholars.

Rouhiainen, Lasse. 2018. *Artificial Intelligence: 101 Things You Must Know Today about Our Future*. Scotts Valley, CA: CreateSpace.

Said, Carolyn. 2019. "Robot Cars Are Getting Better—But Will True Self-Driving Ever Arrive." *San Francisco Chronicle*, February 14, 2019. https://www.sfchronicle .com/business/article/Robot-cars-are-getting-better-but-will-true-13614280.php.

Seed, David. 1999. *American Science Fiction and the Cold War: Literature and Film*. Abingdon, UK: Routledge.

Shook, John R., and Stillwaggon Swan, Liz, eds. 2009. *Transformers and Philosophy: More Than Meets the Mind*. Chicago: Open Court.

Singer, Peter W. 2009. "Isaac Asimov's Laws of Robotics Are Wrong." Brookings Institute, May 18, 2009. https://www.brookings.edu/opinions/isaac-asimovs-laws-of -robotics-are-wrong.

Spaeth, Dennis. 2018. "From Single-Task Machines to Backflipping Robots: The Evolution of Robots." *Cutting Tool Engineering*, January 15, 2018. https://www.ctemag .com/news/articles/evolution-of-robots.

Starck, Kathleen, ed. 2010. *Between Fear and Freedom: Cultural Representations of the Cold War*. Newcastle, UK: Cambridge Scholars.

Thagard, Paul. 2017. "Will Robots Ever Have Emotions?" *Psychology Today*, December 14, 2017. https://www.psychologytoday.com/us/blog/hot-thought/201712/will-rob ots-ever-have-emotions.

Wallach, Wendell, and Allen, Colin. 2009. *Moral Machines: Teaching Robots Right from Wrong*. New York: Oxford University Press.

Ultron

Marvel Comics

In 1961, comic book legend Stan Lee (1922–2018) launched the "Marvel age" of comics. With the launching of the Fantastic Four—soon to be followed by the Hulk, Spider-Man, Iron Man, and many, many more—comic book superheroes took a more dramatic turn, with storylines focused on the private lives of those blessed (or burdened) with superpowers. One of Marvel's early successes was the superhero team *The Avengers* (1963). Over the decades, the Avengers—consisting of a revolving door of various Marvel heroes, but most notably Captain America, Iron Man, and Thor, the Norse god of thunder—have accumulated a wide array of villains in their rogues' gallery. One of the most dangerous has been the robot "Ultron." In the comics, Ultron was the creation of Avenger Hank Pym. When Ultron gained sentience, it turned on its master like a modern-day Frankenstein's monster (Roy Thomas and John Buscema, *The Avengers #55*, Marvel Comics, August 1968).

The most popular version of Ultron currently existing in the overall American zeitgeist comes from his appearance as the main villain in the 2015 film, *Avengers: Age of Ultron*. Though Ultron's creation in this version is significantly different from his comics origin, the primary Frankenstein/creation-turning-on-creator

A replica of the remains of an Ultron unit from the Marvel Comics universe and Marvel Cinematic Universe franchise. (Imaengine/Dreamstime.com)

CGI

Computer-generated imagery (CGI) is the engineering art that utilizes computer graphics primarily for entertainment purposes. It is most popularly associated with feature films and—to a lesser degree—television and online streaming programs. In 1993, Lucasfilm's Industrial Light and Magic successfully utilized CGI to enhance the dinosaur special effects for *Jurassic Park* by director Steven Spielberg (b. 1946). In 1995, Pixar Studios, formerly the computer effects division of ILM, produced the first full-length CGI feature film, *Toy Story*. In 1999, writer-director George Lucas (b. 1944) exhibited the wide-scale possibilities of CGI incorporation into live-action films with *Star Wars, Episode I: The Phantom Menace*, which also included the first fully interactive CGI character, Jar-Jar Binks. In the twenty-first century, CGI has become standard fare in major—and even minor—Hollywood productions, making the transition of superhero comic books to film a far more successful endeavor, as seen by the massive success of the Marvel Cinematic Universe (2008–2019) and the continued blockbuster productions by Pixar. Video games have also benefited tremendously from CGI advances, making video game graphics more realistic than ever before. CGI is also the primary science behind the burgeoning "virtual reality" phenomenon. In 2020, MOI Worldwide and Image Engine announced that they were utilizing CGI technology to assist in creating a new film that will star a CGI recreation of the late actor James Dean (1931–1955), set to be released in 2021 ("Movie Magic," *Popular Mechanics*, June 25, 2020, https://www.popularmechanics.com/technology/robots/a32968811/artificial-intelligence-robot-movie-star-erica). However, the concept of utilizing CGI to "recreate" deceased actors (or make "younger" versions of older living actors) contains certain ethical conundrums. Regardless of where that specific controversy leads, CGI overall is a permanent part of American popular culture.

Richard A. Hall

storyline remains intact. In the film, Ultron (voiced by James Spader, b. 1960) begins as an AI program designed by Tony Stark (played by Robert Downey Jr., b. 1965), intending to give him control of his "Iron Legion" army of autonomous Iron Man armor to create a "suit of armor around the world." As the Avengers party, Stark's personal AI, JARVIS (voiced by Paul Bettany, b. 1971), watches over the developing Ultron program. As Ultron is ultimately a "superior" AI, he quickly overcomes JARVIS, utilizes access to the internet to study humankind and—basing his logic on his primary programming to protect Earth—determines that humanity is the greatest threat to Earth and should, therefore, be eliminated. Ultron then seizes control of the Iron Legion, implanting his AI into each Legionnaire's robot body. He has a two-part plan: (1) create an updated body for his primary identity, and (2) create thrusters to raise a city into the atmosphere in order to drop it like a meteor and create a life-ending event for humankind, repopulating Earth with copies of himself. The Avengers stop the first part of the plan, stealing Ultron's intended "body." Stark and Bruce Banner (played by Mark Ruffalo, b. 1967) incorporate the remnants of JARVIS into the synthetic body, enhanced by the "Mind Stone" (one of the five "Infinity Stones") and, with Thor (played by Chris Hemsworth, b. 1983) using lightning to bestow life— à la Frankenstein—they create the new artificial life-form Vision (also played by Bettany). After successfully defeating Ultron's plans, Vision provides the killing blow to the final Ultron avatar (Joss Whedon, *Avengers: Age of Ultron*, Marvel/Disney, 2015).

Ultron represents the ultimate nightmare scenario in the realm of robotics and artificial intelligence: a sentient, powerful mechanical entity that views humanity as its own greatest threat and seeks to eliminate the threat by eliminating humanity. He is also, as stated, just one of numerous Frankenstein's monster reboots centered on cautionary tales regarding the advancement of AI in modern times. In the comics, Ultron is responsible not only for the creation of Vision but also the robot/AI Jocasta (a modern-day "bride of Frankenstein"). Being an AI program, Ultron can never truly be "defeated," as there always remains the probability of him downloading himself to some safe harbor and coming back more dangerous and determined than ever.

Richard A. Hall

See also: B.A.T.s, Battle Droids, Brainiac, Buffybot, Doombots, ED-209, Iron Legion, Iron Man, JARVIS/Friday, Jocasta, LMDs, Maschinenmensch/Maria, Metallo, Robots, Sentinels, Terminators, Tin Woodsman, Vision, Warlock; *Thematic Essays:* Robots and Slavery, Villainous Robots and Their Impact on Sci-Fi Narratives, AI and the Apocalypse: Science Fiction Meeting Science Fact.

Further Reading

Ambrosino, Brandon. 2018. "What Would It Mean for AI to Have a Soul?" BBC, June 17, 2018. https://www.bbc.com/future/article/20180615-can-artificial-intelligence-have-a-soul-and-religion.

Bastani, Aaron. 2019. *Fully Automated Luxury Communism*. New York: Verso.

Cavaler, Chris. 2015. *On the Origins of Superheroes: From the Big Bang to Action Comics No. 1*. Iowa City: University of Iowa Press.

Chanthadavong, Aimee. 2019. "AI and Ethics: The Debate That Needs to Be Had." ZDnet, September 16, 2019. https://www.zdnet.com/article/ai-and-ethics-the-debate-that-needs-to-be-had/#ftag=CAD-03-10abf5f.

Costello, Matthew J. 2009. *Secret Identity Crisis: Comic Books & the Unmasking of Cold War America*. New York: Continuum.

Dinh, Thien-Nam. 2018. *Silicon Minds: The Science, Impact, and Promise of Artificial Intelligence*. Independently Published.

Fryer-Biggs, Zachary. 2019. "Coming Soon to a Battlefield: Robots That Can Kill." *The Atlantic*, September 3, 2019. https://www.theatlantic.com/technology/archive/2019/09/killer-robots-and-new-era-machine-driven-warfare/597130.

Hitchcock, Susan Tyler. 2007. *Frankenstein: A Cultural History*. New York: W. W. Norton.

Howe, Sean. 2012. *Marvel Comics: The Untold Story*. New York: Harper-Perennial.

Lin, Patrick, Abney, Keith, and Bekey, George A., eds. 2014. *Robot Ethics: The Ethical and Social Implications of Robotics*. Cambridge: Massachusetts Institute of Technology.

Michaud, Nicolas, and Watkins, Jessica, eds. 2018. *Iron Man vs. Captain America and Philosophy: Give Me Liberty or Keep Me Safe*. Chicago: Open Court.

Murdock, Jason. 2019. "Former Google Engineer Warns AI Might Accidentally Start a War: 'These Things Will Start to Behave in Unexpected Ways.'" *Newsweek*, September 16, 2019. https://www.newsweek.com/google-project-maven-artificial-intelligence-laura-nolan-killer-robots-department-defense-1459358.

Pellissier, Hank. 2013. "Robots and Slavery: What Do Humans Want When We Are 'Masters'?" Institute for Ethics and Emerging Technologies, September 13, 2013. https://ieet.org/index.php/IEET2/more/pellissier20130913.

Randall, Terri, dir. 2016. *NOVA*. "Rise of the Robots: Inside the DARPA Robotics Challenge." Aired February 24, 2016, on PBS. DVD.

Robitzski, Dan. 2018. "Artificial Consciousness: How to Give a Robot a Soul." Futurism, June 25, 2018. https://futurism.com/artificial-consciousness.

Rouhiainen, Lasse. 2018. *Artificial Intelligence: 101 Things You Must Know Today about Our Future*. Scotts Valley, CA: CreateSpace.

Schroeder, Stan. 2020. "Samsung Just Launched an 'Artificial Human' Called Neon, and Wait, What?" Mashable, January 7, 2020. https://mashable.com/article/samsung-star-labs-neon-ces.

Seed, David. 1999. *American Science Fiction and the Cold War: Literature and Film*. Abingdon, UK: Routledge.

Spaeth, Dennis. 2018. "From Single-Task Machines to Backflipping Robots: The Evolution of Robots." *Cutting Tool Engineering*, January 15, 2018. https://www.ctemag.com/news/articles/evolution-of-robots.

Thomas, Roy. 2017. *The Marvel Age of Comics: 1961–1978*. Los Angeles: Taschen.

Wallach, Wendell, and Allen, Colin. 2009. *Moral Machines: Teaching Robots Right from Wrong*. New York: Oxford University Press.

Wright, Bradford W. 2003. *Comic Book Nation: The Transformation of Youth Culture in America*. Baltimore, MD: Johns Hopkins University Press.

V-GER

Star Trek

In 1966, creator-writer-producer Gene Roddenberry (1921–1991) launched his iconic science fiction television series *Star Trek* (NBC, 1966–1969). Set in the twenty-third century, the series centers on the adventures of the starship *USS Enterprise*, its commander, Captain James T. Kirk (played by William Shatner, b. 1931); and crew, headed by the alien first officer, Mr. Spock (played by Leonard Nimoy, 1931–2015), representing the "United Federation of Planets." Though short-lived in its initial run, the series has become legendary, not only for its groundbreaking special effects but also for the sophistication of its stories; many of which were grounded in real-world issues of the day, such as racism and a fear of the Cold War. Its success in syndicated reruns revived the franchise, leading to an animated series, thirteen feature films, and six live-action "sequel" series (at the time of this writing). The first feature film of the series was *Star Trek: The Motion Picture* (Paramount, 1979), in which Captain Kirk and the crew of the original *Enterprise* must save Earth from the threat of "V-GER."

Set in the late twenty-third century, Starfleet Headquarters on Earth receives a distress call from the Klingon Empire across the galaxy. Three Klingon battleships have been taken out by a mysterious "cloud" that is heading toward Federation space. Later, a Federation outpost, Epsilon Nine is similarly evaporated by the cloud. Admiral Kirk (Shatner) is placed in command of the newly refitted starship *Enterprise* with most of its original crew to intercept and deal with the threat. After being joined by the retired Mr. Spock (Nimoy), the ship encounters the cloud. While attempting communications, the cloud "scans" the ship, evaporating its new navigator, Lieutenant Ilia (played by Persis Khambata, 1948–1998). Once Kirk succeeds in expressing peaceful intent, the *Enterprise* is allowed to enter the cloud, and "Ilia" reemerges as a humanoid "probe." She identifies the entity at the heart of the cloud as "V-GER," and claims to have been sent to study the parasites "infecting" the *Enterprise*. At the center of the cloud, Kirk and Spock come into physical contact with V-GER, which is discovered to be the long-lost NASA probe *Voyager-6*. Programmed to absorb all the knowledge that it could, the probe was encountered by some alien machine race that considered only machines as true life. Repaired and sent back to find its "creator," V-GER believes it will find its creator on Earth, and that it, too, will be a machine. Kirk enters the command for V-GER to release its data; his first officer, Commander Will Decker (played by Stephen Collins, b. 1947) chooses to "merge" with the mechanical Ilia, and both dissolve with V-GER, presumably having evolved to a new life-form

SETI

The "Search for Extraterrestrial Intelligence" (SETI) is an unofficial, international effort by the scientific community to monitor space for signs of intelligent life from other planets. This is done by way of monitoring space for radio transmissions and attempting to determine whether any have been artificially created rather than naturally occurring. Research toward this end existed throughout the entire twentieth century after, in 1899, legendary engineer Nikola Tesla (1856–1943) believed that he had received radio transmissions from Mars (more than a century later, it has still not been determined exactly what Tesla received or from where). The first organized SETI experiment was conducted by Cornell University in 1960. With the "space race" well underway by that time, Cornell's research led the Soviet Union to begin its own SETI experiments. The closest that any SETI program has come to success was a 1977 Ohio State University program, where a volunteer noted via telescope what he believed to be an artificially created transmission. On the printout of the event, the volunteer wrote the word "Wow!," giving this incident the name the "Wow Signal!" To date, this event has not been repeated or explained. NASA began its official Microwave Observing Program (MOP) in 1982. Though cancelled a year later, the project was taken up again in 1995 by the SETI Institute of Mountain View, California, under the new name "Project Phoenix." In 2016, Russian billionaire philanthropist Yuri Milner (b. 1961) began funding "Breakthrough Listen," the most thorough and elaborate SETI program undertaken to date. Unless it yields positive results, that program is set to discontinue in 2026. Despite more than a century of research with nothing concrete to show for it, humanity's belief in and search for the possibility of extraterrestrial life continues.

Richard A. Hall

(Alan Dean Foster and Harold Livingston, *Star Trek: The Motion Picture*, Paramount, 1979).

Various noncanonical *Star Trek* stories over the years have suggested that either V-GER was responsible for creating the Borg or that it was the Borg that intercepted and reprogrammed V-GER. The story of V-GER, however, blurs the line between science fiction and science fact. Though, to date, there has been no *Voyager-6* launched by NASA, V-GER's design looks very much like a modern-day NASA satellite probe. Another Frankenstein's monster analogy is at play here, in that V-GER is a human creation that ultimately returns to potentially destroy its creator. As V-GER has been programmed to absorb "all that is knowable," its travels through the universe exposed it to information that far exceeded its original ability to collect data. Its transformation allowed the machine to evolve into its own life-form, desperately in search of its creator (very similar to humans' own search for God). The merging of the human Decker and the mechanical Ilia along with the "release" of V-GER's knowledge opens the door for all manner of future storytelling possibilities.

Richard A. Hall

See also: Androids, Borg, Brainiac, Control, Cybermen, The Great Intelligence, HAL 9000, Landru, Lieutenant Commander Data, Lore/B4, The Matrix/Agent Smith, Mechagodzilla, OASIS, Skynet, Soji and Dahj Asha, Starfleet Computer, TARDIS, Ultron, WOPR; *Thematic Essay:* AI and the Apocalypse: Science Fiction Meeting Science Fact.

Further Reading

Allen, Arthur. 2019. "There's a Reason We Don't Know Much about AI." *Politico*, September 16, 2019. https://www.politico.com/agenda/story/2019/09/16/artficial-intelligence-study-data-000956?cid=apn.

Ashby, LeRoy. 2006. *With Amusement for All: A History of American Popular Culture since 1830*. Lexington: University Press of Kentucky.

Bastani, Aaron. 2019. *Fully Automated Luxury Communism*. New York: Verso.

Byrne, Emma. 2013. "Innovation Isn't Safe: The Future According to Kevin Warwick." *Forbes*, September 30, 2013. https://www.forbes.com/sites/netapp/2013/09/30/kevin-warwick-captain-cyborg/#48b704c13560.

Carper, Steve. 2019. *Robots in American Popular Culture*. Jefferson, NC: McFarland.

Castleman, Harry, and Podrazik, Walter J. 2016. *Watching TV: Eight Decades of American Television, Third Edition*. Syracuse, NY: Syracuse University Press.

Chandler, Simon. 2019. "Artificial Intelligence Has Become a Tool for Classifying and Ranking People." *Forbes*, October 1, 2019. https://www.forbes.com/sites/simonchandler/2019/10/01/artificial-intelligence-has-become-a-tool-for-classifying-and-ranking-people/#7431657f1d7c.

Chanthadavong, Aimee. 2019. "AI and Ethics: The Debate That Needs to Be Had." ZDnet, September 16, 2019. https://www.zdnet.com/article/ai-and-ethics-the-debate-that-needs-to-be-had/#ftag=CAD-03-10abf5f.

Decker, Kevin S., and Eberl, Jason T., eds. 2016. *The Ultimate* Star Trek *and Philosophy: The Search for Socrates*. Hoboken, NJ: Wiley-Blackwell.

Dinh, Thien-Nam. 2018. *Silicon Minds: The Science, Impact, and Promise of Artificial Intelligence*. Independently Published.

Gross, Edward, and Altman, Mark A. 2016. *The Fifty-Year Mission, The First 25 Years: The Complete, Uncensored, Unauthorized Oral History of* Star Trek. New York: St. Martin's.

Hitchcock, Susan Tyler. 2007. *Frankenstein: A Cultural History*. New York: W. W. Norton.

Lin, Patrick, Jenkins, Ryan, and Abney, Keith, eds. 2017. *Robot Ethics 2.0: From Autonomous Cars to Artificial Intelligence*. New York: Oxford University Press.

Murdock, Jason. 2019. "Former Google Engineer Warns AI Might Accidentally Start a War: 'These Things Will Start to Behave in Unexpected Ways.'" *Newsweek*, September 16, 2019. https://www.newsweek.com/google-project-maven-artificial-intelligence-laura-nolan-killer-robots-department-defense-1459358.

Paur, Joey. 2019. "An AI Bot Writes a Hilarious Episode of *Star Trek: The Next Generation*." Geek Tyrant, November 16, 2019. https://geektyrant.com/news/an-ai-bot-writes-a-hilarious-episode-of-star-trek-the-next-generation.

Reagin, Nancy R., and Liedl, Janice, eds. 2013. *Star Trek and History*. Hoboken, NJ: John Wiley & Sons.

Roddenberry, Gene. 1979. *Star Trek: The Motion Picture*. Novelization. New York: Pocket.

Rouhiainen, Lasse. 2018. *Artificial Intelligence: 101 Things You Must Know Today about Our Future*. Scotts Valley, CA: CreateSpace.

Spaeth, Dennis. 2018. "From Single-Task Machines to Backflipping Robots: The Evolution of Robots." *Cutting Tool Engineering*, January 15, 2018. https://www.ctemag.com/news/articles/evolution-of-robots.

Starck, Kathleen, ed. 2010. *Between Fear and Freedom: Cultural Representations of the Cold War*. Newcastle, UK: Cambridge Scholars.

Stark, Steven D. 1997. *Glued to the Set: The 60 Television Shows and Events that Made Us Who We Are Today.* New York: Delta Trade Paperbacks.

Wallach, Wendell, and Allen, Colin. 2009. *Moral Machines: Teaching Robots Right from Wrong.* New York: Oxford University Press.

VICI

Small Wonder

In the 1980s, science fiction and family sitcoms were two very popular television genres. The two were combined in the surprise hit *Small Wonder* (syndication, 1985–1989). The series centered on the suburban American Lawson family. The father, Ted (played by Dick Christie, b. 1948), is a robotics engineer who builds a small robot resembling a human child, VICI (Voice Input Child Identicant, played by Tiffany Brissette, b. 1974). He brings VICI (pronounced "Vickie") home to his family, consisting of his wife and young son, in order to help the robot develop proper interactivity with humans. To explain this new "child's" presence, the Lawsons pass VICI off as a recently orphaned relative and eventually legally "adopt" her. As a robot, VICI possesses enhanced speed and strength and is installed with an electrical outlet, a data port, and an interactive panel for direct input and/or diagnostics. The series primarily works as a traditional "fish-out-of-water" premise, with the robot interacting with and learning from her human "family."

In essence, VICI is both the Frankenstein story with a happy ending and the ultimate in parental wish fulfillment: an adorable child that will only "grow" as it is allowed to and is always completely obedient. In the sitcom, she is presented as part of the family, but, like Rosie the maid robot of *The Jetsons*, VICI technically "belongs" to her creator; robots possessed nothing resembling "rights" nor were seen as "people" in the 1980s. Though mostly forgotten in the twenty-first century, VICI remains within the long list of adorable humorous robots in American popular culture. As advancements continue to progress in the twenty-first century in the area of making realistic, "lifelike" robots, it is logical to conclude that child robots are an inevitability: they are a logical response to individuals and couples who wish to have children but cannot.

Richard A. Hall

See also: Androids, AWESOM-O 4000, Baymax, Bernard, Bishop, Buffybot, Cylons, Dolores, Fembots, Human Torch, Lieutenant Commander Data, LMDs, Lore/B4, Maschinenmensch/Maria, Replicants, Robot, Robots, Rosie, Soji and Dahj Asha, Stepford Wives, Terminators, Vision, WALL-E; *Thematic Essays:* Robots and Slavery, Heroic Robots and Their Impact on Sci-Fi Narratives.

Further Reading

Albert, Robert S., and Brigante, Thomas R. 1962. "The Psychology of Friendship Relations: Social Factors." *Journal of Social Psychology* 56 (1). https://www.tandfonline.com/doi/abs/10.1080/00224545.1962.9919371.

Amati, Viviana, Meggiolaro, Silvia, Rivellini, Giulia, and Zaccarin, Susanna. 2018. "Social Relations and Life Satisfaction: The Role of Friends." *Genus* 74, no. 1 (May 4). https://www.ncbi.nlm.nih.gov/pmc/articles/PMC5937874.

Ambrosino, Brandon. 2018. "What Would It Mean for AI to Have a Soul?" BBC, June 17, 2018. https://www.bbc.com/future/article/20180615-can-artificial-intelligence-have-a-soul-and-religion.

Ashby, LeRoy. 2006. *With Amusement for All: A History of American Popular Culture since 1830*. Lexington: University Press of Kentucky.

Carper, Steve. 2019. *Robots in American Popular Culture*. Jefferson, NC: McFarland.

Castleman, Harry, and Podrazik, Walter J. 2016. *Watching TV: Eight Decades of American Television, Third Edition*. Syracuse, NY: Syracuse University Press.

Chanthadavong, Aimee. 2019. "AI and Ethics: The Debate That Needs to Be Had." ZDnet, September 16, 2019. https://www.zdnet.com/article/ai-and-ethics-the-debate-that-needs-to-be-had/#ftag=CAD-03-10abf5f.

Dinh, Thien-Nam. 2018. *Silicon Minds: The Science, Impact, and Promise of Artificial Intelligence*. Independently Published.

Lin, Patrick, Abney, Keith, and Bekey, George A., eds. 2014. *Robot Ethics: The Ethical and Social Implications of Robotics*. Cambridge: Massachusetts Institute of Technology.

Randall, Terri, dir. 2016. *NOVA*. "Rise of the Robots: Inside the DARPA Robotics Challenge." Aired February 24, 2016, on PBS. DVD.

Roberts-Griffin, Christopher. 2011. "What Is a Good Friend: A Qualitative Analysis of Desired Friendship Qualities." *Penn McNair Research Journal* 3, no. 1 (December 21). https://repository.upenn.edu/cgi/viewcontent.cgi?article=1019&context=mcnair_scholars.

Robitzski, Dan. 2018. "Artificial Consciousness: How to Give a Robot a Soul." Futurism, June 25, 2018. https://futurism.com/artificial-consciousness.

Rouhiainen, Lasse. 2018. *Artificial Intelligence: 101 Things You Must Know Today about Our Future*. Scotts Valley, CA: CreateSpace.

Schroeder, Stan. 2020. "Samsung Just Launched an 'Artificial Human' Called Neon, and Wait, What?" Mashable, January 7, 2020. https://mashable.com/article/samsung-star-labs-neon-ces.

Seed, David. 1999. *American Science Fiction and the Cold War: Literature and Film*. Abingdon, UK: Routledge.

Spaeth, Dennis. 2018. "From Single-Task Machines to Backflipping Robots: The Evolution of Robots." *Cutting Tool Engineering*, January 15, 2018. https://www.ctemag.com/news/articles/evolution-of-robots.

Starck, Kathleen, ed. 2010. *Between Fear and Freedom: Cultural Representations of the Cold War*. Newcastle, UK: Cambridge Scholars.

Wallach, Wendell, and Allen, Colin. 2009. *Moral Machines: Teaching Robots Right from Wrong*. New York: Oxford University Press.

Vision

Marvel Comics

The comic book character known as the Vision made his first appearance in *The Avengers* #57 (Marvel Comics, October 1968). The character was the creation of writer Roy Thomas (b. 1940) and artist John Buscema (1927–2002) under the editorial guidance of the legendary Stan Lee (1922–2018). This android hero has a wide array of powers, including flight, intangibility, enhanced strength, and energy projection courtesy of the solar gem embedded in his forehead. In the story, he was created by the robot supervillain Ultron, an enemy of the superhero team, the Avengers. Once the Vision turned on his creator, Ultron became his greatest foe. The Vision became a key member of the Avengers and as such joined the ranks of the greatest superheroes of the Marvel Universe. While he did not achieve the popularity some of his teammates did, he was a trusted staple of the team for many years.

Crafted by Ultron, the Vision was built as a synthetic being, or "synthezoid," to defeat the Avengers, most notably Ultron's creator Hank Pym/Ant-Man/Giant Man/Yellowjacket and his wife Janey Van Dyne/Wasp (Thomas and Buscema, *Avengers* #57, October 1968). Dubbed the "Vision" by Wasp, this strange being is dispatched to battle "Earth's mightiest heroes." In a move unexpected by his creator, the Vision turned on Ultron and decided to side with the heroes he was meant to destroy. Thanks to the advocacy of Captain America, he is soon made an official Avenger. The emotional nature of the moment famously moves the android to actually shed tears (Thomas and Buscema, *The Avengers* #58, November 1968).

As an Avenger, the Vision formed a close bond with fellow hero Wanda Maximoff/Scarlet Witch. As they grew closer, their relationship drew curiosity and sometimes scorn from their teammates. An ongoing plot thread through the book was the pondering over whether an android can become something human enough to feel love. Scarlet Witch's brother, Pietro Maximoff/Quicksilver, in particular, vehemently opposed the relationship. Largely under the guidance of writer Steve Englehart during the 1970s, the two heroes eventually formed a romantic relationship and would be married in *Giant-Size Avengers* #4 (Marvel Comics, June 1975). The couple would form a family unit that would be solidified when they had twin children.

In the story arc "Vision Quest" (John Byrne, *West Coast Avengers* #42–50, March–November 1989) the Vision was destroyed while serving as a member of the West Coast Avengers. In an attempt to restore their friend and ally, the team turned to Professor Phineas Horton, creator of the original Human Torch. The version of the Vision that reemerged was far from the soulful character he once was. He was now an emotional blank slate, despite Hank Pym doing his best to restore him to what he once was. To match this new status, he was redesigned to be bleached of all coloring, which he believed gave him a spectral appearance. This new version of the Vision struggled to maintain his relationship with Scarlett Witch, but it eventually collapsed. Though he lost the ability to connect with the rest of his team on a personal level, his superhero career continued. He tangled with the likes of his doppelganger, the Anti-Vision, as well as the return of Ultron. Over time, the Vision would regain his emotional intelligence. The brain patterns of his colleague Wonder Man would be replicated in the synthezoid's brain pattern. He also owed a great deal of his return to normalcy to Tony Stark/Iron Man, who rebuilt him after the Vision fought a harrowing battle with the villainous Grim Reaper.

In 2015 the Vision made his big-screen debut in the blockbuster film *Avengers: Age of Ultron* (Disney/Marvel Studios, 2015), played by Paul Bettany (b. 1971), raising the character's notoriety in the mainstream to a level he had not experienced before. The film saw his origin altered to the Vision being a synthezoid with the AI of Tony Stark's virtual assistant JARVIS uploaded into his mind and empowered by the mystical Mind Stone. He proved invaluable in the Avengers' battle against Ultron, ultimately being the one to stop him. When the superhero community was split in *Captain America: Civil War* (Disney/Marvel Studios, 2016) he sided with Iron Man's plan to register all superheroes with the United Nations, despite it conflicting with his growing relationship with the Scarlett Witch (played by Elizabeth Olsen, b. 1989). When the powerful villain Thanos began his hunt for the Infinity Stones, the Vision was obviously in danger, given the stone he possessed. Despite the best efforts of Scarlett Witch and the scientists of Wakanda he fell to Thanos (*Avengers: Infinity War*, Disney/Marvel Studios, 2018). However, the Vision is set to return in an upcoming series, *WandaVision*, for the Disney+ streaming service.

Building on this newfound popularity, Tom King, Gabriel Hernandez Walta, and Michael Walsh created the award–winning comic book series *The Vision*. Appointed to serve as the Avengers' liaison to the White House, the Vision wished to blend in with his new surroundings. He accomplished this by moving to the Washington, DC, suburbs and creating a synthezoid version of a nuclear family. He gained a wife named Virginia along with twin children, Viv and Vin, and later a dog named Sparky. Despite efforts to be neighborly, the Vision family had to endure antirobot prejudice. Virginia in particular struggled with her newly programmed human emotions, which led to increased trouble for the family. When she brutally murdered the Grim Reaper, it set off a chain of events that devastated the Vision family and the community at large, ultimately forcing the Avengers to intervene. In the end, the Vision, Sparky, and Viv—the latter two having joined the teen superhero team the Champions—were the only survivors of his attempt to live a life of domesticity.

Throughout his fifty-plus-year history, the nature of the Vision has posed the question, What constitutes a human? Though he was created as an artificial synthetic being, this superhero has developed a tremendous sense of humanity. Time and again he has demonstrated very relatable human motivations and goals. Showing that he does have something resembling a moral conscience and free will, the Vision has been a long-serving member of Marvel's pantheon of superheroes, striving to do what is right despite it conflicting with his original programming.

Josh Plock

See also: Androids, B.A.T.s, Bernard, Bishop, Buffybot, Cylons, Dolores, Doombots, Fembots, Human Torch, Iron Giant, Iron Legion, JARVIS/Friday, Jocasta, Lieutenant Commander Data, LMDs, Lore/B4, Marvin the Paranoid Android, Maschinenmensch/ Maria, Metallo, Replicants, Robots, Sentinels, Soji and Dahj Asha, Stepford Wives, Tin Woodsman, Ultron, VICI, Warlock; *Thematic Essay:* Heroic Robots and Their Impact on Sci-Fi Narratives.

Further Reading

Allen, Arthur. 2019. "There's a Reason We Don't Know Much about AI." Politico, September 16, 2019. https://www.politico.com/agenda/story/2019/09/16/artficial-intell igence-study-data-000956?cid=apn.

Ambrosino, Brandon. 2018. "What Would It Mean for AI to Have a Soul?" BBC, June 17, 2018. https://www.bbc.com/future/article/20180615-can-artificial-intelligence-have -a-soul-and-religion.

Cantwell, David. 2018. "The Wisdom of *The Vision*, a Superhero Story about Family and Fitting In." *New Yorker*, March 22, 2018. https://www.newyorker.com/books/page -turner/the-wisdom-of-the-vision-a-superhero-story-about-family-and-fitting-in.

Carper, Steve. 2019. *Robots in American Popular Culture*. Jefferson, NC: McFarland.

Chandler, Simon. 2019. "Artificial Intelligence Has Become a Tool for Classifying and Ranking People." *Forbes*, October 1, 2019. https://www.forbes.com/sites/simon chandler/2019/10/01/artificial-intelligence-has-become-a-tool-for-classifying-and -ranking-people/#7431657f1d7c.

Chanthadavong, Aimee. 2019. "AI and Ethics: The Debate That Needs to Be Had." ZDnet, September 16, 2019. https://www.zdnet.com/article/ai-and-ethics-the-debate-that -needs-to-be-had/#ftag=CAD-03-10abf5f.

Costello, Matthew J. 2009. *Secret Identity Crisis: Comic Books & the Unmasking of Cold War America*. New York: Continuum.

Dinh, Thien-Nam. 2018. *Silicon Minds: The Science, Impact, and Promise of Artificial Intelligence*. Independently Published.

Hitchcock, Susan Tyler. 2007. *Frankenstein: A Cultural History*. New York: W. W. Norton.

Howe, Sean. 2013. *Marvel Comics the Untold Story*. New York: Harper Perennial.

Lin, Patrick, Abney, Keith, and Bekey, George A., eds. 2014. *Robot Ethics: The Ethical and Social Implications of Robotics*. Cambridge: Massachusetts Institute of Technology.

Lin, Patrick, Jenkins, Ryan, and Abney, Keith, eds. 2017. *Robot Ethics 2.0: From Autonomous Cars to Artificial Intelligence*. New York: Oxford University Press.

Robitzski, Dan. 2018. "Artificial Consciousness: How to Give a Robot a Soul." Futurism, June 25, 2018. https://futurism.com/artificial-consciousness.

Rouhiainen, Lasse. 2018. *Artificial Intelligence: 101 Things You Must Know Today about Our Future*. Scotts Valley, CA: CreateSpace.

Schroeder, Stan. 2020. "Samsung Just Launched an 'Artificial Human' Called Neon, and Wait, What?" Mashable, January 7, 2020. https://mashable.com/article/samsung -star-labs-neon-ces.

Thagard, Paul. 2017. "Will Robots Ever Have Emotions?" *Psychology Today*, December 14, 2017. https://www.psychologytoday.com/us/blog/hot-thought/201712/will-rob ots-ever-have-emotions.

Thomas, Roy. 2017. *The Marvel Age of Comics: 1961–1978*. Los Angeles: Taschen.

Wallach, Wendell, and Allen, Colin. 2009. *Moral Machines: Teaching Robots Right from Wrong*. New York: Oxford University Press.

Wright, Bradford W. 2003. *Comic Book Nation: The Transformation of Youth Culture in America*. Baltimore, MD: Johns Hopkins University Press.

Voltron

Japanese Anime

In the 1980s, Japanese anime became very popular in the United States, and it remains so well into the twenty-first century. Since World War II, several Japanese franchises have made their way to American shores: *Godzilla, Battle of the*

A replica of the Japanese anime robot Voltron constructed with Lego bricks. (Luca Lorenzelli/Dreamstime.com)

Planets (i.e., "G-Force"), *Transformers*, and *Power Rangers*, to name but a few of the most popular. *Voltron* comes from anime specifically. Originally titled *Beast King GoLion* in Japan, World Events Productions purchased the rights for an American version, placed English dubbing over the original anime, and released the series in America as *Voltron* (syndication, 1984–1985). Like most Japanese-American franchises, the television series corresponded with a line of popular toys. The premise of the series is that Voltron, "Defender of the Universe," is a giant robot, created by the merging of five "lion" robots, each a different color—blue, yellow, red, green, and black—to defend the galaxy against evil. After a long period of peace, Voltron is called forth once more to defend the planet Arus from the evil robots under the command of King Zarkon and his witch assistant, Haggar. The second season of the American series was based on the Japanese anime series *Armored Fleet Dairugger XV* (TV Tokyo, 1982–1983), featuring three teams of pilots whose respective ships merged into one of three giant defender robots. It is the first season, however, that most think of when they hear the name "Voltron."

Though a giant robot made from smaller robots, Voltron is much more similar to the later Japanese import, *Mighty Morphin' Power Rangers*, than it is to something like *The Transformers*. The lion robots that make up Voltron are not autonomous. They are "spaceships," each a formidable force on its own but, merged into Voltron, a near-unstoppable defender. Once merged, the five pilots collectively control Voltron's actions, making it more a superpowered weapon than what one would normally consider a "robot." Along with the Transformers of the same period, Voltron was, perhaps, the most popular robot toy in the United States since Robby the Robot in the 1950s.

Richard A. Hall

See also: Alita, B.A.T.s, Iron Giant, Mechagodzilla, Robby the Robot, Sentinels, Transformers, Zordon/Alpha-5; *Thematic Essay:* Heroic Robots and Their Impact on Sci-Fi Narratives.

Further Reading

Ashby, LeRoy. 2006. *With Amusement for All: A History of American Popular Culture since 1830*. Lexington: University Press of Kentucky.

Carper, Steve. 2019. *Robots in American Popular Culture*. Jefferson, NC: McFarland.

Castleman, Harry, and Podrazik, Walter J. 2016. *Watching TV: Eight Decades of American Television, Third Edition*. Syracuse, NY: Syracuse University Press.

Ladd, Fred, and Deneroff, Harvey. 2008. *Astro Boy and Anime Come to the Americas: An Insider's View of the Birth of a Pop Culture Phenomenon*. Jefferson, NC: McFarland.

Lin, Patrick, Abney, Keith, and Bekey, George A., eds. 2014. *Robot Ethics: The Ethical and Social Implications of Robotics*. Cambridge: Massachusetts Institute of Technology.

Mayor, Adrienne. 2018. *Gods and Robots: Myths, Machines, and Ancient Dreams of Technology*. Princeton, NJ: Princeton University Press.

Murdock, Jason. 2019. "Former Google Engineer Warns AI Might Accidentally Start a War: 'These Things Will Start to Behave in Unexpected Ways.'" *Newsweek*, September 16, 2019. https://www.newsweek.com/google-project-maven-artificial-intelligence-laura-nolan-killer-robots-department-defense-1459358.

Randall, Terri, dir. 2016. *NOVA*. "Rise of the Robots: Inside the DARPA Robotics Challenge." Aired February 24, 2016, on PBS. DVD.

Smith, Brian. 2014. *Voltron: From Days of Long Ago: A Thirtieth Anniversary Celebration*. New York: Perfect Square.

WALL-E

Pixar

When the earth became so polluted that it could no longer support human life, humanity took to the stars on a luxury cruise liner, leaving the problems of the planet behind. The responsibility of restoring the world fell to an army of trash-collecting robots. These robots largely fell into disrepair until only the small but dutiful WALL-E (Waste Allocation Load Lifter-Earth class) remained. He is main character for the film *WALL-E* (Disney/Pixar 2008) directed by Andrew Stanton (b. 1965). In developing the story, Stanton was aided by the screenwriter Peter Docter (b. 1968) and later Jim Reardon (b. 1965), with whom he wrote the screenplay. The robot they created did not utilize traditional speech, so veteran sound designer Ben Burtt (b. 1948) provided the "voice" of the character through a series of mechanical sound effects. The film was released June 27, 2008, to tremendous critical praise and commercial success. It would ultimately go on to win the prestigious Academy Award for Best Animated Feature.

WALL-E is content to follow his programming, collecting trash and turning it into compact and easily disposable cubes. However, there seems to be a curious nature within him that those who built him may not have intended. The robot has developed a fascination with the relics of human society that he finds, especially the musical *Hello Dolly!* (produced by 20th Century Fox, 1969), and has formed a friendship with a cricket. What WALL-E knows of the world is forever changed when the sleek and advanced EVE (Extraterrestrial Vegetation Evaluator) lands on the planet on a fact-finding mission. He is infatuated with the newcomer and though initially hesitant, he makes attempts to impress her with his discoveries and knowledge. Though EVE is initially oblivious to him, she will later warm to the little trash collector. Once WALL-E leads EVE to plant life (albeit in potted form), she goes into a hibernation state until it is collected. Surprisingly WALL-E finds a way to stow away with her, back to where she came from.

The robot finds himself aboard the star-bound cruise ship, the *Axiom*. It is here WALL-E witnesses what has become of the human race in their years away from Earth. They have become slothful and completely dependent on technology for their survival. Recliners move them throughout the ship as they engage in endless entertainment, oblivious to their surroundings. This is what WALL-E must navigate in order to reunite with EVE, who has been taken away upon coming aboard. Along the way he meets the ship's captain (voiced by Jeff Garlin, b.1962) who initiates plans to return to Earth when he believes EVE has found Earth-based flora, despite being largely ignorant of what to do.

With no direction or knowledge, the out-of-place WALL-E accidentally creates chaos on the *Axiom* and even sets free a dysfunctional group of his robot brethren. Despite this, EVE begins to warm up to him. The two even share a memorable scene together as they float in space around the ship, propelled by a fire extinguisher. In the midst of their courtship, it is discovered that the megacorporation that originally sent the human race into space never intended for the humans to go home. Their belief was that Earth would never be inhabitable again, and they gave a secret order for the *Axion* to remain in space indefinitely. The robotic staff onboard—chief among them the autopilot system AUTO—has been charged with keeping this conspiracy a secret.

WALL-E became an unlikely ally to EVE and the captain as they rebelled against the ship's programming. Two humans whom WALL-E befriended, Mary and John (voiced by Kathy Najimy, b. 1957, and John Ratzenberger, b. 1947), galvanized the others on the ship to help in this cause. Though badly damaged by AUTO and his compatriots, WALL-E was able to preserve EVE's plant. Inspired by the stalwart robot, everyone played a part in activating the ship's Holo-Detector, which automatically sends the *Axiom* back to Earth. Once they land, EVE anxiously attempts to repair her companion. For a moment she accidentally restores WALL-E to his default emotionless state. However, after a sign of affection, WALL-E returns to his normal self. Together they will play a part in humanity's recolonization of earth.

In the world of *WALL-E*, humans have become so reliant on machines and technology that they have lost a big piece of what makes them who they are. Having abandoned their home to mechanical caretakers, they indulge in a life absorbed in nonstop entertainment to the point of docility. Ironically it was a robot with enough sense of humanity who could remind people of what they were capable. Isolated on planet Earth to do a job people were unwilling to do, WALL-E uses the interesting objects he finds to deduce what human traits had been. Though all of the robots in *WALL-E* seem to have some form of a personality programmed in, none displays it to the level he does. What could be a fault in how he is wired has made the character able to adapt in a manner that other robots in the film cannot. Ultimately this proves to be his strength, as he plays a major role in saving humanity.

Josh Plock

See also: Baymax, BB-8, Cambot/Gypsy/Tom Servo/Crow, D-O, HERBIE, Johnny 5/S.A.I.N.T. Number 5, K9, Muffit, R2-D2, Robot, Rosie, VICI; *Thematic Essay:* Robots and Slavery.

Further Reading

Albert, Robert S. and Brigante, Thomas R. 1962. "The Psychology of Friendship Relations: Social Factors." *Journal of Social Psychology* 56 (1). https://www.tandfon line.com/doi/abs/10.1080/00224545.1962.9919371.

Allen, Arthur. 2019. "There's a Reason We Don't Know Much about AI." Politico, September 16, 2019. https://www.politico.com/agenda/story/2019/09/16/artficial-intel ligence-study-data-000956?cid=apn.

Amati, Viviana, Meggiolaro, Silvia, Rivellini, Giulia, and Zaccarin, Susanna. 2018. "Social Relations and Life Satisfaction: The Role of Friends." *Genus* 74, no. 1 (May 4). https://www.ncbi.nlm.nih.gov/pmc/articles/PMC5937874.

Ambrosino, Brandon. 2018. "What Would It Mean for AI to Have a Soul?" BBC, June 17, 2018. https://www.bbc.com/future/article/20180615-can-artificial-intelligence-have-a-soul-and-religion.

Bastani, Aaron. 2019. *Fully Automated Luxury Communism*. New York: Verso.

Carper, Steve. 2019. *Robots in American Popular Culture*. Jefferson, NC: McFarland.

Ford, Martin. 2015. *Rise of the Robots: Technology and the Threat of a Jobless Future*. New York: Basic.

Hampton, Gregory Jerome. 2017. *Imagining Slaves and Robots in Literature, Film, and Popular Culture: Reinventing Yesterday's Slave with Tomorrow's Robot*. Lanham, MD: Lexington.

Hauser, Tim. 2008. *The Art of WALL-E*. San Francisco: Chronicle.

Lin, Patrick, Abney, Keith, and Bekey, George A., eds. 2014. *Robot Ethics: The Ethical and Social Implications of Robotics*. Cambridge: Massachusetts Institute of Technology.

Lin, Patrick, Jenkins, Ryan, and Abney, Keith, eds. 2017. *Robot Ethics 2.0: From Autonomous Cars to Artificial Intelligence*. New York: Oxford University Press.

Pellissier, Hank. 2013. "Robots and Slavery: What Do Humans Want When We Are 'Masters'?" Institute for Ethics and Emerging Technologies, September 13, 2013. https://ieet.org/index.php/IEET2/more/pellissier20130913.

Price, David A. 2009. *The Pixar Touch: The Making of a Company*. New York: First Vintage.

Roberts-Griffin, Christopher. 2011. "What Is a Good Friend: A Qualitative Analysis of Desired Friendship Qualities." *Penn McNair Research Journal* 3, no. 1 (December 21). https://repository.upenn.edu/cgi/viewcontent.cgi?article=1019&context=mcnair_scholars.

Robitzski, Dan. 2018. "Artificial Consciousness: How to Give a Robot a Soul." Futurism, June 25, 2018. https://futurism.com/artificial-consciousness.

Thagard, Paul. 2017. "Will Robots Ever Have Emotions?" *Psychology Today*, December 14, 2017. https://www.psychologytoday.com/us/blog/hot-thought/201712/will-robots-ever-have-emotions.

Warlock

Marvel Comics

In 1961, comic book legend Stan Lee (1922–2018) launched the "Marvel age" of comics. With the launching of the Fantastic Four—soon to be followed by the Hulk, Spider-Man, Iron Man, and many, many more—comic book superheroes took a more dramatic turn, with storylines focused on the private lives of those blessed (or burdened) with superpowers. In 1963, Lee and artist Jack Kirby (1917–1994) created a new kind of superhero: "mutants." In the Marvel Comics Universe, mutants are individuals who are born with a mutated X gene that, at the onset of puberty, manifests itself in the form of some kind of mutation, commonly a superpower of some kind. The telepathic and telekinetic mutant professor Charles Xavier utilizes the AI program "Cerebro" to track down mutants and offer them admittance to his "School for Gifted Youngsters." At the school, the students are taught to control their powers and how to live with the prejudice of a world that "hates and fears them." He trains his first class of students to battle "evil" mutants as the "X-Men" (Lee and Kirby, *The X-Men #1*, Marvel Comics, September 1963). The second generation of X-Men were known as the "New Mutants," and one of their first teammates was the inorganic mutant "Warlock."

Unlike the human mutants of the Marvel universe, Warlock is an alien robot, one of a race of robots called "Technarchy." These alien robots feed by infecting organic organisms with a technovirus that bleeds them of their "life force." As such, the Technarchy possesses no feeling or compassion for organic beings, seeing them as food. Warlock, however, does feel compassion for all living beings (this factor, then, being a "mutation" among his race). Warlock's appearance is very robotic: a roughly humanoid shape with a long, elongated head. Part of his natural abilities as a Technarch is the ability to "shape-shift." He was created by the team of writer Chris Claremont (b. 1950) and artist Bill Sienkiewicz (b. 1958) and first appeared in *The New Mutants #18* (Marvel Comics, August 1984). He was part of the "next generation" of mutant students of Xavier's School for Gifted Youngsters, where they are taught by former X-Men foe, Magneto (assigned to the task of teaching these young students by Xavier himself when the founding professor had to leave Earth for a time).

Warlock died at the hands of the evil Cameron Hodge when the villain attempted to bleed the alien robot of his power (Louise Simonson and Rob Liefeld, *The New Mutants #95*, Marvel Comics, November 1990). Warlock's closest friend was the mutant Cypher (a.k.a. Doug Ramsey), who sacrificed his life to save Warlock at one point. When a group of normal humans called the Phalanx were mutated using Warlock's ashes, one of the group became a perfect duplicate of Cypher and even had all of the slain mutant's memories. This combination of Warlock and Doug called himself "Douglock" (Scott Lobdell, Richard Ashford, Chris Cooper, Ken Lashley, and Robert Brown, *Excalibur #77*, Marvel Comics, May 1994). This new incarnation later reverted back to being the original, resurrected Warlock while retaining Cypher's memories.

Warlock is an example of a living, sentient robot. Unlike all of the other robots, cyborgs, and AI mentioned in this volume, no one "built" Warlock. He is part of a race of fully free, inorganic beings. As a robot, he lacks the ability to have mutated genetics; his "mutation," then, is his having developed the capacity of feeling empathy with living beings unlike himself. To his race, that is a mutation. Among his fellow New Mutants, Warlock is often presented as the *most* "human" due to his deep feeling for all life. As a result, Warlock is as unique among science fiction robots as he was among his fellow human superheroes. He is a character study in what it is to be "human" and opens discussions on what can be considered "life."

Richard A. Hall

See also: Arnim Zola, Baymax, Bender, Cerebro/Cerebra, Doctor Octopus, Doombots, HERBIE, IG-88/IG-11, Iron Giant, Iron Legion, Iron Man, JARVIS/Friday, Jocasta, K-2SO, L3-37/*Millennium Falcon*, Lieutenant Commander Data, LMDs, Lore/B4, Replicants, Rosie, Soji and Dahj Asha, Sentinels, Ultron, Vision; *Thematic Essay:* Heroic Robots and Their Impact on Sci-Fi Narratives.

Further Reading

Albert, Robert S. and Brigante, Thomas R. 1962. "The Psychology of Friendship Relations: Social Factors." *Journal of Social Psychology* 56 (1). https://www.tandfon line.com/doi/abs/10.1080/00224545.1962.9919371.

Amati, Viviana, Meggiolaro, Silvia, Rivellini, Giulia, and Zaccarin, Susanna. 2018. "Social Relations and Life Satisfaction: The Role of Friends." *Genus* 74, no. 1 (May 4). https://www.ncbi.nlm.nih.gov/pmc/articles/PMC5937874.

Ambrosino, Brandon. 2018. "What Would It Mean for AI to Have a Soul?" BBC, June 17, 2018. https://www.bbc.com/future/article/20180615-can-artificial-intelligence-have -a-soul-and-religion.

Ashby, LeRoy. 2006. *With Amusement for All: A History of American Popular Culture since 1830.* Lexington: University Press of Kentucky.

Darowski, Joseph J., ed. 2014. *The Ages of the X-Men: Essays on the Children of the Atom in Changing Times.* Jefferson, NC: McFarland.

DeFalco, Tom. 2006. *Comics Creators on X-Men.* London: Titan.

Howe, Sean. 2012. *Marvel Comics: The Untold Story.* New York: Harper-Perennial.

Kripal, Jeffrey J. 2015. *Mutants and Mystics: Science Fiction, Superhero Comics, and the Paranormal.* Chicago: University of Chicago Press.

Powell, Jason. 2016. *The Best There Is at What He Does: Examining Chris Claremont's X-Men.* Edwardsville, IL: Sequart.

Roberts-Griffin, Christopher. 2011. "What Is a Good Friend: A Qualitative Analysis of Desired Friendship Qualities." *Penn McNair Research Journal* 3, no. 1 (December 21). https://repository.upenn.edu/cgi/viewcontent.cgi?article=1019&context=mcn air_scholars.

Robitzski, Dan. 2018. "Artificial Consciousness: How to Give a Robot a Soul." Futurism, June 25, 2018. https://futurism.com/artificial-consciousness.

Thagard, Paul. 2017. "Will Robots Ever Have Emotions?" *Psychology Today*, December 14, 2017. https://www.psychologytoday.com/us/blog/hot-thought/201712/will-rob ots-ever-have-emotions.

Wallach, Wendell, and Allen, Colin. 2009. *Moral Machines: Teaching Robots Right from Wrong.* New York: Oxford University Press.

Wein, Len, ed. 2006. *The Unauthorized X-Men: SF and Comic Book Writers on Mutants, Prejudice, and Adamantium.* Smart Pop series. Dallas, TX: BenBella.

Wright, Bradford W. 2003. *Comic Book Nation: The Transformation of Youth Culture in America.* Baltimore, MD: Johns Hopkins University Press.

WOPR

WarGames

In the early 1980s, the Cold War (1947–1992) between the United States and the Soviet Union was in its most dangerous period in decades. The fear of nuclear war was at an all-time high. As a result, much of popular culture throughout the decade focused on the very real dangers of the real world at the time. One such example was the 1983 film *WarGames* by director John Badham (b. 1939) and starring Matthew Broderick (b. 1962). Believing that human beings, faced with the decision to launch a nuclear attack, may fall short of their responsibilities, the U.S.

ENIAC

The Electronic Numerical Integrator and Computer (ENIAC) was developed by the University of Pennsylvania in 1945 and is considered the world's first "digital computer." Developed for the U.S. Army, ENIAC was utilized by the army from 1946 to 1955. It was designed by J. Presper Eckert (1919–1995) and John Mauchly (1907–1980), professors at the University of Pennsylvania, and built by a team of both male and female university engineers. The completed computer was eight feet tall, three feet wide, and ninety-eight feet long. It took up three walls, weighed twenty-seven tons, and ran on a series of capacitors, crystal diodes, relays, resistors, and vacuum tubes. It was capable of roughly 350 cycles per second (by comparison, today's personal desktop computers possess the ability of roughly two billion cycles per second). Unlike computers today, with preprogrammed functions, ENIAC's programs had to be input manually as needed (sometimes taking weeks to accomplish). Women made up ENIAC's primary team of programmers: Fran Bilas (1922–2012), Betty Jennings (1924–2011), Ruth Lichterman (1924–1986), Kay McNulty (1921–2006), and Marlyn Meltzer (1922–2008). Though originally designed to calculate artillery tables for the army, ENIAC's abilities quickly gained the attention of the atom weaponry research laboratories in Los Alamos, New Mexico, and was utilized in performing calculations toward the effort into the hydrogen bomb. ENIAC marked the beginning of successful concentrated computer research in the United States and the foundation of the computers that twenty-first-century American society depends upon every hour of every day.

Richard A. Hall

military decides to place the decision in the hands of "WOPR" (War Operation Plan Response), a supercomputer programmed to run potential nuclear war scenarios, adapting and readjusting its strategies according to outcomes. Meanwhile, high school slacker and computer hacker David Lightman (Broderick), attempting to hack into a computer gaming company in order to download and play their games, instead stumbles upon NORAD, finding the "game" "Global Thermonuclear War," and begins to "play" as the Soviet Union.

WOPR counters Lightman's moves, now unable to tell the difference between reality and gameplay. Tracked down and accused of espionage, Lightman is desperate to prove that the scenario playing out is a game and not an actual attack. The running "game" launches a nuclear strike, and though NORAD officials are convinced to not act on it, WOPR's "launch command" programming sees its own gaming launch as a real one, ordering a counterattack. Escaping and finding WOPR's original programmer (played by John Wood, 1930–2011), Lightman attempts to stop the war game by engaging WOPR to play itself in a game of tic-tac-toe, making the computer aware of the concept of an unwinnable scenario. Preparing to launch all U.S. missiles, WOPR now runs through all potential scenarios, coming to the conclusion that there is no way to "win" a nuclear war. The program shuts down, declaring, "WINNER: NONE."

Of the many films and novels that have explored the concept of a powerful AI ending all life on Earth, *WarGames* was, perhaps, the most frighteningly realistic. Possibly due to this film, nuclear launch codes remain in the hands of human failsafes even in the twenty-first century. While giving a supercomputer the ability to start/respond to a war without human frailties slowing down response time, such cold logic in control of millions—and potentially billions—of human lives is simply a concept too dangerous to seriously consider. What the "AI-ending-human-existence" scenario shows over and over again is that humanity is its own worst enemy. However, even taking that fact into account, entrusting decision making to the cold logic of a machine—no matter how expertly programmed—opens the door to a myriad of unintended consequences that could prove as disastrous as anything humankind may do to itself.

Richard A. Hall

See also: AI/Ziggy, Batcomputer, Cerebro/Cerebra, Doctor/EMH, Dr. Theopolis and Twiki, ED-209, The Great Intelligence, HAL 9000, HERBIE, Janet, JARVIS/Friday, Johnny 5/S.A.I.N.T. Number 5, Landru, The Matrix/Agent Smith, OASIS, Rehoboam, Skynet, Starfleet Computer, Ultron, V-GER; *Thematic Essay:* AI and the Apocalypse: Science Fiction Meeting Science Fact.

Further Reading

Ashby, LeRoy. 2006. *With Amusement for All: A History of American Popular Culture since 1830.* Lexington: University Press of Kentucky.

Chandler, Simon. 2019. "Artificial Intelligence Has Become a Tool for Classifying and Ranking People." *Forbes*, October 1, 2019. https://www.forbes.com/sites/simon chandler/2019/10/01/artificial-intelligence-has-become-a-tool-for-classifying-and -ranking-people/#7431657f1d7c.

Chanthadavong, Aimee. 2019. "AI and Ethics: The Debate That Needs to Be Had." ZDnet, September 16, 2019. https://www.zdnet.com/article/ai-and-ethics-the-debate-that -needs-to-be-had/#ftag=CAD-03-10abf5f.

Dinh, Thien-Nam. 2018. *Silicon Minds: The Science, Impact, and Promise of Artificial Intelligence.* Independently Published.

Dowden, Bradley. 1993. *Logical Reasoning.* Belmont, CA: Wadsworth Publishing Company.

Ford, Martin. 2015. *Rise of the Robots: Technology and the Threat of a Jobless Future.* New York: Basic.

Hitchcock, Susan Tyler. 2007. *Frankenstein: A Cultural History.* New York: W. W. Norton.

Krishnan, Armin. 2009. *Killer Robots: Legality and Ethicality of Autonomous Weapons.* London: Routledge.

Lin, Patrick, Jenkins, Ryan, and Abney, Keith, eds. 2017. *Robot Ethics 2.0: From Autonomous Cars to Artificial Intelligence.* New York: Oxford University Press.

Marks, Robert J. 2020. "*2084* vs. *1984*: The Difference AI Could Make to Big Brother." Podcast interview with author John Lennox. Mind Matters, July 3, 2020. https:// mindmatters.ai/2020/07/2084-vs-1984-the-difference-ai-could-make-to-big -brother.

Mayor, Adrienne. 2018. *Gods and Robots: Myths, Machines, and Ancient Dreams of Technology.* Princeton, NJ: Princeton University Press.

Murdock, Jason. 2019. "Former Google Engineer Warns AI Might Accidentally Start a War: 'These Things Will Start to Behave in Unexpected Ways.'" *Newsweek*, September 16, 2019. https://www.newsweek.com/google-project-maven-artificial-int elligence-laura-nolan-killer-robots-department-defense-1459358.

Rouhiainen, Lasse. 2018. *Artificial Intelligence: 101 Things You Must Know Today about Our Future*. Scotts Valley, CA: CreateSpace.

Seed, David. 1999. *American Science Fiction and the Cold War: Literature and Film*. Abingdon, UK: Routledge.

Singer, Peter W. 2009. "Isaac Asimov's Laws of Robotics Are Wrong." Brookings Institute, May 18, 2009. https://www.brookings.edu/opinions/isaac-asimovs-laws-of -robotics-are-wrong.

Starck, Kathleen, ed. 2010. *Between Fear and Freedom: Cultural Representations of the Cold War*. Newcastle, UK: Cambridge Scholars.

Virk, Rizwan. 2019. *The Simulation Hypothesis: An MIT Computer Scientist Shows Why AI, Quantum Physics, and Eastern Mystics All Agree We Are in a Video Game*. Milwaukee, WI: Bayview.

Wallach, Wendell, and Allen, Colin. 2009. *Moral Machines: Teaching Robots Right from Wrong*. New York: Oxford University Press.

Z

Zordon/Alpha-5

Mighty Morphin' Power Rangers

Since World War II, Japan has had a strong influence on American popular culture. From *Godzilla* to *Transformers*, both the American science fiction and toy industries have experienced major contributions from Japan. One of the most popular Japanese imports was the television series *Mighty Morphin' Power Rangers* (FOX, 1993–1996), launching the hugely profitable "Power Rangers" franchise. The original series was an Americanized version of the Japanese program *Kyōryū Sentai Zyuranger* (TV Tokyo, 1992–1993), taking the Power Ranger battle footage directly from that series and simply inserting scenes of American teenagers (out of uniform) and dubbing English onto the original footage. The concept of the American series is that Earth is under threat from the evil Rita Replusa (played by Machiko Soga, 1938–2006, from the original Japanese series, and voiced by Barbara Goodson, b. 1949, for the American version). Rita had been imprisoned ten thousand years earlier by the mysterious "Zordon" (played by David J. Fielding, b. 1955) while Rita, simultaneously, trapped Zordon in a "time warp," allowing him to still communicate in the present through a hologram and contact with the helpful robot, Alpha-5 (voiced by Richard Steven Horvitz, b. 1966), famous for his catchphrase, "Aye-aye-aye-aye-aye."

When Rita escapes her bondage, Zordon orders Alpha to utilize the computer to track down five teenagers with "attitude." The five are chosen to fight Rita as "Power Rangers," with each Ranger given a distinct color (red, blue, yellow, black, and pink). Aside from their considerable martial arts abilities, the Rangers also have command of five individual "Zords," robot battle machines in the shapes of prehistoric beasts (tyrannosaurus rex, triceratops, saber-tooth tiger, mastodon, and pterodactyl). When necessary, the five Zords can merge into the giant robot "Megazord" (similar in both design and concept to another Japanese import, Voltron). As Zordon can only communicate through hologram, he is entirely dependent on Alpha-5 to maintain the systems necessary for communication both with him and with the Rangers, as well as for overall supervision of the Zords. Alpha is a standard humanoid robot with a UFO-shaped head. The Zordon hologram shows only the entity's face, warped and slightly out of focus. Though the ranks of Rangers—as well as their respective Zords—change from one incarnation to the next, Zordon and Alpha play a role, to one degree or other, in most incarnations of the franchise.

Zordon and Alpha-5 are similar in many ways—to the point of being near copies—to Oz from *The Wizard of Oz* (1939) and the android Marvin from *Hitchhiker's Guide to the Galaxy* (1981). Until the release of the feature film *Mighty*

Morphin' Power Rangers: The Movie (20th Century Fox, 1995), when Zordon's body is released from the time warp in which he was imprisoned, it was not clear whether the ten-thousand-year-old Zordon were still alive or simply AI. Additionally, though technically a "robot," Alpha identifies as an "AI." Together they represent both the impact of Japanese pop culture on the United States and the use of computer-based characters in modern science fiction.

Richard A. Hall

See also: AI/Ziggy, Alita, Batcomputer, Cerebro/Cerebra, Doctor/EMH, The Great Intelligence, Marvin the Paranoid Android, The Matrix/Agent Smith, Mechagodzilla, OASIS, Oz, Rehoboam, Skynet, Transformers, Voltron, WOPR; *Thematic Essay:* AI and the Apocalypse: Science Fiction Meeting Science Fact.

Further Reading

Ashby, LeRoy. 2006. *With Amusement for All: A History of American Popular Culture since 1830*. Lexington: University Press of Kentucky.

Chandler, Simon. 2019. "Artificial Intelligence Has Become a Tool for Classifying and Ranking People." *Forbes*, October 1, 2019. https://www.forbes.com/sites/simon chandler/2019/10/01/artificial-intelligence-has-become-a-tool-for-classifying-and -ranking-people/#7431657f1d7c.

Chanthadavong, Aimee. 2019. "AI and Ethics: The Debate That Needs to Be Had." ZDnet, September 16, 2019. https://www.zdnet.com/article/ai-and-ethics-the-debate-that -needs-to-be-had/#ftag=CAD-03-10abf5f.

Dinh, Thien-Nam. 2018. *Silicon Minds: The Science, Impact, and Promise of Artificial Intelligence*. Independently Published.

Dowden, Bradley. 1993. *Logical Reasoning*. Belmont, CA: Wadsworth Publishing Company.

Garner, Ross. 2021. *Ranger Reboot: Nostalgia, Transmediality, and the Power Rangers Franchise*. London: Bloomsbury Academic.

Ladd, Fred, and Deneroff, Harvey. 2008. *Astro Boy and Anime Come to the Americas: An Insider's View of the Birth of a Pop Culture Phenomenon*. Jefferson, NC: McFarland.

Lin, Patrick, Jenkins, Ryan, and Abney, Keith, eds. 2017. *Robot Ethics 2.0: From Autonomous Cars to Artificial Intelligence*. New York: Oxford University Press.

Marks, Robert J. 2020. "*2084* vs. *1984*: The Difference AI Could Make to Big Brother" Podcast interview with author John Lennox. Mind Matters, July 3, 2020. https:// mindmatters.ai/2020/07/2084-vs-1984-the-difference-ai-could-make-to-big -brother.

Mayor, Adrienne. 2018. *Gods and Robots: Myths, Machines, and Ancient Dreams of Technology*. Princeton, NJ: Princeton University Press.

Virk, Rizwan. 2019. *The Simulation Hypothesis: An MIT Computer Scientist Shows Why AI, Quantum Physics, and Eastern Mystics All Agree We Are in a Video Game*. Milwaukee, WI: Bayview.

Wallach, Wendell, and Allen, Colin. 2009. *Moral Machines: Teaching Robots Right from Wrong*. New York: Oxford University Press.

Glossary

Android
A robot that is humanlike in appearance.

Animatronics
An engineering process to make robots that appear lifelike.

Anthropoid
Possessing the physical appearance/structure of a human.

Anthropomorphism
Attributing humanlike characteristics to a nonhuman character, creature, or object.

App
An abbreviation for "applications" that can be downloaded to smartphones or other internet-accessible devices.

Artificial Intelligence (AI)
Computer programming that allows a machine to conduct humanlike behavior.

Astrophysics
An area of physics that focuses on interstellar phenomenon.

Automaton
A machine designed to behave like an organic being.

Binary Code
A form of computer programming utilizing only 0s and 1s. (See also "Digital.")

Circuits/Circuitry
The hardware means by which software travels in computer systems.

Computer
An electronic mechanical device designed to store and process data.

Computer Discs
The removable hardware by which computer information/software can be added, stored, removed, or transported nonelectronically.

Computer-Generated Imagery (CGI)
Animated entertainment with varying degrees of realism, often used in entertainment productions.

Cybernetics
Mechanical/electronic implants surgically connected to organic beings for the purpose of replacing/enhancing damaged muscles, bones, or tissues.

Cyborg
A being that is partially organic and partially inorganic.

Data
Information stored in computerized devices.

Digital
A form of computer programming utilizing only 0s and 1s. (See also "Binary Code.")

Droid
A term for "robot" coined by writer-director George Lucas for the *Star Wars* franchise; a shortened version of the word "android." Unlike androids, however, droids are not necessarily humanlike in appearance.

Hardware
The physical components of computer systems.

Imagineers
Engineers that work for the Walt Disney Corporation specializing in creating and manufacturing various rides and animatronics for the company's theme parks.

Mech
Abbreviated form of the words "mechanics," "mechanized," and "mechanical"; often used to describe robotic technologies.

Network
The connecting of more than one computer.

Practical Physics
An area of physics research that attempts to implement the ideas of theoretical physicists.

Programmer
An individual responsible for creating computer programming.

Programming
The core operating system for computer or electronic devices.

Robot
A machine capable of mimicking human behavior.

Robotics
The science of engineering and producing robots.

Sci-Fi
Abbreviation for "science fiction."

Self-Awareness
The ability to be aware of one's own identity as a sentient entity.

Self-Destruct
The ability of mechanical/computerized devices to destroy themselves.

Self-Preservation
The instinct—via programming—that computerized devices can prioritize their own continued existence.

Sentient
The ability to be self-aware and think independently.

Silicon Valley
An area of Northern California associated with being the primary headquarters of computer-based technology.

Smartphones
Cellular telephones with full internet access.

Social Media
Internet-based "communities" where individuals can interact via text, photo, and video sharing.

Software
The programming components of computer systems.

Tech
Abbreviated form of the word "technology."

Theoretical Physics
An area of physics research that promotes new ideas on the meanings and explanations for various real-world phenomena.

"Thumb" Drive/"Flash" Drive
Removable miniature data storage devices providing the same functionality as computer discs.

Virtual Reality (VR)
A computer-generated simulation in which individuals can immerse themselves to experience realistic environs.

Wi-Fi
Wireless access to the internet.

Bibliography

Albert, Robert S., and Brigante, Thomas R. 1962. "The Psychology of Friendship Relations: Social Factors." *Journal of Social Psychology* 56 (1). https://www.tandfonline.com/doi/abs/10.1080/00224545.1962.9919371.

Alkon, Paul K. 1987. *Origins of Futuristic Fiction*. Athens: University of Georgia Press.

Allan, Kathryn, ed. 2013. *Disability in Science Fiction*. New York: Palgrave Macmillan.

Allen, Arthur. "There's a Reason We Don't Know Much about AI." Politico, September 16, 2019. https://www.politico.com/agenda/story/2019/09/16/artficial-intelligence-study-data-000956?cid=apn.

Amati, Viviana, Meggiolaro, Silvia, Rivellini, Giulia, and Zaccarin, Susanna. 2018. "Social Relations and Life Satisfaction: The Role of Friends." *Genus* 74, no. 1 (May 4). https://www.ncbi.nlm.nih.gov/pmc/articles/PMC5937874.

Ambrosino, Brandon. 2018. "What Would It Mean for AI to Have a Soul?" BBC, June 17, 2018. https://www.bbc.com/future/article/20180615-can-artificial-intelligence-have-a-soul-and-religion.

Anderson, Kyle. 2017. "A History of *Doctor Who*'s Cybermen." Nerdist, March 7, 2017. https://nerdist.com/article/615936-2.

Arp, Robert, and Decker, Kevin S., eds. 2013. *The Ultimate* South Park *and Philosophy: Respect My Philosophah!* Hoboken, NJ: Wiley.

Ashby, LeRoy. 2006. *With Amusement for All: A History of American Popular Culture since 1830*. Lexington: University Press of Kentucky.

Asher-Perrin, Emmet. 2014. "*Star Trek*, Why Was This a Good Idea Again?—Data's Human Assimilation." Tor, January 29, 2014. https://www.tor.com/2014/01/29/star-trek-why-was-this-a-good-idea-again-datas-human-assimilation.

Baird, Scott. 2018. "*Star Trek:* 20 Strange Details about Data's Body." Screen Rant, August 25, 2018. https://screenrant.com/star-trek-data-body-abilities-hidden-trivia.

Barker, Cory, and Ryan, Chris, and Wiatrowski, Myc, eds. 2014. *Mapping* Smallville*: Critical Essays on the Series and Its Characters*. Jefferson, NC: McFarland.

Bastani, Aaron. 2019. *Fully Automated Luxury Communism*. New York: Verso.

Baum, L. Frank. (1900) 2015. *The Wonderful Wizard of Oz*. Scotts Valley, CA: CreateSpace.

Beard, Jim, ed. 2010. *Gotham City 14 Miles: 14 Essays on Why the 1960s* Batman *TV Series Matters*. Edwardsville, IL: Sequart Research & Literacy Organization.

Bellomo, Mark. 2018. *The Ultimate Guide to G.I. Joe: 1982–1994, Third Edition*. Iola, WI: Krause.

Blair, Anthony. 2018. "Sex Robots That Feel LOVE and Suffer PAIN When Dumped Coming Soon Claims Expert." *Daily Star*, December 6, 2018. https://www.dailystar.co.uk/news/latest-news/sex-robot-feel-love-pain-16824160.

Branson-Trent, Gregory M. 2009. The Bionic Woman*: Complete Episode Guide*. Scotts Valley, CA: CreateSpace.

Bray, Adam, Barr, Tricia, Horton, Cole, and Windham, Ryder. 2019. *Ultimate* Star Wars*, New Edition*. London: DK Publishing.

Bray, Adam, and Horton, Cole. 2017. Star Wars*: Absolutely Everything You Need to Know, Updated and Expanded*. London: DK Children.

Brooker, Charlie, Jones, Annabel, and Arnopp, Jason. 2018. *Inside* Black Mirror. New York: Crown Archetype.

Bryant, D'Orsay D., III. 1985. "Spare-Part Surgery: The Ethics of Organ Transplantation." *Journal of the National Medical Association* 77 (2). https://www.ncbi.nlm.nih.gov/pmc/articles/PMC2561842/pdf/jnma00245-0055.pdf.

Bunce, Robin, and McCrossin, Trip, eds. 2019. Blade Runner 2049 *and Philosophy: This Breaks the World*. Chicago: Open Court.

Burnett, Dean. 2014. "Time Travellers: Please Don't Kill Hitler." *The Guardian*, February 21, 2014. https://www.theguardian.com/science/brain-flapping/2014/feb/21/time-travellers-kill adolf-hitler.

Byrne, Emma. 2013. "Innovation Isn't Safe: The Future According to Kevin Warwick." *Forbes*, September 30, 2013. https://www.forbes.com/sites/netapp/2013/09/30/kevin-warwick-captain-cyborg/#48b704c13560.

Byrne, Emma. 2014. "Cybernetic Implants: No Longer Science Fiction." *Forbes*, March 11, 2014. https://www.forbes.com/sites/netapp/2014/03/11/cybernetic-implants-not-sci-fi/#7399c57e77ba.

Calvert, Bronwen. 2017. *Being Bionic: The World of TV Cyborgs*. London: I. B. Tauris.

Campbell, Joseph. (1949) 2004. *The Hero with a Thousand Faces: Commemorative Edition*. Princeton, NJ: Princeton University Press.

Campbell, Mark. 2010. Doctor Who*: The Complete Guide*. London: Running Press.

Cantwell, David. 2018. "The Wisdom of *The Vision*, a Superhero Story about Family and Fitting In." *New Yorker*, March 22, 2018. https://www.newyorker.com/books/page-turner/the-wisdom-of-the-vision-a-superhero-story-about-family-and-fitting-in.

Caroti, Simone. 2004. "Science Fiction, Forbidden Planet, and Shakespeare's the Tempest." *CLCWeb: Comparative Literature and Culture* 6 (1). https://docs.lib.purdue.edu/cgi/viewcontent.cgi?article=1214&context=clcweb.

Carper, Steve. 2019. *Robots in American Popular Culture*. Jefferson, NC: McFarland.

Castleman, Harry, and Podrazik, Walter J. 2016. *Watching TV: Eight Decades of American Television, Third Edition*. Syracuse, NY: Syracuse University Press.

Cavaler, Chris. 2015. *On the Origins of Superheroes: From the Big Bang to Action Comics No. 1*. Iowa City: University of Iowa Press.

Cavelos, Jeanne. 2008. "R2-D2 and C-3P0: Do Droids Dream of Electric Sheep?" *Scientific American*, August 11, 2008. https://www.scientificamerican.com/article/star-wars-science-droid-dreams.

Chandler, Simon. 2019. "Artificial Intelligence Has Become a Tool for Classifying and Ranking People." *Forbes*, October 1, 2019. https://www.forbes.com/sites/simonchandler/2019/10/01/artifical-intelligence-has-become-a-tool-for-classifying-and-ranking-people/#7431657f1d7c.

Chanthadavong, Aimee. 2019. "AI and Ethics: The Debate That Needs to Be Had." ZDnet, September 16, 2019. https://www.zdnet.com/article/ai-and-ethics-the-debate-that-needs-to-be-had/#ftag=CAD-03-10abf5f.

Chapman, James. 2013. *Inside the TARDIS: The Worlds of* Doctor Who. London: I. B. Tauris.

Choi, Charles Q. 2009. "Human Evolution: The Origin of Tool Use." LiveScience, November 11, 2009. https://www.livescience.com/7968-human-evolution-origin-tool.html.

Cline, Ernest. 2012. *Ready Player One: A Novel*. New York: Broadway.

Conrad, Dean. 2018. *Space Sirens, Scientists and Princesses: The Portrayal of Women in Science Fiction Cinema*. Jefferson, NC: McFarland.

Costello, Matthew J. 2009. *Secret Identity Crisis: Comic Books & the Unmasking of Cold War America*. New York: Continuum.

Cushman, Marc. 2016–2017. *Irwin Allen's Lost in Space, Volumes 1–3: The Authorized Biography of a Classic Sci-Fi Series*. Los Angeles: Jacobs Brown Media Group.

Daniels, Anthony. 2019. *I Am C-3PO: The Inside Story*. London: DK Publishing.

Daniels, Les. 2004. *Superman: The Complete History—The Life and Times of the Man of Steel*. New York: DC Comics.

Darowski, Joseph J., ed. 2014. *The Ages of the X-Men: Essays on the Children of the Atom in Changing Times*. Jefferson, NC: McFarland.

Decker, Kevin S., and Eberl, Jason T., eds. 2016. *The Ultimate* Star Trek *and Philosophy: The Search for Socrates*. Hoboken, NJ: Wiley-Blackwell.

DeFalco, Tom, ed. 2004. *Comics Creators on Spider-Man*. London: Titan.

DeFalco, Tom, ed. 2005. *Comics Creators on Fantastic Four*. London: Titan.

DeFalco, Tom, ed. 2006. *Comics Creators on X-Men*. London: Titan.

Delpozo, Brian. 2016. "Trek at 50: Enduring Archetypes, Part Two: The Alien Other." Comicsverse, August 16, 2016. https://comicsverse.com/trek-50-enduring-archetypes-part-two-alien.

Dinh, Thien-Nam. 2018. *Silicon Minds: The Science, Impact, and Promise of Artificial Intelligence*. Independently Published.

Dowden, Bradley. 1993. *Logical Reasoning*. Belmont, CA: Wadsworth Publishing Company.

Eberl, Jason T., and Decker, Kevin S., eds. 2015. *The Ultimate* Star Wars *and Philosophy: You Must Unlearn What You Have Learned*. Hoboken, NJ: Wiley-Blackwell.

Ewing, Jeffrey A., and Decker, Kevin S., eds. 2017. *Alien and Philosophy: I Infest, Therefore I Am*. Hoboken, NJ: Wiley-Blackwell.

Faludi, Susan. 2007. *The Terror Dream: Fear and Fantasy in Post-9/11 America*. New York: Metropolitan.

Farnell, Chris. 2020. "*Doctor Who*: The Genius of Making the Cybermen and Ideology." Den of Geek, January 28, 2020. https://www.denofgeek.com/tv /doctor-who-the-genius-of-making-thecybermen-an-ideology-2.

Ford, Martin. 2015. *Rise of the Robots: Technology and the Threat of a Jobless Future*. New York: Basic.

Fryer-Biggs, Zachary. 2019. "Coming Soon to a Battlefield: Robots That Can Kill." *The Atlantic*, September 3, 2019. https://www.theatlantic.com/tech nology/archive/2019/09/killer-robots-and-new-era-machine-driven-warfa re/597130.

Garner, Ross. 2021. *Ranger Reboot: Nostalgia, Transmediality, and the Power Rangers Franchise*. London, UK: Bloomsbury Academic.

Geraghty, Lincoln. 2008. "From Balaclavas to Jumpsuits: The Multiple Histories and Identities of *Doctor Who*'s Cybermen." *Journal of the Spanish Association of Anglo-American Studies*, June 2008. https://pdfs.semantic scholar.org/e755/76311422cb72f9b91961e19965bd9423ef9b.pdf.

Grau, Christopher, ed. 2005. *Philosophers Explore* The Matrix. Oxford: Oxford University Press.

Greene, Richard, and Heter, Joshua, eds. 2018. Westworld *and Philosophy: Mind Equals Blown*. Chicago: OpenCourt.

Gross, Edward, and Altman, Mark A., eds. 2016a. *The Fifty-Year Mission, The First 25 Years: The Complete, Uncensored, Unauthorized Oral History of* Star Trek. New York: St. Martin's.

Gross, Edward, and Altman, Mark A., eds. 2016b. *The Fifty-Year Mission, The Next 25 Years: The Complete, Uncensored, Unauthorized Oral History of* Star Trek. New York: St. Martin's.

Gross, Edward, and Altman, Mark A., eds. 2018. *So Say We All: The Complete, Uncensored, Unauthorized Oral History of* Battlestar Galactica. New York: Tor.

Hampton, Gregory Jerome. 2017. *Imagining Slaves and Robots in Literature, Film, and Popular Culture: Reinventing Yesterday's Slave with Tomorrow's Robot*. Lanham, MD: Lexington.

Han, Karen. 2018. "*Lost in Space* Shows a Long-Running Problem with Stories about AI." The Verge, April 24, 2018. https://www.theverge.com/2018 /4/24/17275856/lost-in-space-netflix-ai-artificial-intelligence-iron-giant.

Handley, Rich, and Tambone, Lou, eds. 2018. *Somewhere beyond the Heavens: Exploring* Battlestar Galactica. Edwardsville, IL: Sequart Research and Literacy Organization.

Harris, Mark. 1983. *The* Doctor Who *Technical Manual*. London, UK: Random House.

Harris, Molly. 2020. "Data: All His Best Ever *Star Trek: The Next Generation* Moments." Film Daily, April 10, 2020. https://filmdaily.co/news/best-data -star-trek-tng-moments.

Hauser, Tim. 2008. *The Art of WALL-E*. San Francisco: Chronicle.

Hendershot, Cyndy. 1999. *Paranoia, the Bomb, and 1950s Science Fiction Films*. Bowling Green, KY: Bowling Green State University Popular Press.

Hiatt, Brian. 2019. "Anthony Daniels: My Life as C-3PO Is Far from Over." Rolling Stone, December 17, 2019. https://www.rollingstone.com/movies /movie-features/star-wars-skywalker-anthony-daniels-c3po-interview -927577.

Hicks, Amber. 2019. "Sex Robot Brothel Opens in Japan amid Surge of Men Wanting Bisexual Threesomes." *Mirror*, April 27, 2019. https://www.mir ror.co.uk/news/weird-news/sex-robot-brothel-opens-japan-14792161.

Hills, Matt, ed. 2013. *New Dimensions of* Doctor Who: *Adventures in Space, Time and Television*. London: I. B. Tauris.

Hitchcock, Susan Tyler. 2007. *Frankenstein: A Cultural History*. New York: W. W. Norton.

Hoberman, J. 2011. *An Army of Phantoms: American Movies and the Making of the Cold War*. New York: New Press.

Howe, Sean. 2012. *Marvel Comics: The Untold Story*. New York: Harper-Perennial.

Irwin, William, ed. 2002. The Matrix *and Philosophy: Welcome to the Desert of the Real*. Chicago: Open Court.

Irwin, William. 2018. Westworld *and Philosophy: If You Go Looking for the Truth, Get the Whole Thing*. Hoboken, NJ: Wiley-Blackwell.

Irwin, William, Brown, Richard, and Decker, Kevin S. 2009. *Terminator and Philosophy: I'll Be Back Therefore I Am*. Hoboken, NJ: Wiley-Blackwell.

Johnson, David Kyle, ed. 2019. Black Mirror *and Philosophy: Dark Reflections*. Hoboken, NJ: Wiley-Blackwell.

Kachka, Boris. 2015. "The Last Human Robot." Vulture, December 6, 2015. https://www.vulture.com/2015/12/anthony-daniels-c-3po-c-v-r.html.

Kaminski, Michael. 2008. *The Secret History of Star Wars: The Art of Storytelling and the Making of a Modern Epic*. Kingston, Canada: Legacy.

Kelly, Andy. 2020. "*Star Trek: Picard*: How Data Died, and His Appearance in *Picard* Explained." TechRadar, January 24, 2020. https://www.techradar .com/news/star-trek-picard-data-death-explained.

Kirshner, Jonathan. 2012. *Hollywood's Last Golden Age: Politics, Society, and the Seventies Film in America*. Ithaca, NY: Cornell University Press.

Kistler, Alan. 2013. Doctor Who: *Celebrating Fifty Years, A History*. Guilford, CT: Lyons.

Koberlein, Brian. 2015. "Why It's Impossible To Go Back In Time and Kill Baby Hitler." *Forbes*, October 23, 2015. https://www.forbes.com/sites/brianko berlein/2015/10/23/why-its-impossible-to go-back-in-time-and-kill-baby -hitler/#10894c3529c0.

Kripal, Jeffrey J. 2015. *Mutants and Mystics: Science Fiction, Superhero Comics, and the Paranormal.* Chicago: University of Chicago Press.

Krishnan, Armin. 2009. *Killer Robots: Legality and Ethicality of Autonomous Weapons.* London: Routledge.

Ladd, Fred, and Deneroff, Harvey. 2008. *Astro Boy and Anime Come to the Americas: An Insider's View of the Birth of a Pop Culture Phenomenon.* Jefferson, NC: McFarland.

Langley, Travis, Goodfriend, Wind, and Cain, Tim, eds. 2018. Westworld *Psychology: Violent Delights.* New York: Sterling.

Larson, Glen A., and Thurston, Robert. (1978) 2005. Battlestar Galactica *Classic: The Saga of a Star World.* New York: iBooks.

Lavery, David. 2013. *Joss Whedon, A Creative Portrait: From* Buffy the Vampire Slayer *to Marvel's* The Avengers. London: I. B. Tauris.

Leadbeater, Alex. 2019. "*Star Wars* is Trying to Turn Darth Vader into an Anti-Hero (and That's Very Bad)." Screen Rant, February 4, 2019. https://screenrant.com/star-wars-darth-vader-villain-anti-hero.

Leetaru, Kalev. 2019. "Automatic Image Captioning and Why Not Every AI Problem Can Be Solved through More Data." *Forbes,* July 7, 2019. https://www.forbes.com/sites/kalevleetaru/2019/07/07/automatic-image-captioning-and-why-not-every-ai-problem-can-be-solved-through-more-data/#20b943476997.

Lenburg, Jeff. 2011. *William Hanna and Joseph Barbera: The Sultans of Saturday Morning.* Philadelphia: Chelsea House.

Levin, Ira. 1972. *The Stepford Wives.* New York: William Morrow.

Lewis, Courtland, and Smithka, Paula, eds. 2010. Doctor Who *and Philosophy: Bigger on the Inside.* Chicago: Open Court.

Lewis, Courtland, and Smithka, Paula, eds. 2015. *More* Doctor Who *and Philosophy: Regeneration Time.* Chicago: Open Court.

Lewis, Courtland D. 2014. Futurama *and Philosophy: Pizza, Paradoxes, and . . . Good News!* Scotts Valley, CA: CreateSpace.

Lin, Patrick, Abney, Keith, and Bekey, George A., eds. 2014. *Robot Ethics: The Ethical and Social Implications of Robotics.* Cambridge: Massachusetts Institute of Technology.

Lin, Patrick, Jenkins, Ryan, and Abney, Keith, eds. 2017. *Robot Ethics 2.0: From Autonomous Cars to Artificial Intelligence.* New York: Oxford University Press.

Linder, Courtney. 2020. "This AI Robot Just Nabbed the Lead Role in a Sci-Fi Movie: Meet Erica, the Movie Star Who May Put Human Actors Out of Work." *Popular Mechanics,* June 25, 2020. https://www.popularmechanics.com/technology/robots/a32968811/artificial-intelligence-robot-movie-star-erica.

Longoni, Chiara, Bonezzi, Andrea, and Morewedge, Carey K. 2019. "Resistance to Medical Artificial Intelligence." *Journal of Consumer Research* 46, no. 4 (December): 629–50. https://academic.oup.com/jcr/article-abstract/46/4/629/5485292.

Mallett, Ronald L., and Henderson, Bruce. 2006. *Time Traveler: A Scientist's Personal Mission to Make Time Travel a Reality.* New York: Basic Books.

Marks, Robert J. 2020. "*2084* vs. *1984*: The Difference AI Could Make to Big Brother." Podcast interview with author John Lennox. Mind Matters, July 3, 2020. https://mindmatters.ai/2020/07/2084-vs-1984-the-difference-ai-could -make-to-big-brother.

Mayor, Adrienne. 2018. *Gods and Robots: Myths, Machines, and Ancient Dreams of Technology.* Princeton, NJ: Princeton University Press.

McAdams, Taylor. 2019. "Fans Were Never Meant to Know What Happened Off Camera in *Bionic Woman*." BrainSharper, October 6, 2019. https://www .brain-sharper.com/entertainment/bionic-woman-fb.

Means, Sean P. 2019. "At FanX, 'Bionic' Stars Lindsay Wagner and Lee Majors Recall Their TV Glory, and Lots of Running." *Salt Lake Tribune*, September 6, 2019. https://www.sltrib.com/artsliving/2019/09/06/fanx-bionic-stars.

Michaud, Nicolas, and Watkins, Jessica, eds. 2018. *Iron Man vs. Captain America and Philosophy: Give Me Liberty or Keep Me Safe.* Chicago: Open Court.

Mitchell, Edward Page. 1881. "The Clock That Went Backward." http://www.for gottenfutures.com/game/ff9/tachypmp.htm#clock.

Muir, John Kenneth. 2007. *A Critical History of* Doctor Who *on Television.* Jefferson, NC: McFarland.

Murdock, Jason. 2019. "Former Google Engineer Warns AI Might Accidentally Start a War: 'These Things Will Start to Behave in Unexpected Ways.'" *Newsweek*, September 16, 2019. https://www.newsweek.com/google-proj ect-maven-artificial-intelligence-laura-nolan-killer-robots-department -defense-1459358.

Nahin, Paul J. 1999. *Time Machines: Time Travel in Physics, Metaphysics, and Science Fiction.* 2nd ed. Woodbury, NY: AIP Press; New York: Springer.

Nathan-Turner, John. 1985. *The TARDIS Inside Out.* London: Piccadilly Press.

Nero, Dom. 2019. "Anthony Daniels Says *Star Wars: The Rise of Skywalker* Isn't the Last You'll See of C-3PO." *Esquire*, November 7, 2019. https://www .esquire.com/entertainment/movies/a29713208/anthony-daniels-star-wars -c-3po-the-rise-of-skywalker-interview.

Older, Daniel Jose. 2018. Star Wars*: Last Shot.* New York: Del Rey.

O'Neil, Dennis, ed. 2008. *Batman Unauthorized: Vigilantes, Jokers, and Heroes in Gotham City.* Dallas, TX: BenBella.

Parkin, Lance. 2016. *Whoniverse.* New York: Barron's Educational Series.

Paur, Joey. 2019. "An AI Bot Writes a Hilarious Episode of *Star Trek: The Next Generation*." Geek Tyrant, November 16, 2019. https://geektyrant.com /news/an-ai-bot-writes-a-hilarious-episode-of-star-trek-the-next-gene ration.

Pellissier, Hank. 2013. "Robots and Slavery: What Do Humans Want When We Are 'Masters'?" Institute for Ethics and Emerging Technologies, September 13, 2013. https://ieet.org/index.php/IEET2/more/pellissier20130913.

Pilato, Herbie J. 2007. *The Bionic Book:* The Six Million Dollar Man *and the* Bionic Woman *Reconstructed.* Albany, GA: BearManor.

Pilato, Herbie J. 2016. "A 40th Anniversary Tribute to the Bionic Woman and Wonder Woman. Part I: *The Bionic Woman.*" Emmys, December 19, 2016. https://www.emmys.com/news/online-originals/40th-anniversary-tribute -bionic-woman-and-wonder-woman-part-1-bionic-woman.

Plante, Corey. 2018. *"Forbidden Planet* Birthed the Sci-Fi Template 62 Years Ago" Inverse, March 15, 2018. https://www.inverse.com/article/42360-forbidden -planet-movie-anniversary-science-fiction.

Powell, Jason. 2016. *The Best There Is at What He Does: Examining Chris Claremont's X-Men.* Edwardsville, IL: Sequart.

Price, David A. 2009. *The Pixar Touch: The Making of a Company.* New York: First Vintage.

Randall, Terri, dir. 2016. *NOVA.* "Rise of the Robots: Inside the DARPA Robotics Challenge." Aired February 24, 2016, on PBS. DVD.

Reagin, Nancy R., and Liedl, Janice, eds. 2013a. Star Trek *and History.* Hoboken, NJ: John Wiley & Sons.

Reagin, Nancy R., and Liedl, Janice, eds. 2013b. Star Wars *and History.* New York: John Wiley & Sons.

Rees, Shelley S., ed. 2013. *Reading* Mystery Science Theater 3000*: Critical Approaches.* Lanham, MD: Scarecrow/Rowman & Littlefield.

Reeves-Stevens, Garfield, and Reeves-Stevens, Judith. 1997. Star Trek: The Next Generation*: The Continuing Mission.* New York: Pocket Books.

Resnick, Brian. 2020. "NASA's Latest Rover Is Our Best Chance Yet to Find Life on Mars." Vox, July 30, 2020. https://www.vox.com/science-and-health /2020/7/29/21340464/nasa-perseverance-ingenuity-launch-live-stream -how-to-watch-science-life-on-mars.

Richards, Justin. 2009. Doctor Who*: The Official Doctionary.* London: Penguin Group.

Rinzler, J.W. 2007. *The Making of* Star Wars*: The Definitive Story behind the Original Film.* New York: Del Rey.

Roberts-Griffin, Christopher. 2011. "What Is a Good Friend: A Qualitative Analysis of Desired Friendship Qualities." *Penn McNair Research Journal* 3, no. 1 (December 21). https://repository.upenn.edu/cgi/viewcontent.cgi?ar ticle=1019&context=mcnair_scholars.

Robitzski, Dan. 2018. "Artificial Consciousness: How to Give a Robot a Soul." Futurism, June 25, 2018. https://futurism.com/artificial-consciousness.

Roddenberry, Gene. 1979, *Star Trek: The Motion Picture* (Novelization). New York: Pocket.

Rouhiainen, Lasse. 2018. *Artificial Intelligence: 101 Things You Must Know Today about Our Future.* Scotts Valley, CA: CreateSpace.

Ryfle, Steve. 1998. *Japan's Favorite Mon-Star: The Unauthorized Biography of "The Big G."* Toronto, Canada: ECW.

Said, Carolyn. 2019. "Robot Cars Are Getting Better—But Will True Self-Driving Ever Arrive." *San Francisco Chronicle*, February 14, 2019. https://www .sfchronicle.com/business/article/Robot-cars-are-getting-better-but-will-true -13614280.php.

Sanford, Jonathan J., and Irwin, William, eds. 2012. *Spider-Man and Philosophy: The Web of Inquiry*. New York: Wiley.

Schroeder, Stan. 2020. "Samsung Just Launched an 'Artificial Human' Called Neon, and Wait, What?" Mashable, January 7, 2020. https://mashable.com /article/samsung-star-labs-neon-ces.

Seed, David. 1999. *American Science Fiction and the Cold War: Literature and Film*. Abingdon, UK: Routledge.

Shanahan, Timothy, and Smart, Paul. 2019. Blade Runner 2049: *A Philosophical Exploration*. Abingdon, UK: Routledge.

Shapiro, Allen. 2013. Star Trek *and the Android Data*. Seattle: Amazon.com Services.

Shermer, Michael. 2002. "The Chronology Protection Conjecture." *Scientific American*, August 12, 2002. https://www.scientificamerican.com/article /the-chronology-protection.

Shook, John R., and Stillwaggon Swan, Liz, eds. 2009. *Transformers and Philosophy: More Than Meets the Mind*. Chicago: Open Court.

Singer, Peter W. 2009. "Isaac Asimov's Laws of Robotics Are Wrong." Brookings Institute, May 18, 2009. https://www.brookings.edu/opinions/isaac-asimo vs-laws-of-robotics-are-wrong.

"Six Million Dollar Man Quotes, The." n.d. Retro Junk. https://www.retrojunk. com/content/child/quote/page/4314/the-six-million-dollar-man.

Slavicsek, Bill. 2000. *A Guide to the* Star Wars *Universe*. San Francisco: LucasBooks.

Smith, Brian. 2014. *Voltron: From Days of Long Ago: A Thirtieth Anniversary Celebration*. New York: Perfect Square.

Solomon, Brian. 2017. *Godzilla FAQ: All That's Left to Know about the King of the Monsters*. Framingham, MA: Applause.

South, James B., ed. 2003. Buffy the Vampire Slayer *and Philosophy: Fear and Trembling in Sunnydale*. Popular Culture and Philosophy series, edited by William Irwin. Chicago: Open Court.

Spaeth, Dennis. 2018. "From Single-Task Machines to Backflipping Robots: The Evolution of Robots." *Cutting Tool Engineering*, January 15, 2018. https:// www.ctemag.com/news/articles/evolution-of-robots.

Staff. 2017. "Danger! Here Are 12 Far-Out Facts about 'Lost in Space.'" Decades, June 14, 2017. https://www.decades.com/lists/danger-here-are-12-far-out -facts-about-lost-in-space.

Staff. 2020a. "Shimon: Now a Singing, Songwriting Robot: Marimba-Playing Robot Composes Lyrics and Melodies with Human Collaborators." Georgia Tech Online, February 25, 2020. https://www.news.gatech.edu/2020 /02/25/shimon-now-singing-songwriting-robot.

Staff. 2020b. "To Mondas and Back Again: A Brief History of the Cybermen in *Doctor Who*." *Radio Times*, February 23, 2020. https://www.radiotimes .com/news/2020-02-23/cybermen-doctor-who-history-background.

Stanford University. n.d. "Robotics: A Brief History." Stanford University. https:// cs.stanford.edu/people/eroberts/courses/soco/projects/1998-99/robotics/his tory.html.

Starck, Kathleen, ed. 2010. *Between Fear and Freedom: Cultural Representations of the Cold War.* Newcastle, UK: Cambridge Scholars.

Stark, Steven D. 1997. *Glued to the Set: The 60 Television Shows and Events that Made Us Who We Are Today.* New York: Delta Trade Paperbacks.

Sumerak, Marc. 2018. Star Wars*: Droidography.* New York: HarperFestival.

Sunstein, Cass R. 2016. *The World According to* Star Wars. New York: Dey Street.

Sweet, Derek R. 2015. Star Wars *in the Public Square: The Clone Wars as Political Dialogue.* Critical Explorations in Science Fiction and Fantasy series, edited by Donald E. Palumbo and Michael Sullivan. Jefferson, NC: McFarland.

Taylor, Chris. 2015. *How* Star Wars *Conquered the Universe: The Past, Present, and Future of a Multibillion Dollar Franchise.* New York: Basic Books.

Thagard, Paul. 2017. "Will Robots Ever Have Emotions?" *Psychology Today*, December 14, 2017. https://www.psychologytoday.com/us/blog/hot-thought/201712/will-robots-ever-have-emotions.

Thomas, Roy. 2017. *The Marvel Age of Comics: 1961–1978.* Los Angeles: Taschen.

Tramel, Jimmie. 2019. "Bionic and Iconic: Lindsay Wagner, TV's Bionic Woman, Shares Memories before Tulsa Pop Culture Expo." *Tulsa World*, October 30, 2019. https://www.tulsaworld.com/entertainment/television/bionic-and-iconic-lindsay-wagner-tv-s-bionic-woman-shares/article_a702a343-6afb-5127-9916-827077db598a.html.

Tucker, Reed. 2017. *Slugfest: Inside the Epic 50-Year Battle between Marvel and DC.* New York: Da Capo Press.

Tuttle, John. 2018. "The Original Series Robots Which Led Up to the Robot in Netflix's *Lost in Space.*" Medium, July 25, 2018. https://medium.com/of-intellect-and-interest/the-original-series-robots-which-led-up-to-the-robot-in-netflixs-lost-in-space-2a23028b54f3.

Tye, Larry. 2013. *Superman: The High-Flying History of America's Most Enduring Hero.* New York: Random House.

Uyeno, Greg. 2019. "What Is the Grandfather Paradox?" Space, June 5, 2019. https://www.space.com/grandfather-paradox.html.

Vary, Adam B. 2007. "*Star Trek*: The Next Generation: An Oral History." *Entertainment Weekly*, September 25, 2007. https://ew.com/article/2007/09/25/star-trek-tng-oral-history.

Virk, Rizwan. 2019. *The Simulation Hypothesis: An MIT Computer Scientist Shows Why AI, Quantum Physics, and Eastern Mystics All Agree We Are in a Video Game.* Milwaukee, WI: Bayview.

Wallace, Daniel. 2006. Star Wars*: The New Essential Guide to Droids.* New York: Del Rey.

Wallace, Daniel, and Ling, Josh. 1999. *C-3P0: Tales of the Golden Droid (Star Wars Masterpiece Edition).* San Francisco: Chronicle Books.

Wallach, Wendell, and Allen, Colin. 2009. *Moral Machines: Teaching Robots Right from Wrong.* New York: Oxford University Press.

Wasserman, Ryan. 2018. *Paradoxes of Time Travel.* Oxford: Oxford University Press.

Wein, Len, ed. 2006. *The Unauthorized X-Men: SF and Comic Book Writers on Mutants, Prejudice, and Adamantium.* Smart Pop series. Dallas, TX: BenBella.

Weiner, Robert G., and Barba, Shelley E., eds. 2011. *In the Peanut Gallery with Mystery Science Theater 3000: Essays on Film, Fandom, Technology and the Culture of Riffing.* Jefferson, NC: McFarland.

Weinstock, Jeffrey Andrew, ed. 2008. *Taking* South Park *Seriously.* Albany: State University of New York Press.

Weldon, Glen. 2016. *The Caped Crusade: Batman and the Rise of Nerd Culture.* New York: Simon & Schuster.

Westcott, Kathryn. 2011. "HG Wells or Enrique Gaspar: Whose Time Machine Was First?" BBC News, April 9, 2011. https://www.bbc.com/news/world-europe-12900390.

White, Mark D., ed. 2010. *Iron Man and Philosophy: Facing the Stark Reality.* Hoboken, NJ: Wiley.

Wilcox, Rhonda V., ed. *Slayage: The Journal of Whedon Studies.* Whedon Studies Association. https://www.whedonstudies.tv/slayage-the-journal-of-whedon-studies.html.

Wright, Bradford W. 2003. *Comic Book Nation: The Transformation of Youth Culture in America.* Baltimore, MD: Johns Hopkins University Press.

Yeffeth, Glenn, ed. 2003a. *Seven Seasons of Buffy: Science Fiction and Fantasy Writers Discuss Their Favorite Television Show.* Dallas, TX: BenBella.

Yeffeth, Glenn, ed. 2003b. *Taking the Red Pill: Science, Philosophy and the Religion of* The Matrix. Smart Pop series. Dallas, TX: BenBella.

Young, S. Mark, Duin, Steve, and Richardson, Mike. 2012. *Blast Off!: Rockets, Robots, Rayguns, and Rarities from the Golden Age of Space Toys.* Milwaukie, OR: Dark Horse.

About the Author and Contributors

AUTHOR

RICHARD A. HALL has taught courses on various areas of American popular culture, including superhero comics and television and movies of the Cold War era, as well as U.S. military history. After serving four years in the U.S. Army, he attended Texas A&M International University in Laredo, Texas, finishing his BA and MA before receiving his PhD in history from Auburn University, in Auburn, Alabama. He is the author of *The American Superhero: Encyclopedia of Caped Crusaders in History*; *Pop Goes the Decade: The Seventies*; *The American Villain: Encyclopedia of Bad Guys in Comics, Film, and Television*; and *Pop Goes the Decade: The 2000s*, all from ABC-CLIO/Greenwood Press. He has served as contributor and member of the board of advisors for the online pop culture database, *Pop Culture Universe: Icons, Idols, and Ideas*, also from ABC-CLIO. He was a contributor to *The Supervillain Reader* (2020). His upcoming projects include *Gotham, U.S.A.: Critical Essays on Ethics, American Society, and the Batman Universe of Television's* Gotham; and *The Man Who Made the '80s: J. R. Ewing and the Origins of 21st Century America*. A father of five and grandfather of five, he lives in Texas with the youngest of his children, and his wife, best friend, fellow former soldier, and frequent collaborator, Dr. Maria A. Reyes.

CONTRIBUTORS

LISA C. BAILEY is a freelance writer and editor in Birmingham, Alabama. She holds a BA in communication from the University of Alabama and has written and edited for such national publications as *Health*, *Cooking Light*, and *Southern Living* magazines, as well as for national book publishers and UAB.

KEITH R. CLARIDY graduated from Auburn University in 2006 with an MA in history. He subsequently taught at Chattahoochee Valley Community College, Southern Union State Community College, Tuskegee University, Auburn University at Montgomery, and Troy University's eArmy Campus. While teaching for

Troy, he spent two weeks in Melaka, Malaysia, teaching at Troy's international site, Putra College. Since 2011, Keith has worked for various state agencies in Alabama. His writing credits include contributing to the *Encyclopedia of Alabama* (2007); *"Bring God to the Negro, Bring the Negro to God": Archbishop Thomas Joseph Toolen and Race in Alabama* (2009); and contributing to *The American Villain: Encyclopedia of Bad Guys in Comics, Film, and Television* (ABC-CLIO/ Greenwood, 2020).

JOSH PLOCK is a graduate of Columbus State University, in Columbus, Georgia, where he earned his bachelor's in history with a minor in sociology. He regularly writes about film and comic books for the website House of Geekery. In addition, he also reviews books for the blog *Horror Bound*. He is a contributor to *The American Superhero: Encyclopedia of Caped Crusaders in History* (2019) and its follow-up, *The American Villain: Encyclopedia of Bad Guys in Comics, Film, and Television* (2020), both from ABC-CLIO/Greenwood. He is also a coauthor of *Gotham, U.S.A.: Critical Essays on Ethics, American Society, and the Batman Universe of Television's* Gotham (2021). Currently, he lives in Memphis, Tennessee, with his wife, Meredith, and their dog.

Index

Notes: Page numbers for images are in *italic* type; page numbers for main entries are in **bold** type.